Fuzzy Logic and Expert Systems Applications

Neural Network Systems
Techniques and Applications

Edited by **Cornelius T. Leondes**

Fuzzy Logic and Expert Systems Applications

Edited by

Cornelius T. Leondes
Professor Emeritus
University of California
Los Angeles, California

V O L U M E   **6**   O F

Neural Network Systems
Techniques and Applications

ACADEMIC PRESS
San Diego   London   Boston   New York   Sydney   Tokyo   Toronto

This book is printed on acid-free paper. ∞

Academic Press
a division of Harcourt Brace & Company
525 B Street, Suite 1900, San Diego, California 92101-4495, USA
http://www.apnet.com

Academic Press Limited
24-28 Oval Road, London NW1 7DX, UK
http://www.hbuk.co.uk/ap/

Library of Congress Card Catalog Number: 97-80441

International Standard Book Number: 0-12-443866-0

PRINTED IN THE UNITED STATES OF AMERICA
97 98 99 00 01 02 ML 9 8 7 6 5 4 3 2 1

Contents

Fuzzy Neural Networks Techniques and Their Applications

Hisao Ishibuchi and Manabu Nii

Implementation of Fuzzy Systems

Chu Kwong Chak, Gang Feng, and Marimuthu Palaniswami

Neural Networks and Rule-Based Systems

Aldo Aiello, Ernesto Burattini, and Guglielmo Tamburrini

Construction of Rule-Based Intelligent Systems

Graham P. Fletcher and Chris J. Hinde

Expert Systems in Soft Computing Paradigm

Sankar K. Pal and Sushmita Mitra

Mean-Value-Based Functional Reasoning Techniques in the Development of Fuzzy-Neural Network Control Systems

Keigo Watanabe and Spyros G. Tzafestas

Fuzzy Neural Network Systems in Model Reference Control Systems

Yie-Chien Chen and Ching-Cheng Teng

Wavelets in Identification

A. Juditsky, Q. Zhang, B. Delyon, P.-Y. Glorennec, and A. Benveniste

Contributors

Numbers in parentheses indicate the pages on which the authors' contributions begin.

Aldo Aiello (123), Istituto di Cibernetica C.N.R., I-80072 Arco Felice, Italy

A. Benveniste (315), Institut de Recherche en Informatique et Systemes Aleatoires (IRISA), Campus Universitaire de Beaulieu, 35042 Rennes Cedex, France

Ernesto Burattini (123), Istituto di Cibernetica C.N.R., I-80072 Arco Felice, Italy

Chu Kwong Chak (57), Department of Electrical and Electronic Engineering, University of Melbourne, Parkville, 3052 Victoria, Australia

Yie-Chien Chen (285), Department of Control Engineering, National Chiao-Tung University, Hsinchu, Taiwan

B. Delyon (315), Institut de Recherche en Informatique et Systemes Aleatoires (IRISA), Campus Universitaire de Beaulieu, 35042 Rennes Cedex, France

Gang Feng (57), Department of Systems and Control, School of Electrical Engineering, University of New South Wales, Sydney, New South Wales 2052, Australia

Graham P. Fletcher (175), Department of Computer Sciences, University of Glamorgan, Wales CF37 1DL, United Kingdom

P.-Y. Glorennec (315), Institut de Recherche en Informatique et Systemes Aleatoires (IRISA), Campus Universitaire de Beaulieu, 35042 Rennes Cedex, France

Chris J. Hinde (175), Department of Computer Sciences, University of Glamorgan, Wales CF37 1DL, United Kingdom

Hisao Ishibuchi (1), Department of Industrial Engineering, Osaka Prefecture University, Sakai, Osaka 593, Japan

A. Juditsky (315), Institut de Recherche en Informatique et Systemes Aleatoires (IRISA), Campus Universitaire de Beaulieu, 35042 Rennes Cedex, France

Sushmita Mitra (211), Machine Intelligence Unit, Indian Statistical Institute, Calcutta 700 035, India

Manabu Nii (1), Department of Industrial Engineering, Osaka Prefecture University, Sakai, Osaka 593, Japan

Sankar K. Pal (211), Machine Intelligence Unit, Indian Statistical Institute, Calcutta 700 035, India

Marimuthu Palaniswami (57), Department of Electrical and Electronic Engineering, University of Melbourne, Parkville, 3052 Victoria, Australia

Guglielmo Tamburrini (123), Istituto di Cibernetica C.N.R., I-80072 Arco Felice, Italy

Ching-Cheng Teng (285), Department of Control Engineering, National Chiao-Tung University, Hsinchu, Taiwan

Spyros G. Tzafestas (243), Department of Electrical and Computer Engineering, Intelligent Robotics and Automation Laboratory, National Technical University of Athens, Athens 157 73, Greece

Keigo Watanabe (243), Department of Mechanical Engineering, Faculty of Science and Engineering, Saga University, Saga 840, Japan

Q. Zhang (315), Institut de Recherche en Informatique et Systemes Aleatoires (IRISA), Campus Universitaire de Beaulieu, 35042 Rennes Cedex, France

Preface

Inspired by the structure of the human brain, artificial neural networks have been widely applied to fields such as pattern recognition, optimization, coding, control, etc., because of their ability to solve cumbersome or intractable problems by learning directly from data. An artificial neural network usually consists of a large number of simple processing units, i.e., neurons, via mutual interconnection. It learns to solve problems by adequately adjusting the strength of the interconnections according to input data. Moreover, the neural network adapts easily to new environments by learning, and can deal with information that is noisy, inconsistent, vague, or probabilistic. These features have motivated extensive research and developments in artificial neural networks. This volume is probably the first rather comprehensive treatment devoted to the broad areas of algorithms and architectures for the realization of neural network systems. Techniques and diverse methods in numerous areas of this broad subject are presented. In addition, various major neural network structures for achieving effective systems are presented and illustrated by examples in all cases. Numerous other techniques and subjects related to this broadly significant area are treated.

The remarkable breadth and depth of the advances in neural network systems with their many substantive applications, both realized and yet to be realized, make it quite evident that adequate treatment of this broad area requires a number of distinctly titled but well-integrated volumes. This is the sixth of seven volumes on the subject of neural network systems and it is entitled *Fuzzy Logic and Expert Systems Applications*. The entire set of seven volumes contains

Volume 1: *Algorithms and Architectures*
Volume 2: *Optimization Techniques*
Volume 3: *Implementation Techniques*
Volume 4: *Industrial and Manufacturing Systems*
Volume 5: *Image Processing and Pattern Recognition*
Volume 6: *Fuzzy Logic and Expert Systems Applications*
Volume 7: *Control and Dynamic Systems*

The first contribution to this volume is "Fuzzy Neural Networks Techniques and Their Applications," by Hisao Ishibuchi and Manabu Nii. Fuzzy logic and neural networks have been combined in various ways. In general, hybrid systems of fuzzy logic and neural networks are often referred to as fuzzy neural networks, which in turn can be classified into several categories. The following list is one example of such a classification of fuzzy neural networks:

1. Fuzzy rule-based systems with learning ability,
2. Fuzzy rule-based systems represented by network architectures,
3. Neural networks for fuzzy reasoning,
4. Fuzzified neural networks,
5. Other approaches.

The classification of a particular fuzzy neural network into one of these five categories is not always easy, and there may be different viewpoints for classifying neural networks. This contribution focuses on fuzzy classification and fuzzy modeling. Nonfuzzy neural networks and fuzzified neural networks are used for these tasks. In this contribution, fuzzy modeling means modeling with nonlinear fuzzy number valued functions. Included in this contribution is a description of how feedforward neural networks can be extended to handle the fuzziness of training data. The many implications of this are then treated sequentially and in detail. A rather comprehensive set of illustrative examples is included which clearly manifest the significant effectiveness of fuzzy neural network systems in a variety of applications.

The next contribution is "Implementation of Fuzzy Systems," by Chu Kwong Chak, Gang Feng, and Marimuthu Palaniswami. The expanding popularity of fuzzy systems appears to be related to its ability to deal with complex systems using a linguistic approach. Although many applications have appeared in systems science, especially in modeling and control, there is no systematic procedure for fuzzy system design. The conventional approach to design is to capture a set of linguistic fuzzy rules given by human experts. This empirical design approach encounters a number of problems, i.e., that the design of optimal fuzzy systems is very difficult because no systematic approach is available, that the performance of the fuzzy systems can be inconsistent because the fuzzy systems depend mainly on the intuitiveness of individual human expert, and that the resultant fuzzy systems lack adaptation capability. Training fuzzy systems by using a set of input–output data captured from the complex systems, via some learning algorithms, is known to generate or modify the linguistic fuzzy rules. A neural network is a suitable tool for achieving this purpose because of its capability for learning from data. This contribution presents an in-depth treatment of the neural network implementation of fuzzy systems for modeling and control. With the new space partitioning techniques and the new structure of fuzzy systems developed in this contribution, radial basis function neural networks and sigmoid function neural networks are successfully applied to implement higher order fuzzy systems that effectively treat the problem of rule explosion. Two new fuzzy neural networks along with learning algorithms, such as the Kalman filter algorithm and some hybrid learning algorithms, are presented in this contribution. These fuzzy neural networks can achieve self-organization and adaptation and hence improve the intelligence of fuzzy systems. Some simulation examples are shown to support the effectiveness of the fuzzy neural network approach. An array of illustrative examples clearly manifests the substantive effectiveness of fuzzy neural network system techniques.

The next contribution is "Neural Networks and Rule-Based Systems," by Aldo Aiello, Ernesto Burattini, and Guglielmo Tamburrini. This contribution presents methods of implementing a wide variety of effective rule-based reasoning processes by means of networks formed by nonlinear thresholded neural units. In particular, the following networks are examined:

1. Networks that represent knowledge bases formed by propositional production rules and that perform forward chaining on them.
2. A network that monitors the elaboration of the forward chaining system and learns new production rules by an elementary chunking process.
3. Networks that perform qualitative forms of uncertain reasoning, such as hypothetical reasoning in two-level casual networks and the application of preconditions in default reasoning.
4. Networks that simulate elementary forms of quantitative uncertain reasoning.

The utilization of these techniques is exemplified by the overall structure and implementation features of a purely neural, rule-based expert system for a diagnostic task and, as a result, their substantive effectiveness is clearly manifested.

The next contribution is "Construction of Rule-Based Intelligent Systems," by Graham P. Fletcher and Chris J. Hinde. It is relatively straightforward to transform a propositional rule-based system into a neural network. However, the transformation in the other direction has proved a much harder problem to solve. This contribution explains techniques that allow neurons, and thus networks, to be expressed as a set of rules. These rules can then be used within a rule-based system, turning the neural network into an important tool in the construction of rule-based intelligent systems. The rules that have been extracted, as well as forming a rule-based implementation of the network, have further important uses. They also represent information about the internal structures that build up the hypothesis and, as such, can form the basis of a verification system. This contribution also considers how the rules can be used for this purpose. Various illustrative examples are included.

The next contribution is "Expert Systems in Soft Computing Paradigm," by Sankar K. Pal and Sushmita Mitra. This contribution is a rather comprehensive treatment of the soft computing paradigm, which is the integration of different computing paradigms such as fuzzy set theory, neural networks, genetic algorithms, and rough set theory. The intent of the soft computing paradigm is to generate more efficient hybrid systems. The purpose of soft computing is to provide flexible information processing capability for handling real life ambiguous situations by exploiting the tolerance for imprecision, uncertainty, approximate reasoning, and partial truth to achieve tractability, robustness, and low cost. The guiding principle is to devise methods of computation which lead to an acceptable solution at low cost by seeking an approximate solution to an imprecisely/precisely formulated problem. Several illustrative examples are included.

The next contribution is "Mean-Value-Based Functional Reasoning Techniques in the Development of Fuzzy-Neural Network Control Systems," by Keigo Watanabe and Spyros G. Tzafestas. This contribution reviews first conventional functional reasoning, simplified reasoning, and mean-value-based functional reasoning methods. Design techniques which utilize these fuzzy reasoning methods based on variable structure systems control theory are presented. Techniques for the design of three fuzzy Gaussian neural networks that utilize, respectively, conventional functional reasoning, simplified reasoning, and mean-value-based functional reasoning methods are presented and compared with each other, particularly with regard to the number of learning parameters to be learned in the result. The effectiveness of the mean-value-based functional reasoning technique is made manifest by an illustrative example in the design and simulation of a nonlearning fuzzy controller for a satellite attitude control system. As another illustrative example, a fuzzy neural network controller based on mean-value-based functional reasoning techniques is developed and utilized for the tracking control problem of a mobile robot with two independent driving wheels.

The next contribution is "Fuzzy Neural Network Systems in Model Reference Control Systems," by Yie-Chien and Ching-Cheng Teng. This contribution presents techniques for model reference control systems which utilize fuzzy neural networks. The techniques presented for system model reference control belong to the class of systems referred to as indirect adaptive control. Techniques for the utilization of fuzzy neural network identifiers (FNNI) to identify a controlled plant are presented. The FNNI approximate the system and provide the sensitivity of the controlled plant for the fuzzy neural network controller (FNNC). The techniques presented can be referred to as a genuine adaptation system that can learn to control complex systems and adapt to a wide variation in system plant parameters. Unlike most other techniques presented for adaptive learning neural controllers, the FNNC techniques presented in this contribution are based not only on the theory of neural network systems, but also on the theory of fuzzy logic techniques. The substantive effectiveness of the techniques presented in this contribution are shown by an illustrative example.

The final contribution to this volume is "Wavelets in Identification," by A. Juditsky, Q. Zhang, B. Deylon, P-Y. Glorennec, and A. Benveniste. This contribution presents a rather spendid self-contained treatment of non-parametric nonlinear system identification techniques utilizing both neural network system methods and fuzzy system theory modeling techniques. Wavelet techniques are introduced and a self-contained presentation of wavelet principles is included. The advantages and limitations of the potentially greatly effective wavelet techniques are presented. Illustrative examples are presented throughout this contribution.

This volume on fuzzy logic and expert systems applications clearly reveals the effectiveness and essential significance of the techniques available and, with further development, the essential role they will play in the future. The authors are all to be highly commended for their splendid contributions to this volume which will provide a significant and unique reference for students, research workers, practitioners, computer scientists, and others on the international scene for years to come.

Cornelius T. Leondes

Fuzzy Neural Networks Techniques and Their Applications

Hisao Ishibuchi
Department of Industrial Engineering
Osaka Prefecture University
Sakai, Osaka 593, Japan

Manabu Nii
Department of Industrial Engineering
Osaka Prefecture University
Sakai, Osaka 593, Japan

I. INTRODUCTION

Fuzzy logic and neural networks have been combined in a variety of ways. In general, hybrid systems of fuzzy logic and neural networks are often referred to as fuzzy neural networks [1]. Fuzzy neural networks can be classified into several categories. The following is an example of one such classification of fuzzy neural networks [2]:

1. Fuzzy rule-based systems with learning ability.
2. Fuzzy rule-based systems represented by network architectures.
3. Neural networks for fuzzy reasoning.
4. Fuzzified neural networks.
5. Other approaches.

The classification of a particular fuzzy neural network into one of these five categories is not always easy, and there may be different viewpoints for classifying fuzzy neural networks.

Fuzzy neural networks in the first category are basically fuzzy rule-based systems where fuzzy if-then rules are adjusted by iterative learning algorithms similar to neural network learning (e.g., the back-propagation algorithm [3, 4]). Adaptive fuzzy systems in [5–8] can be classified in this category. In general, fuzzy if-then rules with n inputs and a single output can be written as follows:

If x_1 is A_{j1} and x_2 is A_{j2} and ... and x_n is A_{jn} then y is B_j,

$$j = 1, 2, \ldots, N, \quad (1)$$

where $\mathbf{x} = (x_1, x_2, \ldots, x_n)$ is an n-dimensional input vector, y is an output variable, and $A_{j1}, \ldots, A_{jn}$ and B_j are fuzzy sets. In the first category of fuzzy neural networks, membership functions of the antecedent fuzzy sets (i.e., $A_{j1}, \ldots, A_{jn}$) and the consequent fuzzy set (i.e., B_j) of each fuzzy if-then rule are adjusted in a similar manner as in neural networks.

Usually linguistic labels such as *small* and *large* are associated with the fuzzy sets in the fuzzy if-then rules. An example of a fuzzy if-then rule with two inputs and a single output is

If x_1 is *small* and x_2 is *large* then y is *small*. $\qquad\qquad$ (2)

In a simplified version [5, 6] of fuzzy if-then rules, a real number is used in the consequent part instead of the fuzzy number B_j in (1). That is, simplified fuzzy if-then rules can be written as follows:

If x_1 is A_{j1} and x_2 is A_{j2} and ... and x_n is A_{jn} then y is b_j,

$$j = 1, 2, \ldots, N, \quad (3)$$

where b_j is a real number. Recently these fuzzy if-then rules have frequently been used because of the simplicity of the fuzzy reasoning and the learning.

In the second category of fuzzy neural networks, fuzzy rule-based systems are represented by network architectures. Thus learning algorithms for neural networks such as the back-propagation algorithm [3, 4] can be easily applied to the learning of fuzzy rule-based systems. Various network architectures [9–24] have been proposed for representing fuzzy rule-based systems. In those architectures, usually the membership function of each antecedent fuzzy set (i.e., $A_{j1}, A_{j2}, \ldots, A_{jn}$) corresponds to the activation function of each unit in the neural networks. When the antecedent part (i.e., the condition) of each fuzzy if-then rule is defined by a fuzzy set $\mathbf{A}_j$ on the n-dimensional input space rather than n fuzzy sets $A_{j1}, A_{j2}, \ldots, A_{jn}$ on the n axes in (1), fuzzy if-then rules can be written as follows:

If $\mathbf{x}$ is $\mathbf{A}_j$ then y is B_j, $\qquad j = 1, 2, \ldots, N,$ $\qquad\qquad$ (4)

for the case of the fuzzy consequent, and

If $\mathbf{x}$ is $\mathbf{A}_j$ then y is b_j, $\qquad j = 1, 2, \ldots, N,$ $\qquad\qquad$ (5)

for the case of the real-number consequent. An example of the membership function of the antecedent fuzzy set $\mathbf{A}_j$ is shown in the two-dimensional input space in Fig. 1 where contour lines of the membership function of A_j are depicted. As we can intuitively realize from Fig. 1, the membership function of the antecedent fuzzy set A_j corresponds to a generalized radial basis function. Thus fuzzy rule-based systems with fuzzy if-then rules in (4) or (5) can be viewed as a kind of radial basis function network [25, 26].

Fuzzy neural networks in the third category are neural networks for fuzzy reasoning. Standard feedforward neural networks with special preprocessing procedures are used for fuzzy reasoning in this category. For example, in Keller and Tahani [27, 28], antecedent fuzzy sets and consequent fuzzy sets are represented by membership values at some reference points, and those membership values are used as inputs and targets for the training of feedforward neural networks. In Fig. 2, we illustrate the learning of a three-layer feedforward neural network by the following fuzzy if-then rule:

$$\text{If } x \text{ is } small \text{ then } y \text{ is } large, \tag{6}$$

where each linguistic label is denoted by membership values at 11 reference points. For example, the linguistic label *small* is denoted by the 11-dimensional real vector $(1, 0.6, 0.2, 0, 0, 0, 0, 0, 0, 0, 0)$. Because both the inputs and the targets in Fig. 2 are real-number vectors, the neural network can be trained by the standard back-propagation algorithm [3, 4] with no modification. Neural-network-based fuzzy reasoning methods in [27–33] may be classified in the third category.

The fourth category of fuzzy neural networks consists of fuzzified neural networks. Standard feedforward neural networks can be fuzzified by using fuzzy

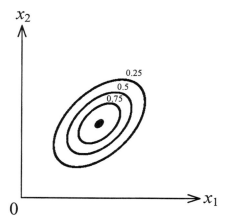

Figure 1 Antecedent fuzzy set on a two-dimensional input space.

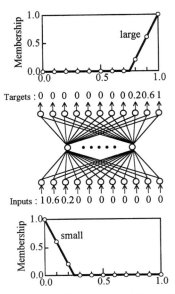

Figure 2   Inputs and targets for the learning from the fuzzy if-then rule: If *x* is *small* then *y* is *large*.

numbers as inputs, targets, and connection weights. This category is clearly distinguished from the other categories because fuzzified neural networks are defined by fuzzy-number arithmetic [34] based on the extension principle of Zadeh [35]. That is, the outputs from fuzzified neural networks are defined by fuzzy arithmetic, whereas other fuzzy neural networks use real-number arithmetic for calculating their outputs. Some examples of fuzzy-number arithmetic are shown in Figs. 3 and 4. The sum and the product of two triangular fuzzy numbers are shown in Fig. 3, and the nonlinear mapping of a fuzzy number by a sigmoidal activation function is shown in Fig. 4. Architectures of fuzzified neural networks and their learning algorithms have been proposed in [36–43]. In Fig. 5, we illustrate the learning of a fuzzified neural network from the fuzzy if-then rule "If *x* is *small* then *y* is *large*." Both the input and the target in Fig. 5 are fuzzy numbers with linguistic labels.

The fifth category of fuzzy neural networks (i.e., other approaches) includes various studies on the combination of fuzzy logic and neural networks. This category includes neural fuzzy point processes by Rocha [44], fuzzy perceptron by Keller and Hunt [45], fuzzy ART (adaptive resonance system) and fuzzy ARTMAP by Carpenter *et al.* [46, 47], max-min neural networks by Pedrycz [48], fuzzy min-max neural networks by Simpson [49, 50], OR/AND neuron by Hirota and Pedrycz [51], and Yamakawa's fuzzy neuron [52].

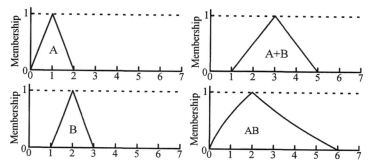

Figure 3   Sum and product of two triangular fuzzy numbers.

In this chapter, we focus our attention on fuzzy classification and fuzzy modeling. Nonfuzzy neural networks and fuzzified neural networks are used for these tasks. In this chapter, *fuzzy modeling* means modeling with nonlinear fuzzy-number-valued functions. This chapter is organized as follows. In Section II, we explain fuzzy classification and fuzzy modeling by nonfuzzy neural networks. In fuzzy classification, an input pattern is not always assigned to a single class. In fuzzy modeling, two nonfuzzy neural networks are trained for realizing an interval-valued function from which a fuzzy-number-valued function is derived. In Section III, interval-arithmetic-based neural networks are explained as the simplest version of fuzzified neural networks. We describe how interval input vectors can be handled in neural networks. Intervals are used for denoting uncertain or missing inputs to neural networks. We also describe the extension of connection weights to intervals, and derive a learning algorithm of the interval connection weights in Section III. Section IV is related to the fuzzification of neural net-

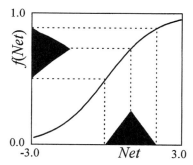

Figure 4   Nonlinear mapping of a triangular fuzzy number by a sigmoidal activation function.

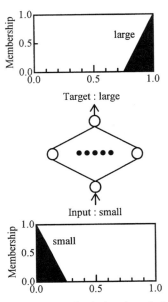

Figure 5   Fuzzy input and fuzzy target for the learning of a fuzzified neural network.

works. Inputs, targets, and connection weights are extended to fuzzy numbers. Fuzzified neural networks are used for the classification of fuzzy inputs, the approximate realization of fuzzy-number-valued functions, the learning of neural networks from fuzzy if-then rules, and the extraction of fuzzy if-then rules from neural networks. Section V concludes this chapter.

II. FUZZY CLASSIFICATION AND FUZZY MODELING BY NONFUZZY NEURAL NETWORKS

A. FUZZY CLASSIFICATION AND FUZZY MODELING

Let us consider a two-class classification problem on the two-dimensional unit cube $[0, 1]^2$ in Fig. 6a where training patterns from Class 1 and Class 2 are denoted by closed circles and open circles, respectively. As we can see from Fig. 6a, the given training patterns are linearly separable. Thus the perceptron learning algorithm [53] can be applied to this problem. On the other hand, training patterns in Fig. 6b are not linearly separable. In this case, we can use a multilayer feedforward neural network. The classification boundary in Fig. 6b was obtained by the

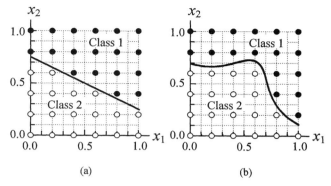

Figure 6 Examples of classification problems: (a) linearly separable classification problem; (b) linearly nonseparable classification problem.

learning of a three-layer feedforward neural network with two input units, three hidden units, and a single output unit.

Theoretically, multilayer feedforward neural networks can generate any classification boundaries because they are universal approximators of nonlinear functions [54–57]. Here let us consider a pattern classification problem in Fig. 7a. Even for such a complicated classification problem, there are neural networks that can correctly classify all the training patterns. In practice, it is not always an appropriate strategy to try to find a neural network with a 100% classification rate for the training patterns because a high classification rate for the training patterns

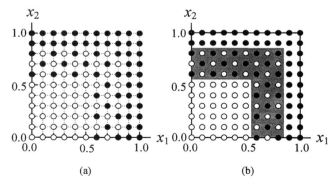

Figure 7 Example of a complicated classification problem with a overlapping region: (a) classification problem; (b) fuzzy boundary.

sometimes leads to poor performance for new patterns (i.e., for test patterns). This observation is known as the overfitting to the training patterns.

In this section, we show how the concept of fuzzy classification is applied to complicated classification problems with overlapping regions such as Fig. 7. Fuzzy classification is also referred to as approximate classification [58, 59]. In the fuzzy classification, we assume that classification boundaries between different classes are not clear but fuzzy. We show an example of the fuzzy classification in Fig. 7b where the dotted area corresponds to the fuzzy boundary between Class 1 and Class 2. The classification of new patterns in the fuzzy boundary is rejected. We can see that the fuzzy boundary in Fig. 7b is intuitively acceptable for the pattern classification problem in Fig. 7a. The fuzzy boundary can be extracted from two neural networks trained by leaning algorithms in Ishibuchi *et al.* [60] based on the concept of possibility and necessity [61]. Those learning algorithms search for the possibility region and the necessity region of each class. Fuzzy classification has also been addressed by Karayiannis and Purushothaman [62–65]. They tackled classification problems similar to Fig. 7, and proposed neural-network-based fuzzy classification methods. The basic idea of their fuzzy classification is similar to ours, but their neural network architectures and learning algorithms are different from those presented in this chapter. Fuzzy classification was also discussed by Archer and Wang in a different manner [66].

The concept of fuzzy data analysis can be introduced to another major application area of neural networks: modeling of nonlinear systems. In general, the input–output relation of an unknown nonlinear system is approximately realized by the learning of a neural network. Let us assume that we have the input–output data in Fig. 8a for an unknown nonlinear system. In this case, we can model the unknown nonlinear system by the learning of a neural network. The nonlinear curve in Fig. 8a is depicted using the output of the neural network trained by the

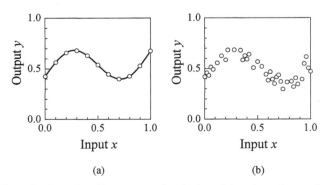

Figure 8   Examples of input–output data for the training of neural networks.

given input–output data. From Fig. 8a, we can see that the input–output relation is well represented by the trained neural network. Now, let us consider the input–output data in Fig. 8b. It does not seem to be an appropriate attempt to represent the input–output data in Fig. 8b by a single nonlinear curve.

For representing such input–output data in an intuitively acceptable way, we use an interval-valued function that approximately covers all the given input–output data. In this section, we describe an identification method [67, 68] of the interval-valued function by two nonfuzzy neural networks in addition to the fuzzy classification. The two neural networks correspond to the lower limit and the upper limit of the interval-valued function, respectively. In this section, we also describe how a fuzzy-number-valued function can be derived from the interval-valued function realized by the two neural networks [68]. Nonlinear modeling by interval-valued functions using neural networks can be viewed as an extension of fuzzy linear regression [69–71] where linear interval models and linear fuzzy models are used for regression analysis (see also [72, 73]).

B. LEARNING FOR FUZZY CLASSIFICATION

In this subsection, we explain the fuzzy classification method in [58–60] based on the concept of possibility and necessity. For simplicity, we start with two-class classification problems. Then we extend the fuzzy classification for two-class problems to the case of multiclass problems.

Let us assume that we have m training patterns $\mathbf{x}_p = (x_{p1}, x_{p2}, \ldots, x_{pn})$, $p = 1, 2, \ldots, m$, from two classes (i.e., Class 1 and Class 2) in an n-dimensional pattern space Ω. In this case, the nonfuzzy pattern classification is to divide the pattern space Ω into two disjoint decision areas Ω_1 and Ω_2. These decision areas satisfy the following relations:

$$\Omega_1 \cup \Omega_2 = \Omega, \tag{7}$$
$$\Omega_1 \cap \Omega_2 = \emptyset, \tag{8}$$

where $\emptyset$ denotes an empty set.

On the other hand, we assume that the class boundary is fuzzy in the fuzzy classification. Thus the pattern space Ω is divided into three disjoint areas for the two-class classification problem:

$$\Omega_1 \cup \Omega_2 \cup \Omega_{FB} = \Omega, \tag{9}$$
$$\Omega_1 \cap \Omega_2 = \emptyset, \qquad \Omega_1 \cap \Omega_{FB} = \emptyset, \qquad \Omega_2 \cap \Omega_{FB} = \emptyset, \tag{10}$$

where Ω_{FB} is the fuzzy boundary between the two classes. The classification of new patterns in the fuzzy boundary is rejected. Figure 7b is an example of the fuzzy boundary.

We use a feedforward neural network with n input units and a single output unit for the two-class pattern classification problem in the n-dimensional pattern space Ω. In the learning of the neural network, we define the target output t_p for each training pattern $\mathbf{x}_p$ as follows:

$$t_p = \begin{cases} 1, & \text{for } \mathbf{x}_p \in \text{Class 1}, \\ 0, & \text{for } \mathbf{x}_p \in \text{Class 2}. \end{cases} \tag{11}$$

The learning of the neural network is to minimize the following cost function:

$$e_p = (t_p - o_p)^2/2, \tag{12}$$

where o_p is the output from the neural network.

Using the output from the trained neural network, we can define the decision area of each class as follows:

$$\Omega_1 = \{\mathbf{x} \mid o(\mathbf{x}) \geq 0.5, \ \mathbf{x} \in \Omega\}, \tag{13}$$
$$\Omega_2 = \{\mathbf{x} \mid o(\mathbf{x}) < 0.5, \ \mathbf{x} \in \Omega\}, \tag{14}$$

where $o(\mathbf{x})$ is the output from the trained neural network for the input vector $\mathbf{x}$. In this manner, we can use the neural network for the two-class classification problem.

The fuzzy classification can be done by slightly modifying the aforementioned procedure. In our fuzzy classification, the cost function is modified for determining the possibility area and the necessity area of each class. For determining the possibility area of Class 1, we use the following cost function:

$$e_p = \begin{cases} (t_p - o_p)^2/2, & \text{if } \mathbf{x}_p \in \text{Class 1}, \\ \omega(u) \cdot (t_p - o_p)^2/2, & \text{if } \mathbf{x}_p \in \text{Class 2}, \end{cases} \tag{15}$$

where u is the number of the iterations of the learning algorithm (i.e., epochs), and $\omega(u)$ is a monotonically decreasing function such that $0 < \omega(u) \leq 1$ and $\omega(u) \to 0$ for $u \to \infty$. For example, we can use the following decreasing function:

$$\omega(u) = 1/\{1 + (u/1000)^2\}. \tag{16}$$

From the definition of the cost function in (15), we can see that the importance of Class 2 patterns is monotonically decreased by the decreasing function $\omega(u)$ during the learning of the neural network. This means that the relative importance of Class 1 patterns is monotonically increased. Thus we can expect that the following relation will hold for Class 1 patterns after the learning of the neural network:

$$o(\mathbf{x}_p) \cong 1 \qquad \text{for } \mathbf{x}_p \in \text{Class 1}. \tag{17}$$

Let us consider a one-dimensional classification problem in Fig. 9 where training patterns from Class 1 and Class 2 are shown by closed circles and open circles,

Figure 9 One-dimensional pattern classification problem.

respectively. For this problem, we used the modified back-propagation algorithm derived from the cost function e_p in (15) with the decreasing function $\omega(u)$ in (16). A three-layer feedforward neural network with five hidden units was trained by iterating the learning algorithm 10,000 times (i.e., 10,000 epochs). In Fig. 10, we show the shape of the output from the neural network. From Fig. 10, we can see that the output from the neural network approached the training patterns from Class 1 (i.e., closed circles in Fig. 10) during the learning. This is because the relative importance of Class 1 patterns was monotonically increased by the decreasing function $\omega(u)$ attached to Class 2 patterns.

From Fig. 10b, we can see that the output from the neural network can be viewed as the possibility grade of Class 1. For example, the output $o(x)$ in Fig. 10b is nearly equal to 1 (full possibility) for the input value $x = 0.35$, whereas the training pattern on $x = 0.35$ belongs to Class 2. We can define the possibility area using the output from the trained neural network. For example,

$$\Omega_1^{\text{Pos}} = \{\mathbf{x} \mid o^{\text{Pos}}(\mathbf{x}) \geq 0.5, \ \mathbf{x} \in \Omega\}, \tag{18}$$

where Ω_1^{Pos} is the possibility area of Class 1 and $o^{\text{Pos}}(\mathbf{x})$ is the output from the neural network trained for the possibility analysis. Input patterns in this possibility area are classified as "having the possibility to belong to Class 1." In Fig. 10b, the possibility area of Class 1 is the interval [0.268, 0.734].

As we can see from Figs. 9 and 10, input patterns around $x = 0.5$ may certainly be classified as Class 1 because there are no Class 2 patterns around $x = 0.5$. For extracting such a certain (i.e., nonfuzzy) decision area, we use a different learning algorithm based on the concept of necessity.

For the necessity analysis of Class 1, we use the following cost function in the learning of the neural network:

$$e_p = \begin{cases} \omega(u) \cdot (t_p - o_p)^2/2, & \text{if } \mathbf{x}_p \in \text{Class 1}, \\ (t_p - o_p)^2/2, & \text{if } \mathbf{x}_p \in \text{Class 2}, \end{cases} \tag{19}$$

where u and $\omega(u)$ are the same as in (15). From (19), we can see that the importance of Class 1 patterns is monotonically decreased by the decreasing function $\omega(u)$ during the learning of the neural network.

For the classification problem in Fig. 9, we used the modified back-propagation algorithm derived from the cost function e_p in (19) with the decreasing function $\omega(u)$ in (16). A three-layer feedforward neural network with five hidden units was trained by iterating the learning algorithm 10,000 times (i.e., 10,000 epochs).

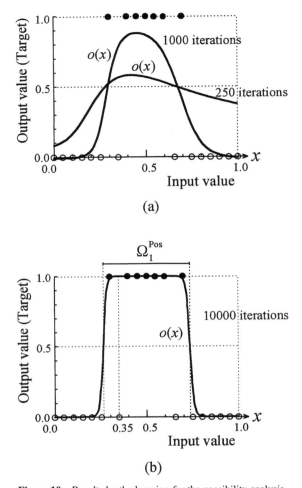

(a)

(b)

Figure 10   Results by the learning for the possibility analysis.

In Fig. 11, we show the shape of the output from the neural network. From Fig. 11, we can see that the output from the neural network approached the training patterns from Class 2 (i.e., open circles in Fig. 11). This is because the relative importance of Class 2 patterns is monotonically increased by the decreasing function $\omega(u)$ attached to Class 1 patterns.

From Fig. 11b, we can see that the output from the neural network can be viewed as the necessity grade of Class 1. For example, the output $o(x)$ is nearly

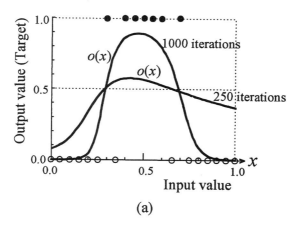

(a)

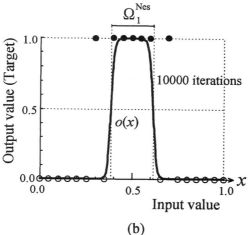

(b)

Figure 11   Results by the learning for the necessity analysis.

equal to 1 (full necessity) for input values around $x = 0.5$. This coincides with our intuition. We can define the necessity area using the output from the trained neural network in the same manner as the possibility area in (18):

$$\Omega_1^{\text{Nes}} = \{\mathbf{x} \mid o^{\text{Nes}}(\mathbf{x}) \geq 0.5, \ \mathbf{x} \in \Omega\}, \tag{20}$$

where Ω_1^{Nes} is the necessity area of Class 1 and $o^{\text{Nes}}(\mathbf{x})$ is the output from the neural network trained for the necessity analysis. Input patterns in this necessity area are classified as "having the necessity to belong to Class 1." In Fig. 11b, the necessity area of Class 1 is the interval $[0.386, 0.607]$. All the input patterns in this interval are certainly classified as Class 1.

The fuzzy boundary is the area that is included in the possibility area but excluded from the necessity area. In the fuzzy boundary, the outputs from the two neural networks trained by the possibility analysis and the necessity analysis are nearly equal to 1 and 0, respectively (see Figs. 10 and 11). Thus the fuzzy boundary can be defined as follows:

$$\Omega_{\text{FB}} = \{\mathbf{x} \mid 0.25 < \mu(\mathbf{x}) < 0.75, \ \mathbf{x} \in \Omega\}, \tag{21}$$

where $\mu(\mathbf{x})$ is a kind of membership grade of $\mathbf{x}$ to Class 1, and defined as follows:

$$\mu(\mathbf{x}) = \frac{o^{\text{Pos}}(\mathbf{x}) + o^{\text{Nes}}(\mathbf{x})}{2}, \tag{22}$$

where $o^{\text{Pos}}(\mathbf{x})$ and $o^{\text{Nes}}(\mathbf{x})$ are the outputs from the neural networks trained for the possibility analysis and the necessity analysis, respectively. The decision area of each class is defined as follows:

$$\Omega_1 = \{\mathbf{x} \mid \mu(\mathbf{x}) \geq 0.75, \ \mathbf{x} \in \Omega\}, \tag{23}$$

$$\Omega_2 = \{\mathbf{x} \mid \mu(\mathbf{x}) \leq 0.25, \ \mathbf{x} \in \Omega\}. \tag{24}$$

In Fig. 12, we show the shape of $\mu(\mathbf{x})$ that was obtained from the outputs of the two neural networks in Figs. 10b and 11b. The fuzzy boundary is also shown in Fig. 12. From Fig. 12, we can see that the fuzzy classification coincides with our intuition.

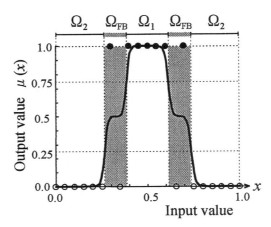

Figure 12 Fuzzy boundary and decision areas obtained by the fuzzy classification method.

For comparison, we show some results by the standard back-propagation algorithm based on the squared error in (12). Figure 13a and b was obtained after 50,000 iterations of the back-propagation algorithm for three-layer feedforward neural networks with five hidden units and ten hidden units, respectively.

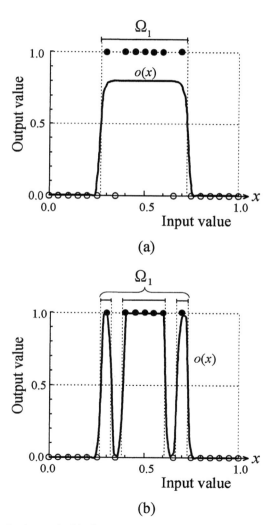

(a)

(b)

Figure 13 Results by the standard back-propagation algorithm: (a) output from the trained neural network with five hidden units; (b) output from the trained neural network with ten hidden units.

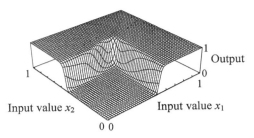

Figure 14   Shape of the output from the neural network trained for the possibility analysis.

In Fig. 13a, the learning seems to be incomplete. On the contrary, the learning in Fig. 13b seems to be the overfitting to the training data. We can see that the result of the fuzzy classification in Fig. 12 coincides very well with our intuition if compared with the results of the standard back-propagation algorithm in Fig. 13.

We also applied our fuzzy classification method to the two-dimensional pattern classification problem in Fig. 7. In Fig. 14, we show the shape of the output from the neural network with five hidden units trained for the possibility analysis. From Fig. 7, we can see that the output from the neural network in Fig. 14 represents the grade of possibility of Class 1 very well. On the other hand, in Fig. 15, we show the grade of necessity of Class 1 obtained for the necessity analysis. We can see from Fig. 7 that Fig. 15 represents the necessity grade of Class 1 very well. The function $\mu(\mathbf{x})$, which is shown in Fig. 16, was obtained by the two outputs in Figs. 14 and 15. In Fig. 17, we show the fuzzy boundary obtained from $\mu(\mathbf{x})$ in Fig. 16. We can see that an intuitively acceptable fuzzy boundary was obtained by the proposed fuzzy classification method in Fig. 17.

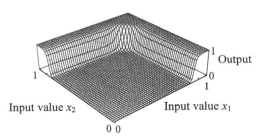

Figure 15   Shape of the output from the neural network trained for the necessity analysis.

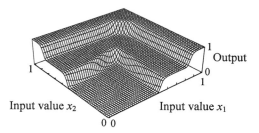

Figure 16 Shape of the function $\mu(\mathbf{x})$.

The fuzzy classification method for two-class problems can be extended to the case of multiclass classification problems. For a c-class classification problem, we divide the pattern space Ω into the following $(c + 1)$ areas:

$$\Omega_1 \cup \Omega_2 \cup \cdots \cup \Omega_c \cup \Omega_{\mathrm{FB}} = \Omega, \tag{25}$$

$$\Omega_h \cap \Omega_k = \emptyset \qquad \text{for } \forall h, k, \ h \neq k. \tag{26}$$

Let us assume that we have m training patterns $\mathbf{x}_p = (x_{p1}, x_{p2}, \dots, x_{pn})$, $p = 1, 2, \dots, m$, from c classes. For this c-class classification problem with the n-dimensional pattern space Ω, we use a feedforward neural network with n input units and c output units. The target vector $\mathbf{t}_p = (t_{p1}, t_{p2}, \dots, t_{pc})$ for the input pattern $\mathbf{x}_p$ is defined as follows:

$$t_{pk} = \begin{cases} 1, & \text{if } \mathbf{x}_p \in \text{Class } k, \\ 0, & \text{otherwise,} \end{cases} \tag{27}$$

for $k = 1, 2, \dots, c$.

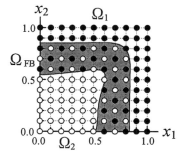

Figure 17 Fuzzy boundary and decision areas obtained by the fuzzy classification method.

$$\xrightarrow{\hspace{6cm}} \mathcal{X}$$

0.0   0.5   1.0

Figure 18   Three-class classification problem on the one-dimensional pattern space [0, 1].

For the possibility analysis, we define the cost function for the input pattern $\mathbf{x}_p$ by the target vector $\mathbf{t}_p = (t_{p1}, t_{p2}, \ldots, t_{pc})$ and the output vector $\mathbf{o}_p = (o_{p1}, o_{p2}, \ldots, o_{pc})$ from the neural network as follows:

$$e_p = \sum_{k=1}^{c} e_{pk}, \tag{28}$$

where e_{pk} is the cost function for the kth output unit, which is defined as

$$e_{pk} = \begin{cases} (t_{pk} - o_{pk})^2/2, & \text{if } \mathbf{x}_p \in \text{Class } k, \\ \omega(u) \cdot (t_{pk} - o_{pk})^2/2, & \text{otherwise.} \end{cases} \tag{29}$$

From the comparison between (15) and (29), we can see that the cost function e_{pk} for the kth output unit in (29) is for the possibility analysis of Class k. Let us consider a three-class classification problem on the one-dimensional pattern space [0, 1] in Fig. 18 where closed circles, open circles, and squares denote the training patterns from Class 1, Class 2, and Class 3, respectively. We applied the modified back-propagation algorithm derived from the cost function in (28) and (29) to this three-class pattern classification problem. We also used the decreasing function $\omega(u)$ in (16). The outputs from the trained neural network with a single input unit, five hidden units, and three output units are shown in Fig. 19. From this

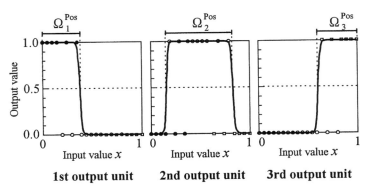

Figure 19   Results of the possibility analysis.

figure, we can see that the output from each output unit represents the possibility grade of the corresponding class very well.

For the necessity analysis, we modify the cost function e_{pk} for the kth output unit as follows:

$$e_{pk} = \begin{cases} \omega(u) \cdot (t_{pk} - o_{pk})^2/2, & \text{if } \mathbf{x}_p \in \text{Class } k, \\ (t_{pk} - o_{pk})^2/2, & \text{otherwise.} \end{cases} \qquad (30)$$

As we can see from the comparison between (19) and (30), the cost function e_{pk} in (30) is for the necessity analysis of Class k. In Fig. 20, we show the results of the learning based on this cost function. From Fig. 20, we can see that the output from each unit represents the necessity grade of the corresponding class very well.

If some region in the pattern space has high possibility grades for at least two classes, such a region can be viewed as a fuzzy boundary. On the contrary, if some region has a high possibility grade for only a single class, such a region can be viewed as the decision area of the corresponding class. To formulate this intuitive discussion, let us define $\mu_k(\mathbf{x})$ for each class as follows:

$$\mu_k(\mathbf{x}) = o_k^{\text{Pos}}(\mathbf{x}) - \max\{o_h^{\text{Pos}}(\mathbf{x}) \mid h = 1, 2, \ldots, c; \ h \neq k\}, \qquad (31)$$

where $o_k^{\text{Pos}}(\mathbf{x})$ is the output from the kth output unit of the neural network trained for the possibility analysis. When $\mu_k(\mathbf{x})$ is large, we can see that the input vector $\mathbf{x}$ has a high possibility grade only for Class k. Thus the input vector $\mathbf{x}$ is classified as Class k. From this idea, the decision area of each class is defined by $\mu_k(\mathbf{x})$ as follows:

$$\Omega_k = \{\mathbf{x} \mid \mu_k(\mathbf{x}) \geq 0.5, \ \mathbf{x} \in \Omega\}, \qquad k = 1, 2, \ldots, c. \qquad (32)$$

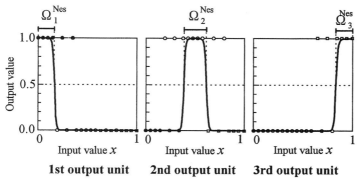

Figure 20 Results of the necessity analysis.

The fuzzy boundary is defined from (25) as follows:

$$\Omega_{FB} = \Omega - \{\Omega_1 \cup \Omega_2 \cup \cdots \cup \Omega_c\}. \tag{33}$$

The decision areas and the fuzzy boundary for the classification problem in Fig. 18 are shown in Fig. 21 together with the shape of $\mu_k(\mathbf{x})$. From this figure, we can see that intuitively acceptable results were obtained by our fuzzy classification method. Our fuzzy classification method is more than the classification with the reject option. Because our method is based on the possibility analysis, we can get the information about the possible classes for each of the rejected input patterns. For example, let us consider an input pattern at $x = 0.3$ in Fig. 21. As is shown in Fig. 21, the classification of this pattern is rejected. Thus we examine the output from the trained neural network for the possibility analysis (i.e., Fig. 19). Because the output corresponding to the input $x = 0.3$ is $(1.00, 1.00, 0.00)$, we can see that the possible classes of this input pattern are Class 1 and Class 2. We can also see that there is no possibility that the input pattern belongs to Class 3.

In order to examine the performance of our fuzzy classification method, we applied it to the well-known iris classification data (see, e.g., Fisher [74]). First we examined the performance for training data by applying our fuzzy classification method to the iris data using all the 150 samples as training patterns. The computer simulation was iterated 20 times using a three-layer feedforward neural network with four input units, two hidden units, and three output units. The average simulation results are summarized in Table I. From this table, we can see that no pattern was misclassified by the fuzzy classification method. The classification of 4.3 patterns was rejected on the average over the 20 trials. From Table I, we can also see that Class 1 patterns are clearly separable from the other patterns (i.e., Class 2 and Class 3 patterns).

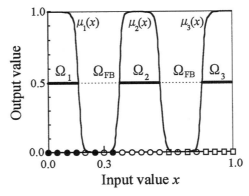

Figure 21   Fuzzy boundary and decision areas obtained by the fuzzy classification method.

Table I

Classification Results by the Fuzzy Classification for Training Data

Correct	Classification results			
class	Class 1	Class 2	Class 3	Boundary
1	50	0	0	0
2	0	47.7	0	2.3
3	0	0	48	2

Next we examined the performance of our fuzzy classification method for test data by the leaving-one-out procedure (see, e.g., Weiss and Kulikowski [75]). In the leaving-one-out procedure, a single pattern was used as a test pattern and the other 149 patterns were used for training. This procedure was iterated 150 times so that every pattern was used as a test pattern just once. In our computer simulation, this leaving-one-out procedure was iterated five times. The average results are summarized in Table II. From this table, we can see that 2.8 patterns (i.e., 1.87%) were misclassified on the average over the five iterations of the leaving-one-out procedure. This error rate is less than almost all the reported results in the literature (e.g., 3.3% by the back-propagation algorithm in [75]). This low error rate was achieved by rejecting the classification of 10.4 patterns (i.e., 6.93%) on the average.

C. LEARNING FOR FUZZY MODELING

Modeling of fuzzy systems has been addressed in the field of fuzzy regression [69–71] where the following fuzzy regression model is used for a fuzzy system with n nonfuzzy inputs and a single fuzzy output:

$$\widetilde{Y}(\mathbf{x}) = \widetilde{A}_0 + \widetilde{A}_1 x_1 + \cdots + \widetilde{A}_n x_n, \tag{34}$$

Table II

Classification Results by the Fuzzy Classification for Test Data

Correct	Classification results			
class	Class 1	Class 2	Class 3	Boundary
1	50	0	0	0
2	0	42.2	1.8	6
3	0	1	44.6	4.4

where $\mathbf{x} = (x_1, x_2, \ldots, x_n)$ is an n-dimensional real-number input vector, $\widetilde{Y}(\mathbf{x})$ is a fuzzy-number output from the fuzzy regression model, and $\widetilde{A}_0, \widetilde{A}_1, \ldots, \widetilde{A}_n$ are fuzzy-number coefficients. Thus the fuzzy regression model maps the nonfuzzy input vector $\mathbf{x} = (x_1, x_2, \ldots, x_n)$ to the fuzzy-number output $\widetilde{Y}(\mathbf{x})$.

The simplest version of the fuzzy regression model is the following interval regression model:

$$Y(\mathbf{x}) = A_0 + A_1 x_1 + \cdots + A_n x_n, \tag{35}$$

where $Y(\mathbf{x})$ is an interval output from the interval regression model, and $A_0, A_1, \ldots, A_n$ are interval coefficients.

Let us assume that m input–output pairs $(\mathbf{x}_p; y_p)$, $p = 1, 2, \ldots, m$, are given as training data where $\mathbf{x}_p = (x_{p1}, x_{p2}, \ldots, x_{pn})$ is an n-dimensional real-number input vector and y_p is a real-number output. The interval coefficients of the interval regression model in (35) are determined by solving the following linear programming problem:

$$\text{Minimize} \sum_{p=1}^{m} w\big(Y(\mathbf{x}_p)\big), \tag{36}$$

$$\text{subject to } y_p \in Y(\mathbf{x}_p), \qquad p = 1, 2, \ldots, m, \tag{37}$$

where $w(\cdot)$ denotes the width of the interval. The objective function (36) is to minimize the sum of the widths of the interval outputs $Y(\mathbf{x}_p)$'s. The constraint condition (37) means that the interval output $Y(\mathbf{x}_p)$ has to include the given output y_p. The given output y_p can be viewed as the target in the learning of neural networks. In Fig. 22, we show an example of the interval regression model with a single input and a single output. From Fig. 22, we can see that all the given input–output pairs are included in the interval regression model.

In this subsection, we extend the linear interval model in (35) to nonlinear models using neural networks. Now let us assume that the input–output pairs in

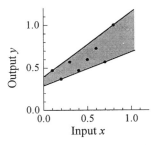

Figure 22   Interval regression model and given input–output data.

Fig. 8b are given. From this figure, we can see that no linear model is appropriate for the given data. In Fig. 23a, we show the output from a three-layer feedforward neural network with five hidden units trained by the standard back-propagation algorithm. An interval function determined by our method, which will be explained in this subsection, is shown in Fig. 23b. From the comparison between Fig. 23a and Fig. 23b, we can see that the interval function in Fig. 23b can represent the given data much better than the nonlinear curve in Fig. 23a.

In our method for determining a nonlinear interval function such as Fig. 23b, we use two feedforward neural networks. One is used for representing the upper bound of the nonlinear interval function, and the other is used for the lower bound.

Let $o_*(\mathbf{x})$ and $o^*(\mathbf{x})$ be the outputs from the two neural networks corresponding to the input vector $\mathbf{x}$. Using the two neural networks, a nonlinear interval function $Y(\mathbf{x})$ can be constructed as follows:

$$Y(\mathbf{x}) = \left[o_*(\mathbf{x}), o^*(\mathbf{x})\right], \tag{38}$$

where $o_*(\mathbf{x})$ and $o^*(\mathbf{x})$ are the lower bound and the upper bound of the interval function $Y(\mathbf{x})$, respectively. The linear programming problem in (36) and (37) is modified for the nonlinear interval function $Y(\mathbf{x})$ as follows:

$$\text{Minimize} \sum_{p=1}^{m} \left|o^*(\mathbf{x}_p) - o_*(\mathbf{x}_p)\right|, \tag{39}$$

$$\text{subject to } o_*(\mathbf{x}_p) \leq y_p \leq o^*(\mathbf{x}_p), \qquad p = 1, 2, \ldots, m, \tag{40}$$

$$o_*(\mathbf{x}) \leq o^*(\mathbf{x}) \qquad \text{for } \forall \mathbf{x}. \tag{41}$$

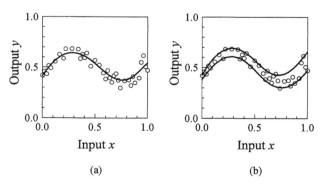

Figure 23 Comparison of two approaches: (a) modeling with a real-number-valued function; (b) modeling with an interval-valued function.

It is not easy to derive a learning algorithm for this nonlinear optimization problem. We show a simple approach for approximately solving this problem [67, 68]. For determining the lower bound $o_*(\mathbf{x})$ of the nonlinear interval function $Y(\mathbf{x})$, we define the following cost function for the input–output pair $(\mathbf{x}_p; y_p)$:

$$e_p = \begin{cases} (y_p - o_*(\mathbf{x}_p))^2/2, & \text{if } y_p < o_*(\mathbf{x}_p), \\ \omega(u) \cdot (y_p - o_*(\mathbf{x}_p))^2/2, & \text{if } o_*(\mathbf{x}_p) \leq y_p, \end{cases} \tag{42}$$

where u and $\omega(u)$ are the same as in the last subsection for the fuzzy classification. That is, u is the number of iterations of the learning algorithm, and $\omega(u)$ is a monotonically decreasing function such that $0 < \omega(u) \leq 1$ and $\omega(u) \to 0$ for $u \to \infty$. From (42), we can see that the squared error is discounted by $\omega(u)$ when the inequality constraint $o_*(\mathbf{x}) \leq y_p$ in (40) is satisfied by the output $o_*(\mathbf{x})$ from the neural network. Because $\omega(u)$ becomes almost zero after enough iterations, the cost function is negligible when the inequality constraint is satisfied. On the contrary, if the inequality constraint $o_*(\mathbf{x}) \leq y_p$ is not satisfied, the cost function is the same as in the standard back-propagation algorithm. In this case, the output $o_*(\mathbf{x})$ from the neural network approaches the given target y_p. In this manner, it is expected that the inequality constraint $o_*(\mathbf{x}) \leq y_p$ is approximately satisfied after enough iterations of the learning algorithm based on the cost function in (42).

Using the input–output data in Fig. 23, we trained a neural network with five hidden units by the modified back-propagation algorithm derived from the cost function in (42). As the decreasing function $\omega(u)$ in (42), we used the following function:

$$\omega(u) = 1/\{1 + (u/2000)^3\}. \tag{43}$$

In Fig. 24, we show the shape of this decreasing function and the shape of the output $o_*(\mathbf{x})$ from the neural network during the learning. From Fig. 24a, we can see that $\omega(u)$ is very small after 5000 iterations, whereas it is relatively large before 2000 iterations. From Fig. 24b, we can see that the output $o_*(\mathbf{x})$ approximately satisfies the inequality constraint $o_*(\mathbf{x}) \leq y_p$ for all the given input–output data after 10,000 iterations.

The upper bound $o^*(\mathbf{x})$ of the nonlinear interval function $Y(\mathbf{x})$ can be also determined by the learning of a neural network. The learning is performed in order to approximately satisfy the inequality constraint $y_p \leq o^*(\mathbf{x})$ in (40). The cost function to be minimized in the learning is defined for the input–output pair $(\mathbf{x}_p; y_p)$ as follows:

$$e_p = \begin{cases} (y_p - o^*(\mathbf{x}_p))^2/2, & \text{if } o^*(\mathbf{x}_p) < y_p, \\ \omega(u) \cdot (y_p - o^*(\mathbf{x}_p))^2/2, & \text{if } y_p \leq o^*(\mathbf{x}_p), \end{cases} \tag{44}$$

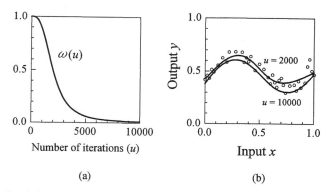

Figure 24 Simulation results by the learning for determining the lower limit of a nonlinear interval function: (a) decreasing function $\omega(u)$; (b) shape of the output from the neural network.

where the squared error is discounted by $\omega(u)$ when the inequality constraint $y_p \leq o^*(\mathbf{x})$ in (40) is satisfied by the output $o^*(\mathbf{x})$ from the neural network. In a similar manner as in Fig. 24b, we trained the neural network with five hidden units by the modified back-propagation algorithm derived from the cost function in (44). In Fig. 25, we show the shape of the output $o^*(\mathbf{x})$ from the neural network during the learning. From Fig. 25, we can see that the output $o^*(\mathbf{x})$ approximately satisfies the inequality constraint $y_p \leq o^*(\mathbf{x})$ after 10,000 iterations.

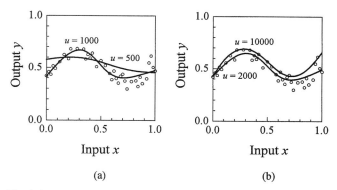

Figure 25 Simulation results by the learning for determining the upper limit of a nonlinear interval function.

In the fuzzy regression analysis [69–73], the fuzzy-number coefficients of the fuzzy regression model in (34) are determined by the following linear programming problem:

$$\text{Minimize } \sum_{p=1}^{m} w\big([\widetilde{Y}(\mathbf{x}_p)]_h\big), \tag{45}$$

$$\text{subject to } y_p \in [\widetilde{Y}(\mathbf{x}_p)]_h, \qquad p = 1, 2, \ldots, m, \tag{46}$$

where $[\cdot]_h$ is the h-level set of a fuzzy number (see Fig. 26). Because the h-level set of a fuzzy number is a closed interval, the linear programming problem in (45) and (46) for the fuzzy regression analysis is basically the same as the problem in (36) and (37) for the interval regression analysis.

Therefore the fuzzy regression model can be derived from the following relation (see Fig. 26):

$$[\widetilde{Y}(\mathbf{x})]_h = Y(\mathbf{x}). \tag{47}$$

For the case of nonlinear models, we can also derive nonlinear fuzzy functions $\widetilde{Y}(\mathbf{x})$'s using the preceding relation from a nonlinear interval function $Y(\mathbf{x})$. Two nonlinear fuzzy functions are shown in Fig. 27 for the case of triangular fuzzy outputs and trapezoidal fuzzy outputs. Figure 27a is depicted from the nonlinear interval function in Fig. 23b by the following relation:

$$[\widetilde{Y}(\mathbf{x})]_{h=0.0} = Y(\mathbf{x}), \tag{48}$$

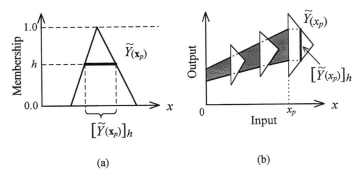

(a)   (b)

Figure 26   Illustration of the h-level set.

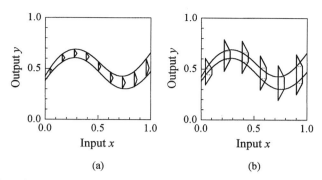

Figure 27   Two fuzzy functions derived from the interval function in Fig. 23b.

where $\widetilde{Y}(\mathbf{x})$ is a symmetric triangular fuzzy number (see Fig. 27a). On the other hand, Fig. 27b is depicted from the same nonlinear interval function by the relation:

$$\left[\widetilde{Y}(\mathbf{x})\right]_{h=1.0} = Y(\mathbf{x}), \qquad (49)$$

where $\widetilde{Y}(\mathbf{x})$ is a symmetric trapezoidal fuzzy number (see Fig. 27b).

III. INTERVAL-ARITHMETIC-BASED NEURAL NETWORKS

A. INTERVAL ARITHMETIC IN NEURAL NETWORKS

In real-world applications, training data may include uncertain inputs or missing inputs. Let us consider a two-class classification problem on the two-dimensional pattern space $[0, 1]^2$. We assume that we have a training pattern $(0.2, ?)$ from Class 1 where "?" denotes the missing input. One of the simplest approaches to the handling of this training pattern with the missing input is to ignore this pattern. Another approach is to substitute the most likely value for the missing input. Now let us assume that we have a new pattern $(?, 0.5)$ to be classified by a trained neural network. In this case, the first approach cannot be used because we have to classify this new pattern. The substitution in the second approach is not easy because the classification of this new pattern is unknown. In this section, we employ an interval-arithmetic-based approach to the handling of missing inputs. The new pattern $(?, 0.5)$ is represented as an interval pattern $([0, 1], 0.5)$ in our approach. This interval representation can be also used for handling uncertain inputs. For example, let us assume that we have the following information about an uncertain pattern on the two-dimensional pattern space $[0, 1]^2$:

  (i) The first input is not more than 0.3 (i.e., $x_1 \leq 0.3$).
  (ii) The second input is not less than 0.8 (i.e., $0.8 \leq x_2$).

From these two pieces of information, we can represent this uncertain pattern as an interval pattern $([0, 0.3], [0.8, 1])$ because the pattern space is the unit square $[0, 1]^2$.

When an interval input pattern is presented to a neural network, interval arithmetic [76, 77] is used for calculating the input–output relation of the neural network. Interval arithmetic is also used when connection weights of neural networks are given as intervals. In this subsection, we briefly describe the interval arithmetic that will be used for the handling of interval input patterns and interval connection weights.

Interval arithmetic is the generalization of ordinary arithmetic on real numbers to closed intervals. In this section, we denote real numbers and closed intervals by lowercase letters (e.g., $a, b, c, \ldots$) and uppercase letters (e.g., $A, B, C, \ldots$), respectively. An interval is also represented by its lower limit and upper limit as

$$A = \left[a^L, a^U \right], \tag{50}$$

where the superscripts "L" and "U" denote the lower limit and the upper limit, respectively.

The inclusion relation between intervals can be defined as

$$A \subseteq B \quad \Leftrightarrow \quad b^L \leq a^L \text{ and } a^U \leq b^U, \tag{51}$$

where $A = [a^L, a^U]$ and $B = [b^L, b^U]$. As a special case of this relation, the inclusion relation between an interval and a real number can be defined as

$$a \in B \quad \Leftrightarrow \quad b^L \leq a \leq b^U. \tag{52}$$

We have already used this inclusion relation in the previous section.

The following addition and multiplication are used in this section for calculating the total input to each unit in interval-arithmetic-based neural networks:

$$A + B = \left[a^L, a^U \right] + \left[b^L, b^U \right] = \left[a^L + b^L, a^U + b^U \right], \tag{53}$$

$$a \cdot B = a \cdot \left[b^L, b^U \right] = \begin{cases} [a \cdot b^L, a \cdot b^U], & \text{if } a \geq 0, \\ [a \cdot b^U, a \cdot b^L], & \text{if } a < 0, \end{cases} \tag{54}$$

$$\begin{aligned} A \cdot B &= \left[a^L, a^U \right] \cdot \left[b^L, b^U \right] \\ &= \big[\min\{ a^L b^L, a^L b^U, a^U b^L, a^U b^U \}, \\ &\quad \max\{ a^L b^L, a^L b^U, a^U b^L, a^U b^U \} \big]. \end{aligned} \tag{55}$$

In the case of $0 \le a^L \le a^U$ (i.e., if A is nonnegative), the preceding product operation on intervals can be simplified as

$$A \cdot B = [a^L, a^U] \cdot [b^L, b^U] = [\min\{a^L b^L, a^U b^L\}, \max\{a^L b^U, a^U b^U\}]. \quad (56)$$

As an example, let us consider a very simple network with two input units (i.e., units 1 and 2) and a single output unit (i.e., unit j) in Fig. 28 where O_{p1} and O_{p2} are interval outputs from the two input units, w_{j1} and w_{j2} are real-number connection weights, θ_j is a real-number bias, and O_{pj} is an interval output from the unit j. The total input to the unit j is calculated by interval arithmetic as follows (see Fig. 28).

$$
\begin{aligned}
\text{Net}_{pj} &= w_{j1} \cdot O_{p1} + w_{j2} \cdot O_{p2} + \theta_j \\
&= -2 \cdot [1, 2] + 1 \cdot [2, 3] + 1 = [-4, -2] + [2, 3] + [1, 1] \\
&= [-1, 2],
\end{aligned}
\quad (57)
$$

where a real number is treated as a special interval whose lower and upper limits are the same.

The sigmoidal function, which is used as an activation function at hidden and output units, is extended to the case of interval inputs as follows:

$$f(\text{Net}) = \{ f(x) \mid x \in \text{Net} \}, \quad (58)$$

where Net is an interval input and

$$f(x) = 1 / \{ 1 + \exp(-x) \}. \quad (59)$$

Because the sigmoidal function in (59) is a strictly increasing function, the interval output $f(\text{Net})$ in (58) can be calculated as

$$f(\text{Net}) = f([\text{net}^L, \text{net}^U]) = [f(\text{net}^L), f(\text{net}^U)]. \quad (60)$$

This is illustrated in Fig. 29.

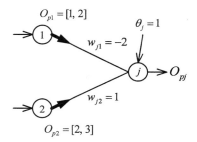

Figure 28 Simple network with two interval inputs and an interval output.

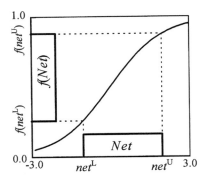

Figure 29 Interval activation function at hidden and output units.

The interval output O_{pj} in Fig. 28 is calculated as follows:

$$O_{pj} = f(\text{Net}_j) = f([-1, 2]) = [f(-1), f(2)] = [0.269, 0.881]. \quad (61)$$

B. NEURAL NETWORKS FOR HANDLING INTERVAL INPUTS

As we have already described, intervals can be used for representing uncertain inputs and missing inputs [78, 79]. Interval representation is also useful for utilizing experts' knowledge in the learning of neural networks [80]. Let us consider a two-class classification problem in the pattern space $[0, 1]^2$. Now we assume that the following two pieces of information are given from domain experts:

(i) If $x_1 \leq 0.5$ and $x_2 \leq 0.5$ then Class 1.
(ii) If $x_1 \geq 0.8$ or $x_2 \geq 0.8$ then Class 2.

These two rules are shown in Fig. 30a. We also assume that we have training patterns in Fig. 30b where closed circles and open circles are training patterns from Class 1 and Class 2, respectively. Our problem is to train a neural network from both experts' knowledge (i.e., the if-then rules in Fig. 30a) and the numerical data in Fig. 30b.

We can denote the first rule as an interval pattern $([0, 0.5], [0, 0.5])$ from Class 1. This interval pattern is shown as the square in Fig. 30a. The second rule can be denoted by two interval patterns $([0.8, 1], [0, 1])$ and $([0, 1], [0.8, 1])$ from Class 2. These two input patterns correspond to the two rectangles in Fig. 30a. Because real numbers can be viewed as a special case of closed intervals whose

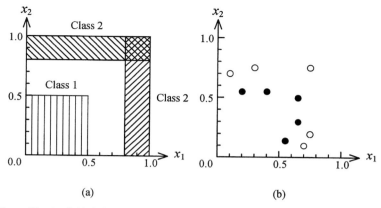

Figure 30 Available information for learning: (a) experts' knowledge; (b) numerical data.

upper and lower limits are the same (e.g., $0.5 = [0.5, 0.5]$), we can represent both the if-then rules and the numerical patterns as a set of interval patterns.

In general, let us assume that we have m interval patterns $\mathbf{X}_p = (X_{p1}, X_{p2}, \ldots, X_{pn})$, $p = 1, 2, \ldots, m$, from c classes. For these interval patterns, we use a standard three-layer feedforward neural network with n input units, n_H hidden units, and c output units. The input–output relation of each unit can be written as follows (see Fig. 31):

Input units: $\quad O_{pi} = X_{pi}, \qquad i = 1, 2, \ldots, n,$ $\qquad\qquad$ (62)

Hidden units: $\quad \text{Net}_{pj} = \displaystyle\sum_{i=1}^{n} w_{ji} \cdot O_{pi} + \theta_j, \qquad j = 1, 2, \ldots, n_H,$ $\quad$ (63)

$$O_{pj} = f(\text{Net}_{pj}), \qquad j = 1, 2, \ldots, n_H,$$ $\qquad\qquad$ (64)

Output units: $\quad \text{Net}_{pk} = \displaystyle\sum_{j=1}^{n_H} w_{kj} \cdot O_{pj} + \theta_k, \qquad k = 1, 2, \ldots, c,$ $\quad$ (65)

$$O_{pk} = f(\text{Net}_{pk}), \qquad k = 1, 2, \ldots, c.$$ $\qquad\qquad$ (66)

We can see that these formulations are the same as the architecture of standard feedforward neural networks except that the input and the output of each unit are intervals. The calculation of the input–output relation of each unit is done by interval arithmetic described in the previous subsection. For example, the input–

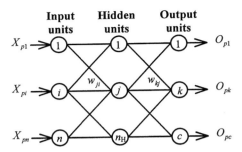

Figure 31 Architecture of interval-arithmetic-based neural networks with interval inputs and real-number connection weights.

output relation of the kth output unit can be rewritten from interval arithmetic as

$$\text{net}^L_{pk} = \sum_{\substack{j=1 \\ w_{kj} \geq 0}}^{n_H} w_{kj} \cdot o^L_{pj} + \sum_{\substack{j=1 \\ w_{kj} < 0}}^{n_H} w_{kj} \cdot o^U_{pj} + \theta_k, \qquad (67)$$

$$\text{net}^U_{pk} = \sum_{\substack{j=1 \\ w_{kj} \geq 0}}^{n_H} w_{kj} \cdot o^U_{pj} + \sum_{\substack{j=1 \\ w_{kj} < 0}}^{n_H} w_{kj} \cdot o^L_{pj} + \theta_k, \qquad (68)$$

$$O_{pk} = \left[o^L_{pk}, o^U_{pk} \right] = \left[f\left(\text{net}^L_{pk} \right), f\left(\text{net}^U_{pk} \right) \right]. \qquad (69)$$

For the learning of the neural network from the interval patterns, we define the target vector $\mathbf{t}_p = (t_{p1}, t_{p2}, \dots, t_{pc})$ corresponding to the interval input pattern $\mathbf{X}_p$ as follows:

$$t_{pk} = \begin{cases} 1, & \text{if } \mathbf{X}_p \in \text{Class } k, \\ 0, & \text{otherwise,} \end{cases} \qquad (70)$$

for $k = 1, 2, \dots, c$. The cost function to be minimized in the learning is defined as follows:

$$e_p = \sum_{k=1}^{c} \left(t_{pk} - o^L_{pk} \right)^2 / 2 + \sum_{k=1}^{c} \left(t_{pk} - o^U_{pk} \right)^2 / 2. \qquad (71)$$

A back-propagation-type learning algorithm can be derived from this cost function for adjusting the connection weights and the biases [78–80].

To illustrate our approach, we first trained a neural network with two input units, five hidden units, and a single output unit by the standard back-propagation algorithm using only the numerical data in Fig. 30b. The classification boundary

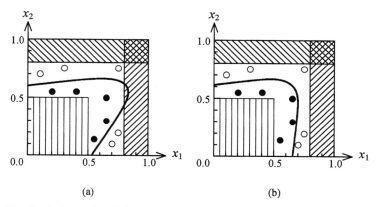

Figure 32 Simulation results: (a) learning from only numerical data; (b) learning from both experts' knowledge and numerical data.

obtained by the learning is shown in Fig. 32a. From this figure, we can see that all the given patterns are correctly classified. Because the experts' knowledge in Fig. 30a was not used in the learning, the classification boundary violates the second if-then rule "If $x_1 \geq 0.8$ or $x_2 \geq 0.8$ then Class 2."

We next trained the same neural network using both the experts' knowledge and the numerical data. That is, the three interval patterns in Fig. 30a and the ten patterns in Fig. 30b were used for the learning of the neural network in our approach. The classification result obtained by the learning is shown in Fig. 32b. From this figure, we can see that the classification boundary is clearly consistent with both the given patterns and the experts' knowledge.

We show another simulation result by our approach in Fig. 33a, which was obtained by the learning using the six interval patterns in this figure. From this figure, we can see that all the interval patterns are correctly classified. For comparison, we applied the standard back-propagation algorithm to this problem using the four vertexes of each interval pattern. We show the simulation result in Fig. 33b. As shown in this figure, all the vertexes are correctly classified but the classification boundary violates an interval pattern.

Our interval-arithmetic-based approach can also be employed when a new pattern has uncertain or missing inputs. In the same manner as in the training patterns, we represent the new pattern with uncertain or missing inputs by an interval pattern $\mathbf{X}_p = (X_{p1}, X_{p2}, \ldots, X_{pn})$. For example, a new pattern $(0.3, ?, 0.8)$ in the three-dimensional pattern space $[0, 1]^3$ is represented as an interval pattern $([0.3, 0.3], [0, 1], [0.8, 0.8])$ where real numbers are also represented as closed intervals. The classification of the interval pattern $\mathbf{X}_p$ is done by presenting this pattern to the trained neural network. As we have already explained, an interval

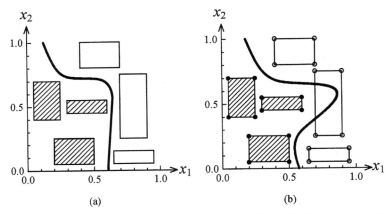

Figure 33 Simulation results: (a) learning from interval data; (b) learning from vertexes of interval data.

output vector $\mathbf{O}_p = (O_{p1}, O_{p2}, \ldots, O_{pc})$ is obtained from the interval input pattern $\mathbf{X}_p$ by interval arithmetic. Now our problem is to assign the interval input pattern $\mathbf{X}_p$ to one of the given c classes based on the interval output vector $\mathbf{O}_p$.

To classify the interval input pattern $\mathbf{X}_p$, we use the following rule [79, 81]:

$$\text{If } o_{pk}^L > o_{ph}^U \text{ for } h = 1, 2, \ldots, c, \ h \neq k \text{ then classify } \mathbf{X}_p \text{ as Class } k. \quad (72)$$

The condition part of this rule means that the following inequality holds:

$$o_{pk} > o_{ph} \quad \text{for } \forall o_{pk} \in O_{pk}, \ \forall o_{ph} \in O_{ph},$$
$$\text{and } h = 1, 2, \ldots, c, \ h \neq k. \quad (73)$$

For example, an interval input pattern is classified as Class 2 if the corresponding interval output vector $\mathbf{O}_p = (O_{p1}, O_{p2}, O_{p3}, O_{p4})$ is as in Fig. 34a. On the other hand, in the case of Fig. 34b, the classification of an interval input pattern is rejected. This is because the condition part of (72) does not hold for any class in Fig. 34b.

For illustration, first we trained a neural network by the standard back-propagation algorithm using the training patterns in Fig. 35a where the classification boundary obtained from the trained neural network is also shown. We presented two interval patterns in Fig. 35b to the trained neural network, and examined the corresponding outputs. One interval input pattern $\mathbf{X}_A$ is ([0.1, 0.4], [0.5, 0.8]), and the other interval input pattern $\mathbf{X}_B$ corresponds to an input pattern (0.8, ?) with a missing input. For the interval input pattern $\mathbf{X}_A$, the interval output vector ([0.00, 0.00], [0.99, 0.99], [0.00, 0.00]) was obtained.

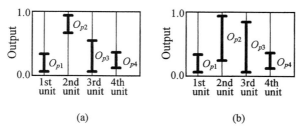

(a)   (b)

Figure 34 Examples of interval outputs: (a) classifiable case; (b) unclassifiable case.

From this interval output vector, we can classify $\mathbf{X}_A$ as Class 2 by the classification rule in (72). On the other hand, the classification of $\mathbf{X}_B$ is rejected because the corresponding interval output vector is ([0.00, 0.99], [0.00, 0.02], [0.00, 0.99]).

If the condition part in (72) holds, any patterns included in $\mathbf{X}_p$ are also classified as the same class. This is because the following inclusion relation holds:

$$X_{qi} \subseteq X_{pi} \text{ for } i = 1, 2, \ldots, n \quad \Rightarrow \quad O_{qk} \subseteq O_{pk} \text{ for } k = 1, 2, \ldots, c, \quad (74)$$

where $\mathbf{X}_p = (X_{p1}, X_{p2}, \ldots, X_{pn})$ and $\mathbf{X}_q = (X_{q1}, X_{q2}, \ldots, X_{qn})$ are interval input patterns, and $\mathbf{O}_p = (O_{p1}, O_{p2}, \ldots, O_{pc})$ and $\mathbf{O}_q = (O_{q1}, O_{q2}, \ldots, O_{qc})$ are the corresponding interval output vectors from the neural network. The relation in (74), which is called "inclusion monotonicity," is one of the basic features of interval arithmetic. For example, from this relation, we can see that any interval

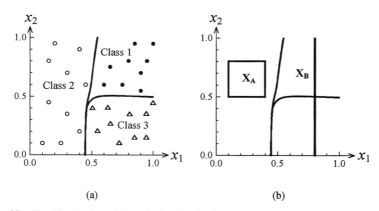

(a)   (b)

Figure 35 Classification boundaries obtained by the trained neural network and interval input vectors: (a) classification boundaries and training data; (b) classification boundaries and new interval input vectors.

(and real number) input patterns included in $\mathbf{X}_A$ in Fig. 35b are always classified as Class 2 because $\mathbf{X}_A$ has already been classified as Class 2.

C. NEURAL NETWORKS WITH INTERVAL WEIGHTS

In the previous subsection, we described how feedforward neural networks can be extended to the case of interval inputs. In this subsection, we extend connection weights to intervals.

Let us start with a feedforward neural network with real-number input vectors and interval connection weights. Such a neural network is used for approximately realizing a nonlinear interval function. For a nonlinear interval function with n inputs and a single output, we use an interval-arithmetic-based neural network that maps an n-dimensional real-number input vector $\mathbf{x}_p = (x_{p1}, x_{p2}, \ldots, x_{pn})$ to an interval output O_p. The input–output relation of each unit of the interval-arithmetic-based neural network with interval connection weights is written for the real-number input vector $\mathbf{x}_p = (x_{p1}, x_{p2}, \ldots, x_{pn})$ as follows:

Input units: $o_{pi} = x_{pi}, \qquad i = 1, 2, \ldots, n,$ \hfill (75)

Hidden units: $\text{Net}_{pj} = \sum_{i=1}^{n} W_{ji} \cdot o_{pi} + \Theta_j, \qquad j = 1, 2, \ldots, n_H,$ \hfill (76)

$O_{pj} = f(\text{Net}_{pj}), \qquad j = 1, 2, \ldots, n_H,$ \hfill (77)

Output unit: $\text{Net}_p = \sum_{j=1}^{n_H} W_j \cdot O_{pj} + \Theta,$ \hfill (78)

$O_p = f(\text{Net}_p).$ \hfill (79)

This interval-arithmetic-based neural network is the same as the standard feedforward neural network except that the connection weights W_{ji}, W_j and the biases Θ_j, Θ are given by intervals. The architecture of this neural network and its example are shown in Fig. 36. As in Fig. 36a, we denote the interval connection weights and the interval biases by their lower and upper limits as

$$W_{ji} = [w_{ji}^L, w_{ji}^U], \qquad W_j = [w_j^L, w_j^U],$$
$$\Theta_j = [\theta_j^L, \theta_j^U], \qquad \Theta = [\theta^L, \theta^U]. \tag{80}$$

The input–output relation of each unit in (75)–(79) is calculated by interval arithmetic. For example, the input–output relation of the output unit can be rewritten

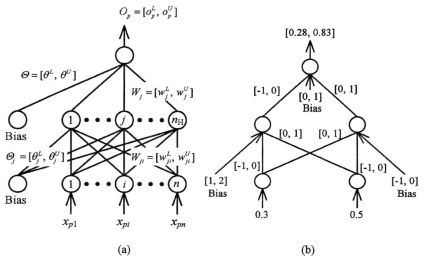

Figure 36 Interval-arithmetic-based neural networks with real-number input vectors and interval connection weights: (a) general architecture; (b) an example.

because the interval outputs O_{pj}'s from the hidden units are always nonnegative [see (60) and Fig. 29]:

$$\text{net}_p^L = \sum_{\substack{j=1 \\ w_j^L \geq 0}}^{n_H} w_j^L \cdot o_{pj}^L + \sum_{\substack{j=1 \\ w_j^L < 0}}^{n_H} w_j^L \cdot o_{pj}^U + \theta^L, \tag{81}$$

$$\text{net}_p^U = \sum_{\substack{j=1 \\ w_j^U \geq 0}}^{n_H} w_j^U \cdot o_{pj}^U + \sum_{\substack{j=1 \\ w_j^U < 0}}^{n_H} w_j^U \cdot o_{pj}^L + \theta^U, \tag{82}$$

$$O_p = \left[o_p^L, o_p^U\right] = \left[f\left(\text{net}_p^L\right), f\left(\text{net}_p^U\right)\right]. \tag{83}$$

Let us assume that we have m input–output pairs $(\mathbf{x}_p; Y_p)$, $p = 1, 2, \ldots, m$, as training data where $\mathbf{x}_p = (x_{p1}, x_{p2}, \ldots, x_{pn})$ is an n-dimensional real-number input vector, and $Y_p = [y_p^L, y_p^U]$ is the corresponding interval output. The given output Y_p is used as a target interval. We show an example of such training data in Fig. 37a for the case of $n = 1$. Our problem here is to train the interval-arithmetic-based neural network using the given training data.

The learning is performed so that the interval output O_p from the neural network becomes approximately equal to the target interval Y_p for all the given

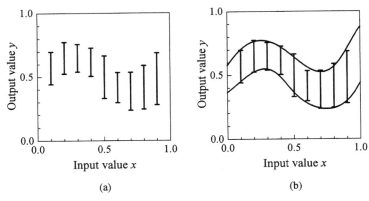

Figure 37   Simulation result: (a) given training data; (b) shape of the output from the trained neural network.

input–output pairs $(\mathbf{x}_p; Y_p)$, $p = 1, 2, \ldots, m$. Thus we define the cost function for the given input–output pair $(\mathbf{x}_p; Y_p)$ as follows:

$$e_p = \left(y_p^L - o_p^L\right)^2/2 + \left(y_p^U - o_p^U\right)^2/2. \tag{84}$$

A back-propagation-type learning algorithm can be derived from this cost function for adjusting the interval connection weights and the interval biases [82, 83]. The adjustment of the interval connection weights and the interval biases is performed by updating their lower and upper limits. For example, the interval connection weight $W_j = [w_j^L, w_j^U]$ is adjusted by updating its lower limit w_j^L and its upper limit w_j^U using the partial derivatives $\partial e_p/\partial w_j^L$ and $\partial e_p/\partial w_j^U$. It should be noted that the inequality $w_j^L \leq w_j^U$ always has to be satisfied.

In Fig. 37b, we show the result of the learning of the interval-arithmetic-based neural network with a single input, five hidden units, and a single output unit. The two curves in this figure correspond to the lower limit and the upper limit of the interval output from the trained neural network.

In the previous section, we described how a nonlinear interval function can be approximately realized by two standard feedforward neural networks. As shown in Fig. 37b, a single interval-arithmetic-based neural network with interval connection weights can also represent a nonlinear interval function. The main difference between these two approaches is that the two standard neural networks are independently trained, whereas the lower and upper limits of the interval connection weights are adjusted with the inequality constraints such as $w_j^L \leq w_j^U$. This difference is clearly demonstrated in Fig. 38. In Fig. 38a, two curves are outputs from two standard neural networks that were independently trained. In this figure,

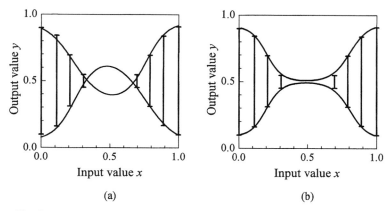

Figure 38 Comparison of two approaches: (a) two standard neural networks; (b) single interval-arithmetic-based neural network.

the output from one neural network for the lower limit is larger than that for the upper limit in some range (i.e., around $x = 0.5$). In Fig. 38b, however, the lower limit of the interval output is always smaller than the upper limit.

Interval-arithmetic-based neural networks with interval connection weights can be trained so as to include all the given training data as shown in Fig. 39a or be included in the target intervals as shown in Fig. 39b.

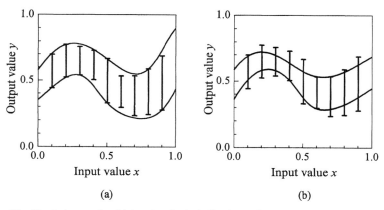

Figure 39 Simulation results: (a) learning for including interval targets; (b) learning for being included in interval targets.

They can also handle real-number targets. For those extensions, see Ishibuchi *et al.* [82]. The most general architecture of interval-arithmetic-based neural networks has interval input vectors, interval connection weights, and interval target vectors. That is, interval-arithmetic-based neural networks for approximately realizing nonlinear interval functions are extended to the case of interval input vectors and multiple output units. The learning of those neural networks is also performed by updating the lower and upper limits of the interval connection weights and biases [83]. Those neural networks are used for approximately realizing nonlinear mappings of interval vectors (i.e., mappings from interval vectors to interval vectors).

IV. FUZZIFIED NEURAL NETWORKS

A. FUZZY ARITHMETIC IN NEURAL NETWORKS

In the previous section, we extended inputs, connection weights, biases, and targets to intervals. Here they are extended to fuzzy numbers for the fuzzification of multilayer feedforward neural networks. As we have already shown in Figs. 3 and 4, fuzzy arithmetic [34] based on the extension principle [35] is used for defining the input–output relation of fuzzified neural networks.

We denote fuzzy numbers by uppercase letters with tildes such as $\widetilde{A}, \widetilde{B}, \widetilde{C}$, etc. A fuzzy number $\widetilde{A}$ is specified by its membership function $\mu_{\widetilde{A}}(\cdot)$ on the real line $\Re$ (i.e., on the set of real numbers). In Fig. 40, we show two examples of fuzzy numbers. They can be interpreted as "about 5" and "about 10," respectively. Linguistic values such as "*small*" and "*large*" are also viewed as fuzzy numbers.

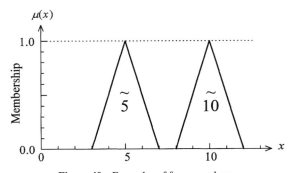

Figure 40   Examples of fuzzy numbers.

Five linguistic values (S: *small*, MS: *medium small*, M: *medium*, ML: *medium large*, and L: *large*) defined on the unit interval [0, 1] are shown in Fig. 41. Fuzzy numbers can be used for representing various linguistic concepts such as *"hot"* water, a *"warm"* day, and a *"tall"* man. Fuzzified neural networks can handle such a linguistic concept as well as numerical data.

We use the following addition, multiplication, and nonlinear mapping of fuzzy numbers in our fuzzified neural networks (see Figs. 3 and 4):

$$\mu_{\widetilde{A}+\widetilde{B}}(z) = \max\{\mu_{\widetilde{A}}(x) \wedge \mu_{\widetilde{B}}(y) \mid z = x + y\}, \tag{85}$$

$$\mu_{\widetilde{A}\cdot\widetilde{B}}(z) = \max\{\mu_{\widetilde{A}}(x) \wedge \mu_{\widetilde{B}}(y) \mid z = x \cdot y\}, \tag{86}$$

$$\mu_{f(\widetilde{\mathrm{Net}})}(z) = \max\{\mu_{\widetilde{\mathrm{Net}}}(x) \mid z = f(x)\}, \tag{87}$$

where $\wedge$ is the minimum operator and $f(x) = 1/\{1 + \exp(-x)\}$.

These fuzzy-number operations are numerically performed by interval arithmetic on level sets of fuzzy numbers. The h-level set of a fuzzy number $\widetilde{A}$ is defined as follows:

$$[\widetilde{A}]_h = \{x \mid \mu_{\widetilde{A}}(x) \geq h, \ x \in \Re\} \qquad \text{for } 0 < h \leq 1. \tag{88}$$

The h-level set $[\widetilde{A}]_h$ is illustrated in Fig. 42a. A fuzzy number can be approximately represented by a collection of its h-level sets for various values of h. In Fig. 42b, a fuzzy number $\widetilde{A}$ is approximately represented by its ten h-level sets for $h = 0.1, 0.2, \ldots, 1.0$.

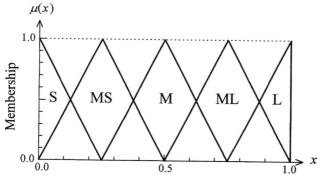

Figure 41 Five linguistic values (**S**: *small*, **MS**: *medium small*, **M**: *medium*, **ML**: *medium large*, and **L**: *large*).

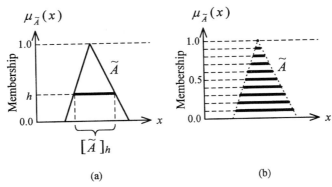

Figure 42 Level sets of a fuzzy number $\tilde{A}$: (a) h-level set; (b) approximate representation of a fuzzy number $\tilde{A}$ by a collection of its level sets.

As we can see from Fig. 42, h-level sets of fuzzy numbers are closed intervals. Thus we use interval arithmetic for approximately calculating the fuzzy input–output relation of each unit of our fuzzified neural networks.

B. NEURAL NETWORKS FOR HANDLING FUZZY INPUTS

In this subsection, we describe how multilayer feedforward neural networks can be extended to the case of fuzzy inputs. Fuzzy inputs may be obtained from uncertain measurement or linguistic knowledge of human experts. For example, let us assume that we have the following linguistic knowledge for a three-class pattern classification problem on the two-dimensional pattern space $[0, 1]^2$:

If x_1 is *small* and x_2 is *small* then Class 1,

If x_1 is *small* and x_2 is *large* then Class 2,

If x_1 is *large* then Class 3,

where "*small*" and "*large*" are fuzzy numbers defined in Fig. 41. These three fuzzy if-then rules are shown in Fig. 43. We also assume that we have numerical data in Fig. 43 where closed circles, open circles, and squares are training patterns from Class 1, Class 2, and Class 3, respectively. Our problem is to train a feedforward neural network using both the linguistic knowledge and the numerical data in Fig. 43.

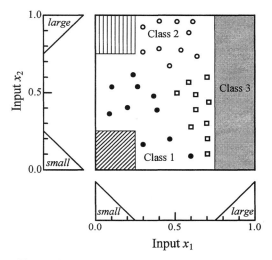

Figure 43   Linguistic information and numerical data.

The previous three fuzzy if-then rules can be viewed as the following fuzzy training patterns because the pattern space is the unit square $[0, 1]^2$:

$$(small, small) \Rightarrow \text{Class 1},$$
$$(small, large) \Rightarrow \text{Class 2},$$
$$(large, [0, 1]) \Rightarrow \text{Class 3}.$$

Numerical data are also handled as fuzzy training patterns in our fuzzified neural networks because real numbers can be viewed as a special case of fuzzy numbers. A real number a can be viewed as a fuzzy number with the following membership function:

$$\mu_a(x) = \begin{cases} 1, & \text{if } x = a, \\ 0, & \text{otherwise.} \end{cases} \tag{89}$$

In this manner, both the expert knowledge and the numerical data are handled as fuzzy training patterns. That is, they are simultaneously utilized in the learning of neural networks.

In general, for a c-class pattern classification problem on the n-dimensional pattern space $[0, 1]^n$, let us assume that we have m fuzzy training patterns $\widetilde{\mathbf{X}}_p = (\widetilde{X}_{p1}, \widetilde{X}_{p2}, \ldots, \widetilde{X}_{pn})$, $p = 1, 2, \ldots, m$. For this pattern classification problem, we use a three-layer feedforward neural network with n input units, n_H hidden units, and c output units. The input–output relation of each unit of this neural network is written for the fuzzy input pattern $\widetilde{\mathbf{X}}_p = (\widetilde{X}_{p1}, \widetilde{X}_{p2}, \ldots, \widetilde{X}_{pn})$ as follows:

Input units: $\quad \widetilde{O}_{pi} = \widetilde{X}_{pi}, \qquad i = 1, 2, \ldots, n,$ (90)

$$\text{Hidden units: } \widetilde{\text{Net}}_{pj} = \sum_{i=1}^{n} w_{ji} \cdot \widetilde{O}_{pi} + \theta_j, \qquad j = 1, 2, \ldots, n_H, \quad (91)$$

$$\widetilde{O}_{pj} = f(\widetilde{\text{Net}}_{pj}), \qquad j = 1, 2, \ldots, n_H, \quad (92)$$

$$\text{Output units: } \widetilde{\text{Net}}_{pk} = \sum_{j=1}^{n_H} w_{kj} \cdot \widetilde{O}_{pj} + \theta_k, \qquad k = 1, 2, \ldots, c, \quad (93)$$

$$\widetilde{O}_{pk} = f(\widetilde{\text{Net}}_{pk}), \qquad k = 1, 2, \ldots, c. \quad (94)$$

The input–output relation of each unit is defined by fuzzy-number arithmetic described in the previous subsection. These formulations are the same as the architecture of standard feedforward neural networks except that the input and the output of each unit are fuzzy numbers. The numerical calculation of the input–output relation is done by interval arithmetic on h-level sets of fuzzy numbers. For example, the input–output relation of the kth output unit can be rewritten for the h-level sets as follows:

$$\left[\widetilde{\text{Net}}_{pk}\right]_h^L = \sum_{\substack{j=1 \\ w_{kj} \geq 0}}^{n_H} w_{kj} \cdot \left[\widetilde{O}_{pj}\right]_h^L + \sum_{\substack{j=1 \\ w_{kj} < 0}}^{n_H} w_{kj} \cdot \left[\widetilde{O}_{pj}\right]_h^U + \theta_k, \quad (95)$$

$$\left[\widetilde{\text{Net}}_{pk}\right]_h^U = \sum_{\substack{j=1 \\ w_{kj} \geq 0}}^{n_H} w_{kj} \cdot \left[\widetilde{O}_{pj}\right]_h^U + \sum_{\substack{j=1 \\ w_{kj} < 0}}^{n_H} w_{kj} \cdot \left[\widetilde{O}_{pj}\right]_h^L + \theta_k, \quad (96)$$

$$[O_{pk}]_h = \left[[O_{pk}]_h^L, [O_{pk}]_h^U\right] = \left[f\left([\widetilde{\text{Net}}_{pk}]_h^L\right), f\left([\widetilde{\text{Net}}_{pk}]_h^U\right)\right], \quad (97)$$

where $[\cdot]_h$ denotes the h-level set of a fuzzy number and $[\cdot]_h^L$ and $[\cdot]_h^U$ denote the lower limit and the upper limit of the h-level set.

For the learning of the neural network from the fuzzy training patterns, we define the target vector $\mathbf{t}_p = (t_{p1}, t_{p2}, \ldots, t_{pc})$ corresponding to the fuzzy input pattern $\widetilde{\mathbf{X}}_p$ as follows:

$$t_{pk} = \begin{cases} 1, & \text{if } \widetilde{\mathbf{X}}_p \in \text{Class } k, \\ 0, & \text{otherwise,} \end{cases} \quad (98)$$

for $k = 1, 2, \ldots, c$. The cost function to be minimized in the learning is defined as follows:

$$e_p = \sum_h \left\{ \sum_{k=1}^{c} (t_{pk} - [\tilde{O}_{pk}]_h^L)^2/2 + \sum_{k=1}^{c} (t_{pk} - [\tilde{O}_{pk}]_h^U)^2/2 \right\}. \qquad (99)$$

A back-propagation-type learning algorithm can be derived from this cost function for adjusting the connection weights and the biases [41, 84].

To illustrate our approach, we first trained a neural network with two input units, five hidden units, and three output units by the standard back-propagation algorithm using only the numerical data in Fig. 43. The classification boundaries obtained by the learning are shown in Fig. 44a. Next we trained the same neural network using both the experts' knowledge and the numerical data. In the learning, we used ten levels (i.e., $h = 0.1, 0.2, \ldots, 1.0$) in the cost function in (99). The classification boundaries obtained by the learning from both the experts' knowledge and the numerical data are shown in Fig. 44b. From this figure, we can see the classification boundaries are clearly consistent with both the experts' knowledge and the numerical data.

Linguistic information from human experts can also be utilized for modeling problems. Let us assume that we have the following linguistic information for the modeling of a single-input and single-output nonlinear system:

If x is *small* then y is *small*,

If x is *large* then y is *large*,

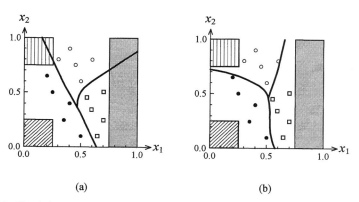

(a)   (b)

Figure 44 Simulation results: (a) learning from only numerical data; (b) learning from both numerical data and linguistic information.

where "*small*" and "*large*" are defined in Fig. 41. These fuzzy if-then rules can be viewed as the following fuzzy training data:

$$\{(\widetilde{X}_p; \widetilde{Y}_p)\} = \{(small; small), (large; large)\}. \tag{100}$$

In general, let us assume that we have m fuzzy input–output pairs $(\widetilde{\mathbf{X}}_p; \widetilde{Y}_p)$, $p = 1, 2, \ldots, m$, from an n-input and single-output nonlinear system, where $\widetilde{\mathbf{X}}_p = (\widetilde{X}_{p1}, \widetilde{X}_{p2}, \ldots, \widetilde{X}_{pn})$. As we have already described for classification problems, nonfuzzy input–output pairs can also be represented in this form. Thus the fuzzy training data $(\widetilde{\mathbf{X}}_p; \widetilde{Y}_p)$, $p = 1, 2, \ldots, m$, may include nonfuzzy input–output pairs as well as fuzzy input–output pairs.

For the modeling of a nonlinear system with n inputs and a single output, we use a neural network with n input units and a single output unit. When the n-dimensional fuzzy vector $\widetilde{\mathbf{X}}_p$ is presented to the neural network, the corresponding fuzzy output $\widetilde{O}_p$ is defined in the same manner as in (90)–(94). The given fuzzy output $\widetilde{Y}_p$ is used as the fuzzy target. The cost function to be minimized in the learning of the neural network is defined as follows:

$$e_p = \sum_h \{([\widetilde{Y}_p]_h^L - [\widetilde{O}_p]_h^L)^2/2 + ([\widetilde{Y}_p]_h^U - [\widetilde{O}_p]_h^U)^2/2\}. \tag{101}$$

A back-propagation-type learning algorithm can be derived from this cost function for adjusting the connection weights and the biases of the neural network [41, 85].

For illustration, we show simulation results in Fig. 45 where numerical data are denoted by closed circles. Figure 45a is the simulation result by the learning from only the numerical data where the standard back-propagation algorithm was

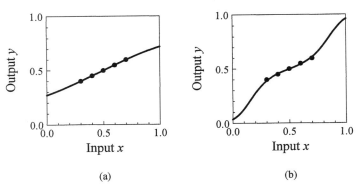

(a)   (b)

Figure 45 Simulation results for a function approximation problem: (a) learning from only numerical data; (b) learning from both numerical data and linguistic information.

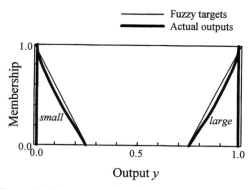

Figure 46 Fuzzy outputs from the trained neural network.

used. In Fig. 45b, both the numerical data and the linguistic information in (100) were used for the learning by our approach. The fuzzy outputs from the trained neural network are shown in Fig. 46 together with the fuzzy targets (i.e., *small* and *large*). From Fig. 46, we can see that a good fit to the fuzzy targets was obtained by the learning of the neural network.

As shown in Fig. 46, our fuzzified neural networks for fuzzy inputs can be used for approximately realizing fuzzy if-then rules. High fitting ability to given fuzzy if-then rules and high interpolation ability of sparse fuzzy if-then rules were demonstrated in [41, 85]. Our approach can also be used for extracting fuzzy if-then rules from trained neural networks [86, 87]. For the rule extraction, a linguistic input vector corresponding to the antecedent part of each fuzzy if-then rule was presented to the trained neural network, and the corresponding fuzzy output was examined to determine the consequent part of the fuzzy if-then rule.

C. NEURAL NETWORKS WITH FUZZY WEIGHTS

Multilayer feedforward neural networks can be fuzzified by extending their connection weights and biases to fuzzy numbers. Fuzzified neural networks with nonfuzzy input vectors are used for the modeling of fuzzy functions [42, 88, 89]. Fuzzified neural networks with fuzzy input vectors are used for approximately realizing fuzzy if-then rules [42, 43, 90].

In this subsection, we describe a general architecture of fully fuzzified three-layer feedforward neural networks [43]. Let us assume that we have m fuzzy input–output pairs $(\widetilde{\mathbf{X}}_p; \widetilde{\mathbf{Y}}_p)$, $p = 1, 2, \ldots, m$, where $\widetilde{\mathbf{X}}_p = (\widetilde{X}_{p1}, \widetilde{X}_{p2}, \ldots, \widetilde{X}_{pn})$ is an n-dimensional fuzzy input vector and $\widetilde{\mathbf{Y}}_p = (\widetilde{Y}_{p1}, \widetilde{Y}_{p2}, \ldots, \widetilde{Y}_{pc})$ is a c-dimensional fuzzy target vector. $\widetilde{\mathbf{X}}_p$ and $\widetilde{\mathbf{Y}}_p$ may be viewed as the antecedent part and the consequent part of a fuzzy if-then rule, respectively. Our problem is to approximately realize a nonlinear fuzzy mapping from $\tilde{\mathbf{X}}_p$ to $\tilde{\mathbf{Y}}_p$. For this problem, we use a fuzzified neural network with fuzzy connection weights and fuzzy biases. The input–output relation of each unit of the fuzzified neural network with n input units, n_H hidden units, and c output units is written as follows (see Fig. 47):

$$\text{Input units:} \quad \tilde{O}_{pi} = \tilde{X}_{pi}, \qquad i = 1, 2, \ldots, n, \tag{102}$$

$$\text{Hidden units:} \quad \widetilde{\text{Net}}_{pj} = \sum_{i=1}^{n} \tilde{W}_{ji} \cdot \tilde{O}_{pi} + \tilde{\Theta}_j, \qquad j = 1, 2, \ldots, n_H, \tag{103}$$

$$\tilde{O}_{pj} = f\left(\widetilde{\text{Net}}_{pj}\right), \qquad j = 1, 2, \ldots, n_H, \tag{104}$$

$$\text{Output units:} \quad \widetilde{\text{Net}}_{pk} = \sum_{j=1}^{n_H} \tilde{W}_{kj} \cdot \tilde{O}_{pj} + \tilde{\Theta}_k, \qquad k = 1, 2, \ldots, c, \tag{105}$$

$$\tilde{O}_{pk} = f\left(\widetilde{\text{Net}}_{pk}\right), \qquad k = 1, 2, \ldots, c. \tag{106}$$

In this formulation, the connection weights $\tilde{W}_{ji}$, $\tilde{W}_{kj}$ and the biases $\tilde{\Theta}_j$, $\tilde{\Theta}_k$ are fuzzy numbers.

As we have already described, the input–output relation of the fuzzified neural network is defined by fuzzy-number arithmetic, and the numerical calculation is

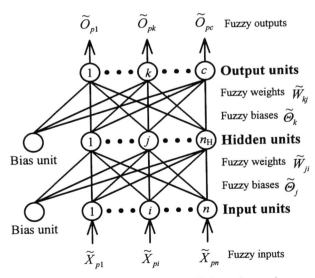

Figure 47 Architecture of fully fuzzified neural networks.

performed by interval arithmetic on h-level sets of fuzzy numbers (see Figs. 3 and 4).

Triangular fuzzy numbers and trapezoidal fuzzy numbers are usually used as the fuzzy connection weights $\widetilde{W}_{ji}$, $\widetilde{W}_{kj}$ and the fuzzy biases $\widetilde{\Theta}_j$, $\widetilde{\Theta}_k$ (see Fig. 48). The learning of the fuzzified neural network is performed by adjusting the fuzzy connection weights and the fuzzy biases. The adjustment of the fuzzy connection weights and the fuzzy biases is done by updating their parameter values. For example, the adjustment of the nonsymmetric triangular fuzzy weight $\widetilde{W}_{kj}$ in Fig. 48a is done by updating its three parameter values, that is, its lower limit w_{kj}^L, center w_{kj}^C, and upper limit w_{kj}^U.

In the learning, the following cost function is used for the fuzzy input–output pair $(\widetilde{\mathbf{X}}_p; \widetilde{\mathbf{Y}}_p)$:

$$e_p = \sum_h \left\{ \sum_{k=1}^{c} ([\widetilde{Y}_{pk}]_h^L - [\widetilde{O}_{pk}]_h^L)^2/2 + \sum_{k=1}^{c} ([\widetilde{Y}_{pk}]_h^U - [\widetilde{O}_{pk}]_h^U)^2/2 \right\}. \quad (107)$$

A back-propagation-type learning algorithm can be derived from this cost function for updating the parameter values of the fuzzy connection weights and the fuzzy biases [42, 43, 90, 91]. For example, the learning algorithm for adjusting the nonsymmetric triangular fuzzy weight $\widetilde{W}_{kj}$ in Fig. 48a can be derived by calculating the partial derivatives $\partial e_p/\partial w_{kj}^L$, $\partial e_p/\partial w_{kj}^C$, and $\partial e_p/\partial w_{kj}^U$.

To illustrate our approach, we show some simulation results. First, we show an example of fuzzy modeling, that is, approximate realization of a nonlinear fuzzy function. Let us assume that we have three input–output pairs for a single-input and single-output nonlinear fuzzy system in Fig. 49a where inputs are real numbers and outputs are trapezoidal fuzzy numbers. Using these three input–output pairs, we trained a fuzzified neural network with nonsymmetric trapezoidal fuzzy numbers as connection weights and biases. Four parameters of the nonsymmetric trapezoidal fuzzy numbers (see Fig. 48b) were adjusted by the learning algorithm derived from the cost function in (107) with ten levels (i.e., $h = 0.1, 0.2, \ldots, 1.0$).

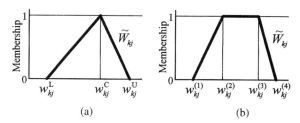

(a)   (b)

Figure 48 Fuzzy connection weight $\widetilde{W}_{kj}$: (a) nonsymmetric triangular fuzzy number; (b) nonsymmetric trapezoidal fuzzy number.

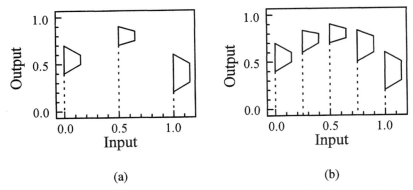

Figure 49 Simulation results: (a) training data; (b) fuzzy outputs from the trained neural network.

The fuzzy outputs from the trained neural network are shown in Fig. 49b. From Fig. 49, we can observe a good fit to the fuzzy targets and a good generalization for new inputs.

We also trained the same fuzzified neural network using the following fuzzy if-then rules:

If x is *small* then y is *small*,

If x is *medium* then y is *medium small* or *medium*,

If x is *large* then y is *medium* or *medium large* or *large*,

where the membership functions of disjunctive combinations of linguistic values are defined by trapezoidal fuzzy numbers as shown in Fig. 50. In the same manner as in the previous example, we trained the fuzzified neural network by the learning algorithm derived from the cost function in (107). The fuzzy outputs from the

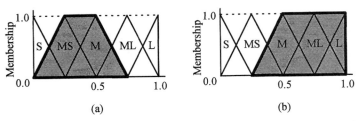

Figure 50 Membership functions of disjunctive combinations of linguistic values: (a) *medium small* or *medium*; (b) *medium* or *medium large* or *large*.

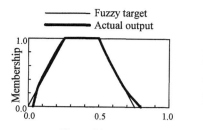
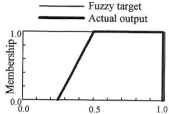

Figure 51   Fuzzy outputs from the trained neural network.

trained neural network are shown in Fig. 51 together with the fuzzy targets. From this figure, we can see that a good fit to the fuzzy targets was realized by the learning of the fuzzified neural network.

V. CONCLUSION

In this chapter, we described how feedforward neural networks can be extended for handling the fuzziness of training data. First, we explained a fuzzy classification method where we assumed that classification boundaries between different classes are not crisp but fuzzy. In our fuzzy classification method, possible classes of an input pattern can be suggested by the trained neural network. Next we explained a fuzzy modeling method by two standard neural networks. One neural network was used for representing the lower limit of a nonlinear interval function, and the other was used for the upper limit. The two neural networks were trained in order that the nonlinear interval function should approximately cover all the given input–output pairs. Then we explained interval-arithmetic-based neural networks where inputs, connection weights, biases, and targets were extended to intervals. Interval-arithmetic-based neural networks can be used for the handling of uncertain or missing inputs. They can also be used for approximately realizing nonlinear interval functions. Finally, we extended inputs, connection weights, biases, and targets to fuzzy numbers in order to fuzzify multilayer feedforward neural networks. Fuzzified neural networks can be used for the handling of linguistic inputs, the learning from fuzzy if-then rules, and the approximation of nonlinear fuzzy functions.

As we have mentioned in this chapter, various architectures have been referred to as "fuzzy neural networks." Most of those architectures have been proposed for control problems. That is, they map real-number input vectors to real numbers. Our fuzzy neural networks in this chapter have the ability to handle the fuzziness in training data. Thus they can be trained from linguistic information as well as numerical information. They can also be used for extracting linguistic knowledge from neural networks trained by numerical information. In this manner, our fuzzy neural networks serve as a bridge between two kinds of information, that is, numerical information and linguistic information.

REFERENCES

[1] S. C. Lee and E. T. Lee. Fuzzy neural networks. *Math. Biosci.* 23:151–177, 1975.

[2] H. Ishibuchi. Development of fuzzy neural networks. In *Fuzzy Modeling: Paradigms and Practice* (W. Pedrycz, Ed.), pp. 185–202. Kluwer Academic, Boston, 1996.

[3] D. E. Rumelhart, G. E. Hinton, and R. J. Williams. Learning representations by back-propagating errors. *Nature* 323:533–536, 1986.

[4] D. E. Rumelhart, J. L. McClelland, and the PDP Research Group. *Parallel Distributed Processing*, Vol. 1. MIT Press, Cambridge, MA, 1986.

[5] H. Ichihashi and T. Watanabe. Learning control by fuzzy models using a simplified fuzzy reasoning. *J. Japan Soc. Fuzzy Theory Systems* 2:429–437, 1990 (in Japanese).

[6] H. Nomura, I. Hayashi, and N. Wakami. A learning method of fuzzy inference rules by descent method. In *Proceedings of the First IEEE International Conference on Fuzzy Systems*, San Diego, pp. 203–210, 1992.

[7] L.-X. Wang and J. M. Mendel. Back-propagation fuzzy system as nonlinear dynamic system identifiers. In *Proceedings of the First IEEE International Conference on Fuzzy Systems*, San Diego, pp. 1409–1418, 1992.

[8] L.-X. Wang. Stable adaptive fuzzy control of nonlinear systems. *IEEE Trans. Fuzzy Systems* 1:146–155, 1993.

[9] C.-T. Lin and C. S. G. Lee. Neural-network-based fuzzy logic control and decision system. *IEEE Trans. Comput.* 40:1320–1336, 1991.

[10] C.-T. Lin and C. S. G. Lee. Reinforcement structure/parameter learning for neural-network-based fuzzy logic control systems. *IEEE Trans. Fuzzy Systems* 2:46–63, 1994.

[11] H. R. Berenji. A reinforcement learning-based architecture for fuzzy logic control. *Internat. J. Approximate Reasoning* 6:267–292, 1992.

[12] H. R. Berenji and P. Khedkar. Learning and tuning fuzzy logic controllers through reinforcements. *IEEE Trans. Neural Networks* 3:724–740, 1992.

[13] J.-S. R. Jang. Fuzzy controller design without domain experts. In *Proceedings of the First IEEE International Conference on Fuzzy Systems*, San Diego, pp. 289–296, 1992.

[14] J.-S. R. Jang. ANFIS: adaptive-network-based fuzzy inference system. *IEEE Trans. Systems Man Cybernet.* 23:665–685, 1993.

[15] S. Horikawa, T. Furuhashi, and Y. Uchikawa. On fuzzy modeling using fuzzy neural networks with the back-propagation algorithm. *IEEE Trans. Neural Networks* 3:801–806, 1992.

[16] T. Hasegawa, S. Horikawa, T. Furuhashi, and Y. Uchikawa. On design of adaptive fuzzy controller using fuzzy neural networks and a description of its dynamical behavior. *Fuzzy Sets Systems* 71:3–23, 1995.

[17] C.-T. Sun. Rule-base structure identification in an adaptive-network-based fuzzy inference system. *IEEE Trans. Fuzzy Systems* 2:64–73, 1994.

[18] C. M. Higgins and R. M. Goodman. Fuzzy rule-based networks for control. *IEEE Trans. Fuzzy Systems* 2:82–88, 1994.

[19] H. K. Kwan and Y. Cai. A fuzzy neural network and its application to pattern recognition. *IEEE Trans. Fuzzy Systems* 2:185–193, 1994.

[20] I. H. Suh and T. W. Kim. Fuzzy membership function based neural networks with applications to the visual servoing of robot manipulators. *IEEE Trans. Fuzzy Systems* 2:203–220, 1994.

[21] C.-L. Chen and W.-C. Chen. Fuzzy controller design by using neural network techniques. *IEEE Trans. Fuzzy Systems* 2:235–244, 1994.

[22] C.-T. Lin, C.-J. Lin, and C. S. G. Lee. Fuzzy adaptive learning control network with on-line neural learning. *Fuzzy Sets Systems* 71:25–45, 1995.

[23] M. L. Presti, R. Poluzzi, and A. M. Zanaboni. Synthesis of fuzzy controllers through neural networks. *Fuzzy Sets Systems* 71:47–70, 1995.

[24] J. J. Shann and H. C. Fu. A fuzzy neural network for rule acquiring on fuzzy control systems. *Fuzzy Sets Systems* 71:345–357, 1995.

[25] J. S. R. Jang and C. T. Sun. Functional equivalence between radial basis function networks and fuzzy inference systems. *IEEE Trans. Neural Networks* 4:156–163, 1993.

[26] J. Nie and D. A. Linkens. Learning control using fuzzified self-organizing radial basis function network. *IEEE Trans. Fuzzy Systems* 1:280–287, 1993.

[27] J. M. Keller and H. Tahani. Backpropagation neural networks for fuzzy logic. *Inform. Sci.* 62:205–221, 1992.

[28] J. M. Keller and H. Tahani. Implementation of conjunctive and disjunctive fuzzy logic rules with neural networks. *Internat. J. Approximate Reasoning* 6:221–240, 1992.

[29] H. Takagi and I. Hayashi. NN-driven fuzzy reasoning. *Internat. J. Approximate Reasoning* 5:191–212, 1991.

[30] I. Hayashi, H. Nomura, H. Yamasaki, and N. Wakami. Construction of fuzzy inference rules by NDF and NDFL. *Internat. J. Approximate Reasoning* 6:241–266, 1992.

[31] S. K. Pal and S. Mitra. Multi-layer perceptron, fuzzy sets and classification. *IEEE Trans. Neural Networks* 3:683–697, 1992.

[32] S. Mitra. Fuzzy MLP based expert system for medical diagnosis. *Fuzzy Sets Systems* 65:285–296, 1994.

[33] S. Mitra and L. I. Kuncheva. Improving classification performance using fuzzy MLP and two-level selective partitioning of the feature space. *Fuzzy Sets Systems* 70:1–13, 1995.

[34] A. Kaufmann and M. M. Gupta. *Introduction to Fuzzy Arithmetic.* Van Nostrand–Reinhold, New York, 1985.

[35] L. A. Zadeh. The concept of a linguistic variable and its application to approximate reasoning. 1, 2, and 3. *Inform. Sci.* 8:199–249, 8:301–357, 9:43–80, 1975.

[36] Y. Hayashi, J. J. Buckley, and E. Czogala. Fuzzy neural network with fuzzy signals and weights. *Internat. J. Intelligent Systems* 8:527–537, 1993.

[37] J. J. Buckley and Y. Hayashi. Fuzzy neural networks: a survey. *Fuzzy Sets Systems* 66:1–13, 1994.

[38] P. V. Krishnamraju, J. J. Buckley, K. D. Reilly, and Y. Hayashi. Genetic learning algorithms for fuzzy neural nets. In *Proceedings of the Third IEEE International Conference on Fuzzy Systems*, Orlando, pp. 1969–1974, 1994.

[39] J. J. Buckley and Y. Hayashi. Neural nets for fuzzy systems. *Fuzzy Sets Systems* 71:265–276, 1995.

[40] J. J. Buckley, K. D. Reilly, and K. V. Penmetcha. Backpropagation and genetic algorithms for training fuzzy neural nets. In *Proceedings of the Fifth IEEE International Conference on Fuzzy Systems*, New Orleans, pp. 2–6, 1996.

[41] H. Ishibuchi, R. Fujioka, and H. Tanaka. Neural networks that learn from fuzzy if-then rules. *IEEE Trans. Fuzzy Systems* 1:85–97, 1993.

[42] H. Ishibuchi, K. Kwon, and H. Tanaka. A learning algorithm of fuzzy neural networks with triangular fuzzy weights. *Fuzzy Sets Systems* 71:277–293, 1995.

[43] H. Ishibuchi, K. Morioka, and I. B. Turksen. Learning of fuzzified neural networks. *Internat. J. Approximate Reasoning* 13:327–358, 1995.

[44] A. F. Rocha. Neural fuzzy point processes. *Fuzzy Sets Systems* 5:127–140, 1981.

[45] J. M. Keller and D. J. Hunt. Incorporating fuzzy membership functions into the perceptron algorithm. *IEEE Trans. Pattern Anal. Machine Intell.* 7:693–699, 1985.

[46] G. A. Carpenter, S. Grossberg, and D. B. Rosen. Fuzzy ART: fast stable learning and categoriza-
     tion of analog patterns by an adaptive resonance system. *Neural Networks* 4:759–771, 1991.
[47] G. A. Carpenter, S. Grossberg, N. Markuzon, J. H. Reynolds, and D. B. Rosen. Fuzzy ARTMAP:
     a neural network architecture for incremental supervised learning of analog multidimensional
     maps. *IEEE Trans. Neural Networks* 3:698–713, 1992.
[48] W. Pedrycz. Neurocomputations in relational systems. *IEEE Trans. Pattern Anal. Machine Intell.*
     13:289–297, 1991.
[49] P. K. Simpson. Fuzzy min-max neural networks. 1: Classification. *IEEE Trans. Neural Networks*
     3:776–786, 1992.
[50] P. K. Simpson. Fuzzy min-max neural networks. 2: Clustering. *IEEE Trans. Fuzzy Systems* 1:32–
     45, 1993.
[51] K. Hirota and W. Pedrycz. OR/AND neuron in modeling fuzzy set connectives. *IEEE Trans.
     Fuzzy Systems* 2:151–161, 1994.
[52] M. Furukawa and T. Yamakawa. The design algorithms of membership functions for a fuzzy
     neuron. *Fuzzy Sets Systems* 71:329–343, 1995.
[53] F. Rosenblatt. The perceptron: a probabilistic model for information strategy and organization in
     the brain. *Psychol. Rev.* 65:386–408, 1958.
[54] K. Funahashi. On the approximate realization of continuous mappings by neural networks. *Neu-
     ral Networks* 2:183–192, 1989.
[55] K. Hornik. Multilayer feedforward networks are universal approximators. *Neural Networks*
     2:359–366, 1989.
[56] H. White. Connectionist nonparametric regression: multilayer feedforward networks can learn
     arbitrary mappings. *Neural Networks* 3:535–549, 1990.
[57] K. Hornik. Approximation capabilities of multilayer feedforward networks. *Neural Networks*
     4:251–257, 1991.
[58] H. Ishibuchi and H. Tanaka. Approximate pattern classification using neural networks. In *Fuzzy
     Logic: State of the Art* (R. Lowen and M. Roubens, Eds.), pp. 225–236. Kluwer Academic,
     Dordrecht, 1993.
[59] H. Ishibuchi, K. Nozaki, and R. Weber. Approximate pattern classification with fuzzy boundary.
     In *Proceedings of the International Joint Conference on Neural Networks*, Nagoya, Japan, pp.
     693–696, 1993.
[60] H. Ishibuchi, R. Fujioka, and H. Tanaka. Possibility and necessity pattern classification using
     neural networks. *Fuzzy Sets Systems* 48:331–340, 1992.
[61] D. Dubois and P. Prade. *Possibility Theory*. Plenum, New York, 1988.
[62] N. B. Karayiannis and G. Purushothaman. Fuzzy pattern recognition using feed-forward neural
     networks with multilevel hidden neurons. In *Proceedings of the IEEE International Conference
     on Neural Networks*, Orlando, pp. 1577–1582, 1994.
[63] G. Purushothaman and N. B. Karayiannis. Feed-forward neural architectures for membership
     estimation and fuzzy classification. In *Intelligent Engineering Systems Through Artificial Neural
     Networks* (C. H. Dagli, B. R. Fernandez, J. Ghosh, and S. R. T. Kumara, Eds.), Vol. 4, pp. 235–
     240. ASME, New York, 1994.
[64] G. Purushothaman and N. B. Karayiannis. On the capability of feed-forward neural networks
     for fuzzy classification. In *Intelligent Engineering Systems Through Artificial Neural Networks*
     (C. H. Dagli, B. R. Fernandez, J. Ghosh, and S. R. T. Kumara, Eds.), Vol. 5, pp. 253–258. ASME,
     New York, 1995.
[65] G. Purushothaman and N. B. Karayiannis. Quantum neural networks (QNNs): inherently fuzzy
     feedforward neural networks. In *Proceedings of the IEEE International Conference on Neural
     Networks*, Washington, DC, pp. 1085–1090, 1996.
[66] N. P. Archer and S. Wang. Fuzzy set representation of neural network classification boundary.
     IEEE Trans. Systems Man Cybernet. 21:735–742, 1991.

[67] H. Ishibuchi and H.Tanaka. Regression analysis with interval model by neural networks. In *Proceedings of the IEEE International Joint Conference on Neural Networks*, Singapore, pp. 1594–1599, 1991.

[68] H. Ishibuchi and H. Tanaka. Fuzzy regression analysis using neural networks. *Fuzzy Sets Systems* 50:257–266, 1992.

[69] H. Tanaka, S. Uejima, and K. Asai. Linear regression analysis with fuzzy model. *IEEE Trans. Systems Man Cybernet.* 12:903–907, 1982.

[70] H. Tanaka. Fuzzy data analysis by possibilistic linear models. *Fuzzy Sets Systems* 24:363–375, 1987.

[71] J. Kacprzyk and M. Fedrizzi, Eds. *Fuzzy Regression Analysis*. Omnitech Press, Warsaw, 1992.

[72] H. Ishibuchi and M. Nii. Fuzzy regression analysis by neural networks with non-symmetric fuzzy number weights. In *Proceedings of the IEEE International Conference on Neural Networks*, Washington, DC, pp. 1191–1196, 1996.

[73] H. Ishibuchi and M. Nii. Fuzzy regression analysis with non-symmetric fuzzy number coefficients and its neural network implementation. In *Proceedings of the Fifth IEEE International Conference on Fuzzy Systems*, New Orleans, pp. 318–324, 1996.

[74] R. A. Fisher. The use of multiple measurements in taxonomic problems. *Ann. Eugenics* 7:179–188, 1936.

[75] S. M. Weiss and C. A. Kulikowski. *Computer Systems That Learn*. Morgan Kaufmann, San Mateo, CA, 1991.

[76] R. E. Moore. *Methods and Applications of Interval Analysis*. SIAM, Philadelphia, 1979.

[77] G. Alefeld and J. Herzberger. *Introduction to Interval Computations*. Academic Press, New York, 1983.

[78] H. Ishibuchi, A. Miyazaki, K. Kwon, and H. Tanaka. Learning from incomplete training data with missing values and medical application. In *Proceedings of the International Joint Conference on Neural Networks*, Nagoya, Japan, pp. 1871–1874, 1993.

[79] H. Ishibuchi, A. Miyazaki, and H. Tanaka. Neural-network-based diagnosis systems for incomplete data with missing inputs. In *Proceedings of the IEEE International Conference on Neural Networks*, Orlando, pp. 3457–3460, 1994.

[80] H. Ishibuchi and H. Tanaka. An extension of the BP-algorithm to interval input vectors—learning from numerical data and expert's knowledge. In *Proceedings of the IEEE International Joint Conference on Neural Networks*, Singapore, pp. 1588–1593, 1991.

[81] H. Ishibuchi and A. Miyazaki. Determination of inspection order for classifying new samples by neural networks. In *Proceedings of the IEEE International Conference on Neural Networks*, Orlando, pp. 2907–2910, 1994.

[82] H. Ishibuchi, H. Tanaka, and H. Okada. An architecture of neural networks with interval weights and its application to fuzzy regression analysis. *Fuzzy Sets Systems* 57:27–39, 1993.

[83] K. Kwon, H. Ishibuchi, and H. Tanaka. Neural networks with interval weights for nonlinear mapping of interval vectors. *IEICE Trans. Inform. Systems* E77-D:409–417, 1994.

[84] H. Ishibuchi, R. Fujioka, and H. Tanaka. An architecture of neural networks for input vectors of fuzzy numbers. In *Proceedings of the First IEEE International Conference on Fuzzy Systems*, San Diego, pp. 1293–1300, 1992.

[85] H. Ishibuchi, H. Tanaka, and H. Okada. Interpolation of fuzzy if-then rules by neural networks. *Internat. J. Approximate Reasoning* 10:3–27, 1994.

[86] H. Ishibuchi and K. Morioka. Classification of fuzzy input patterns by neural networks. In *Proceedings of the IEEE International Conference on Neural Networks*, Perth, Australia, pp. 3118–3123, 1995.

[87] H. Ishibuchi and M. Nii. Generating fuzzy if-then rules from trained neural networks: linguistic analysis of neural networks. In *Proceedings of the IEEE International Conference on Neural Networks*, Washington, DC, pp. 1133–1138, 1996.

[88] A. Miyazaki, K. Kwon, H. Ishibuchi, and H. Tanaka. Fuzzy regression analysis by fuzzy neural networks and its application. In *Proceedings of the Third IEEE International Conference on Fuzzy Systems*, Orlando, pp. 52–57, 1994.

[89] H. Ishibuchi and K. Morioka. Determination of type II membership functions by fuzzified neural networks. In *Proceedings of the Third European Congress on Intelligent Techniques and Soft Computing*, Aachen, Germany, pp. 529–533, 1995.

[90] H. Ishibuchi, K. Morioka, and H. Tanaka. A fuzzy neural network with trapezoid fuzzy weights. In *Proceedings of the Third IEEE International Conference on Fuzzy Systems*, Orlando, pp. 228–233, 1994.

[91] H. Ishibuchi and M. Nii. Learning of fuzzy connection weights in fuzzified neural networks. In *Proceedings of the Fifth IEEE International Conference on Fuzzy Systems*, New Orleans, pp. 373–379, 1996.

Implementation of Fuzzy Systems

Chu Kwong Chak
Department of Electrical and Electronic
Engineering
University of Melbourne
Parkville, 3052 Victoria, Australia

Gang Feng
Department of Systems and Control
School of Electrical Engineering
University of New South Wales
Sydney, New South Wales 2052, Australia

Marimuthu Palaniswami
Department of Electrical and Electronic
Engineering
University of Melbourne
Parkville, 3052 Victoria, Australia

I. INTRODUCTION

In the 1960s, Zadeh [1, 2] developed a linguistic approach to deal with linguistic vague information based on fuzzy sets and fuzzy logic. Since then there have been a number of applications of the approach to a variety of fields including meteorology, engineering, medicine, management, computer science, expert systems, and systems science.

In the field of systems science, many complex plants are difficult to deal with by the conventional approach (precise mathematical equations) because of their nonlinear, time-varying behavior and imprecise measurement information. Nevertheless, human operators can handle these complex plants by their practical experience. They only need imprecise system states and a set of imprecise linguistic if-then rules. The fuzzy system theory developed by Zadeh [3] based on fuzzy sets and fuzzy logic can be used to deal with such complex systems.

Fuzzy systems accept numeric inputs from the outside world and convert these into linguistic values that can be manipulated by using fuzzy logic operations with linguistic if-then rules given by human operators. The linguistic outputs, the

Fuzzy Logic and Expert Systems Applications

result of the fuzzy logic operations, are converted into numeric outputs which are then delivered to the outside world. Thus, fuzzy systems provide a framework of representing human expert rules with fuzzy logic to infer human decision. Based on this ability, fuzzy systems can approximate human reasoning and achieve some intelligence.

Fuzzy systems can be used for different kinds of purposes such as modeling, prediction, classification, and control in the field of systems science. In particular, the possible use of fuzzy systems in modeling and control has generated great attention. Fuzzy systems for modeling and control have emerged as one of the most active and fruitful areas for research in the application of fuzzy set theory. The application was pioneered by Mamdani [4], who successfully carried out a pilot study on a model steam engine using fuzzy systems. His study showed that fuzzy systems may profitably and easily be used by control engineers. A number of successful control applications have also been reported. These included heat exchange process control [5], steam engine control [6, 7], traffic junction control [8], cement kiln control [9], model car parking control [10], automobile speed control [11], robot control [12, 13], aircraft autopilot control [14], camera autofocus control, and automobile transmission control [15].

However, at present there is no systematic procedure for the design of fuzzy systems. Usually the linguistic rules are generated by converting the human operator's experience into linguistic form directly or by summarizing the sampled input–output pairs of the systems to be dealt with. Unfortunately, it is difficult for systems designers to obtain optimal fuzzy rules because these are most likely to be influenced by the intuitiveness of the operators and the systems designers. Moreover, some information will be lost when human operators express their experience by linguistic rules. This results in a set of less than optimal linguistic rules. Therefore, fuzzy systems capable of developing and improving the linguistic rules and structures automatically are highly desired [16–18].

Neural network implementation of fuzzy systems has been proposed as a possible approach for fuzzy systems design [19–29]. The resulting systems, which are sometimes called fuzzy neural networks or neural-network-based fuzzy systems, will possess the advantages of both types of systems and overcome the difficulties of each type of system. In fact, the resulting systems not only support numerical mathematical analysis, hardware implementation, distributed parallel processing, and self-learning but are also capable of dealing with difficulties arising from uncertainty, imprecision, and noise.

Another aim of developing neural-network-based fuzzy systems is to enhance fuzzy systems with higher intelligence. Fuzzy systems simulate human reasoning to achieve intelligence by manipulating a set of heuristic rules given by a human expert. Thus, the intelligence is totally limited by the given set of rules. There will be neither chance for the fuzzy system to improve nor useful rules to be added. To make fuzzy systems more intelligent, fuzzy systems with learning and adaptation are desired.

The fuzzy neural network discussed in this chapter is a hybrid system which functions as a fuzzy system with the processing mechanism realized by a neural network. Thus, the capability of learning imposed upon a fuzzy system can be achieved by the learning algorithm of a neural network. In principle, a fuzzy neural network is a fuzzy system implemented within the framework of neural networks so as to achieve the capability of learning using input–output data which will lead to improvement of the fuzzy rules and fuzzy system intelligence.

In general, there are two approaches to the integration of fuzzy systems and neural networks. In the first approach, one may incorporate the concept of fuzzy logic into the neural network. A fuzzy neuron is designed to function in much the same way as a nonfuzzy neuron, except that it reflects the fuzzy nature and has the ability to cope with fuzzy information [23–26].

The other approach [19–22, 27–29] is to realize the process of fuzzy reasoning by the structure of a neural network and to express the parameters of fuzzy reasoning by the connection weights of the neural network. The resulting fuzzy neural network can automatically identify the fuzzy rules and tune membership functions by modifying the connection weights of the network using some learning algorithm. This second approach is closer to dealing with the problem of fuzzy systems design. This chapter will deal mainly with the second approach to fuzzy neural networks. This approach has been discussed by a number of researchers [19–22].

Horikawa *et al.* [22] described three general structures of fuzzy neural networks in accordance with the structure of the consequences of fuzzy rules. The first type is concerned with the consequence being a crisp constant, the second one with the consequence being a function of input variables, and the third one with the consequence being a fuzzy value. The error back-propagation algorithm was used for training.

Lin and Lee [20, 30] proposed a neural-network-based fuzzy logic control system. This work considered finding centers/widths of membership functions by self-organized clustering and finding fuzzy logic rules by competitive learning. The fuzzy logic control system implemented was of a conventional type, and error back propagation was applied to tune the consequence parameters of output membership functions and premise parameters of input membership functions. The system was enhanced with a reinforcement learning method when obtaining exact training data became expensive [31].

Jang [19] implemented the Sugeno–Takagi fuzzy logic system using an adaptive network (which can be regarded as a neural network) that utilized hybrid learning rules. A gradient descent techniques was applied to tune premise parameters, and the least-squares estimation techniques was used to estimate consequence parameters. The membership functions were chosen to be bell-shaped functions (highly nonlinear functions; e.g., of the Gaussian type). It was shown that the system was functionally equivalent to a radial basis function network [32].

The fuzzy neural networks proposed in the aforementioned papers suffered from the limitation that if the number of input fuzzy partitions is large, the required number of consequence parameters will be very large, and the least-squares estimation algorithm cannot be implemented easily because the calculation of very large matrices is required. Thus, the application of the networks is limited to some low-dimensional systems. Moreover, the learning processes were typically slow.

This chapter discusses the neural network implementation of fuzzy systems based on Takagi–Sugeno fuzzy systems [33] because they have many advantages for modeling and control. Takagi–Sugeno fuzzy systems differ from conventional fuzzy systems in that linear systems instead of fuzzy sets are formed in the consequences of the fuzzy rules. The output of the fuzzy systems is a "fuzzy" combination of a set of linear systems. In what follows, the basic concepts of fuzzy sets, fuzzy logic, and structure of fuzzy systems are presented first, and fuzzy neural network designs are then discussed in the latter part of this chapter.

II. STRUCTURE OF FUZZY SYSTEMS FOR MODELING AND CONTROL

This section gives an insight into the structure of fuzzy systems for modeling and control. Some of the basic vocabulary relating to fuzzy systems is presented, which is required for the development of fuzzy systems and the design of fuzzy neural networks in this chapter.

A. FUZZY SETS AND FUZZY LOGIC

In the real world, objects are often classified into different categories. For such categories as *tall man, high inflation rate, pretty woman* etc., all of them convey linguistic vague information. The concept of membership of an object in such categories is not obvious and not precise. Thus, the application of classical two-valued logic to the real world is limited in some cases. The idea of fuzzy sets proposed by Zadeh [1] aims to deal with such information.

Fuzzy set theory is an extension of classical set theory. In classical set theory, an element either belongs to a set or does not belong to a set. In fuzzy set theory, an element may partially belong to a set. Fuzzy sets have gradations of set membership which is represented by a function referred to as a membership function, and so they resemble the kinds of categories ordinary people use in natural thought or communication. The formal presentation of the fuzzy set theory is as follows:

DEFINITION 1. Let $x \in U$ and let S be a subset of U. $\mu(x) : U \rightarrow [0, 1]$ is called the membership function which represents the degree of x belonging to the subset S. U is called the universe of discourse. Then the fuzzy set A is defined to be a set of ordered pairs $A = \{(x, \mu(x)) \mid x \in S, S \subset U\}$. The membership function is denoted by $\mu_A(x)$ for the fuzzy set A. The support of a fuzzy set A denoted as $A_{\sup}$ is the crisp set of all points x in U such that $\mu_A(x) > 0$. A fuzzy set A whose support $A_{\sup}$ contains a single point x in U with $\mu_A(x) = 1$ is referred to as a fuzzy singleton. A fuzzy set A whose support $A_{\sup}$ is the universe of discourse U with $\mu(x) = 1$ is referred to as a fuzzy universe. It is denoted by Z. If the universe of discourse U is a set of real numbers, the fuzzy sets defined on U are called fuzzy numbers.

The fuzzy set operations are defined via their membership functions.

DEFINITION 2. Let A_1 and A_2 be fuzzy sets in U and let B be a fuzzy set in V.

(i) Union:

$$A_1 \cup A_2 = \{x, \mu_{A_1 \cup A_2}(x) \mid x \in U\}, \qquad \text{where } \mu_{A_1 \cup A_2}(x) = \mu_{A_1}(x) \vee \mu_{A_2}(x);$$

(ii) Intersection:

$$A_1 \cap A_2 = \{x, \mu_{A_1 \cap A_2}(x) \mid x \in U\}, \qquad \text{where } \mu_{A_1 \cap A_2}(x) = \mu_{A_1}(x) \wedge \mu_{A_2}(x);$$

(iii) Complement:

$$\overline{A_1} = \{x, \mu_{\overline{A_1}}(x) \mid x \in U\}, \qquad \text{where } \mu_{\overline{A_1}}(x) = 1 - \mu_{A_1}(x);$$

(iv) Cartesian product:

$$A_1 \times B = \{v, \mu_{A_1 \times B}(v) \mid v = (x_1, x_2) \in W, \ W = U \times V\},$$
$$\text{where } \mu_{A_1 \times B}(v) = \mu_{A_1}(x_1) \wedge \mu_B(x_2).$$

The operators $\wedge$ and $\vee$ can be any kind of triangular norms and triangular conorms, respectively [34], for example, product, sum, max, or min. Refer to [35, 36] for additional fuzzy set operations.

A linguistic variable can be regarded as a variable whose values are defined in linguistic terms (e.g., negative large, negative small, positive small, and positive large). These terms which are imprecise and ill-defined can be represented by fuzzy sets. In fact, the use of fuzzy sets provides a basis for the systematic manipulation of such linguistic variables or such linguistic terms.

Based on linguistic information, human experts can describe the behavior of a system using a set of rules such as "If A then B" in which A and B are fuzzy sets representing linguistic information. Each rule can be expressed as a fuzzy implication. The ideas of fuzzy implication are as follows:

In classical logic, the rule "If A then B" in the form of an implication is written as $A \rightarrow B$ which is equivalent to the relation $R := {\sim}A \vee B$ (not A or B). For fuzzy logic, the fuzzy implication "If A then B" where A and B are fuzzy sets with membership functions μ_A and μ_B, respectively, which represent linguistic variables, is expressed in a different way. Instead of using $R := {\sim}A \vee B$ as its relation, the fuzzy relation R is defined to be a fuzzy set of the product $A \times B$ characterized by a membership function μ_R which is obtained by $\mu_R = \mu_A \wedge \mu_B$. Thus, the fuzzy rule "If A then B" can be expressed as a fuzzy implication denoted by $A \rightarrow B$ using the fuzzy relation R. In the context of fuzzy logic, there are many ways to define a fuzzy implication. In fuzzy control literature, the commonly used fuzzy implication is based on the composition rule of inference for approximate reasoning suggested in [3].

B. BASIC STRUCTURE OF FUZZY SYSTEMS FOR MODELING AND CONTROL

Fuzzy systems for either modeling or control have similar operations. Figure 1 shows the block diagram of the structure of conventional fuzzy systems for modeling and control. The fuzzy system is composed of four function blocks: fuzzification, rule base, inference engine, and defuzzification.

The mechanism of fuzzy systems is as follows: the measurements x of the outside world in the form of crisp data are transformed by fuzzification into linguistic values. Then the linguistic values are processed by the fuzzy rules in the rule base in the form of "if-then" through fuzzy implication. The output expressed in fuzzy sets after fuzzy implication is finally transformed by defuzzification into a nonfuzzy (crisp) output as the output of the system to the outside world.

C. TYPES OF FUZZY SYSTEMS FOR MODELING AND CONTROL

The evolution of the structure of fuzzy systems is mainly affected by the different reasoning methods developed, a better understanding of fuzzy logic, and an ambition of wider application. The evolution is too extensive to be fully discussed. We will restrict our discussion within the context of system modeling and control.

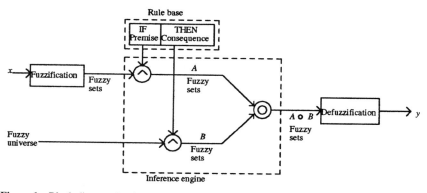

Figure 1 Block diagram for the structure of conventional fuzzy systems for modeling and control.

The first fuzzy system for a control application was developed by Mamdani [4]. In this fuzzy system, one level forward data-driven inference is employed as the inference mechanism. The format of his fuzzy rules is

If x_1 is A_1 and x_2 is A_2 and ... and x_n is A_n, then y is B,

where $A_1, A_2, \ldots, A_n$ and B are fuzzy sets. It is noted that the consequence of implication is a fuzzy set.

His study showed that fuzzy systems may profitably and easily be used by control engineers. A number of successful control applications have been reported in accordance with the structure of the fuzzy system ever since. These include heat exchange process control [5], steam engine control [6, 7], traffic junction control [8], and cement kiln control [9]. The fuzzy system developed by Mamdani is referred to as a conventional fuzzy system (Fig. 1).

In 1985, Takagi and Sugeno [33] modified the consequence of implication from fuzzy sets to linear functions and developed the so-called "Takagi–Sugeno fuzzy systems" which were applied to parking control of a model car [10]. The format of their fuzzy rules is

If x_1 is A_1 and x_2 is A_2 and ... and x_n is A_n, then $y = a_0 + a_1 x_1 + \cdots + a_n x_n$.

The structure of these systems varies significantly from that of the previous ones (the conventional ones). As a consequence of implication, they contain a linear function by which the output can be computed. (It is noted that the term "linear system" may be interchanged with the term "linear function" in the latter part of this chapter.) The aim of the linear function in Takagi–Sugeno fuzzy systems is to describe the local linear behavior of the system. Fuzziness, which appears only in the premise part of the fuzzy rule, indicates the uncertainty about which the output range of the linear function varies.

Takagi–Sugeno fuzzy systems have a number of advantages by their nature. The systems can be easily understood and the local system equations can be directly related to the local behavior of the system. Each local system can be clearly described and the dynamics are separately modeled. Takagi–Sugeno fuzzy systems include two kinds of knowledge: one is the qualitative knowledge represented by the if-then rules, and the other is the quantitative knowledge represented by the local functions. The systems allow us to formulate these two kinds of knowledge into a unified mathematical framework.

In the following subsections, we will discuss the details of each part of a fuzzy system. We will give more precise definitions of the terms which will be used in the latter sections.

D. INPUT DOMAIN AND OUTPUT DOMAIN

Every system has its input and output domains. The input domain and the output domain of a fuzzy system are determined in relation to the input universe of discourse and the output universe of discourse of fuzzy sets in the fuzzy system. When a fuzzy system is designed, the fuzzy sets of fuzzy rules in the universe of discourse should have the input domain and output domain covered while the fuzzy system is operating.

DEFINITION 3. Let $x = [x_1 \quad x_2 \quad \cdots \quad x_n]^T \in R^n$ be the input vector and Ξ_X be the vector space spanned by x. Ξ_X is called the input space. The subset of the space Ξ_X from which a fuzzy system accepts inputs is called the input domain U.

DEFINITION 4. Let $y = [y_1 \quad y_2 \quad \cdots \quad y_m]^T \in R^m$ be the output vector and Ξ_Y be the vector space spanned by y. Ξ_Y is called the output space. The subset of the space Ξ_Y to which a fuzzy system delivers outputs is called the output domain W.

E. RULE BASE

The behavior of a fuzzy system is characterized by a set of linguistic rules which constitutes a rule base. A typical linguistic rule is of the following form:

If (*a set of conditions is satisfied*), then (*a set of consequences can be inferred*).

The premise of a rule is a condition in the input domain U and the consequence is an action to be performed in the output domain W. Because the premises and the consequences of these if-then rules are associated with fuzzy concepts, the rules are expressed as fuzzy rules, for example,

$$\text{If } x_1 \text{ is } A_1 \text{ and } x_2 \text{ is } A_2, \text{ then } y \text{ is } B, \tag{1}$$

where x_1 and x_2 are *scalar* inputs, A_1 and A_2 are input linguistic terms represented by fuzzy sets, and B is an output linguistic term represented by a fuzzy set.

Now consider a general rule base for n-dimensional fuzzy systems whose fuzzy rules are in the form

$$R_l: \text{ If } (x \text{ is } A_1^l \text{ and } x \text{ is } A_2^l \text{ and} \ldots \text{and } x \text{ is } A_{l_1}^l)$$
$$\text{or } (x \text{ is } A_{l_1+1}^l \text{ and } x \text{ is } A_{l_1+2}^l \text{ and} \ldots \text{and } x \text{ is } A_{l_2}^l)$$
$$\vdots$$
$$\text{or } (x \text{ is } A_{l_{K-1}+1}^l \text{ and } x \text{ is } A_{l_{K-1}+2}^l \text{ and} \ldots \text{and } x \text{ is } A_{l_{K_l}}^l),$$
$$\text{then } y \text{ is } B^l, \quad k_l \in \{1, 2, \ldots, K_l\}, \ l = 1, 2, \ldots, L, \tag{2}$$

where x is the input *vector* of the fuzzy systems, the A^l's are fuzzy terms of input (input fuzzy terms) which are represented by fuzzy sets, and B^l is a fuzzy term of output (output fuzzy terms) which is represented by a fuzzy set. Each rule has K_l n-dimensional input fuzzy terms the projection of which into each dimensional is the input linguistic terms. The n-dimensional input fuzzy terms are represented by n-dimensional fuzzy sets A^l. A number of input fuzzy terms can be combined by *AND* and then *OR* to form the premise of a fuzzy rule.

It is noted that the input fuzzy terms and output fuzzy terms used here are fuzzy sets with multidimensional membership functions.

The fuzzy rules given previously are very general so that the fuzzy rules of conventional fuzzy systems can be included. For example, the fuzzy rule in Eq. (1) is equivalent to

$$\text{If } (x_1 \text{ is } A_1 \text{ and } x_2 \text{ is } Z) \text{ and } (x_1 \text{ is } Z \text{ and } x_2 \text{ is } A_2), \text{ then } y \text{ is } B,$$

where Z is a fuzzy universe (Definition 1), or

$$\text{If } (x \text{ is } A_1 \times Z) \text{ and } (x \text{ is } Z \times A_2), \text{ then } y \text{ is } B,$$

where $A_1 \times Z$ and $Z \times A_2$ are two-dimensional fuzzy sets, or

$$\text{If } (x \text{ is } A_1 \times A_2), \text{ then } y \text{ is } B,$$

where $A_1 \times A_2$ is a two-dimensional fuzzy set.

It should be noted that the multidimensional fuzzy sets which represent the input fuzzy terms can be projected into each dimension to obtain the one-dimensional fuzzy sets to represent input linguistic terms.

The fuzzy rules in the rule base in Eq. (2) can be rewritten in the following equivalent rule base:

R'_l: If (x is A^l_1 and x is A^l_2 and...and x is $A^l_{l_1}$), then y is B_l also

R'_{l+1}: If (x is $A^l_{l_1+1}$ and x is $A^l_{l_1+2}$ and...and x is $A^l_{l_2}$), then y is B_l also

$$\vdots$$

R'_{l+K_l}: If (x is $A^l_{n_{K-1}+1}$ and x is $A^l_{n_{K-1}+2}$ and...and x is $A^l_{l_{K_l}}$), then y is B_l also

$$k_l \in \{1, 2, \ldots, K_l\}, \quad l = 1, 2, \ldots, L. \quad (3)$$

From the rule base in Eq. (2) and the rule base in Eq. (3), we see that there are two ways of implementing fuzzy systems. The rule base in Eq. (2) requires more complicated reasoning and implications but less rules, whereas the rule base in Eq. (3) requires more rules but less complicated reasoning and implications. The first one is preferred here because less rule consequences are advantageous for neural network implementation.

It is noted that the same argument as above can also be applied to Takagi–Sugeno fuzzy systems by the replacement of the output labels in Eqs. (1)–(3) with linear functions.

F. INPUT FUZZY PARTITIONS

The input fuzzy partition of the input domain U is related to the interpretation of the premise of fuzzy rules in a rule base. There are a number of input fuzzy terms (fuzzy sets) in the premise of a fuzzy rule. An inferred fuzzy set of the premise of a fuzzy rule can be obtained from the input fuzzy terms (fuzzy sets) of the fuzzy rule. The support of the inferred fuzzy set occupies a subspace of the input space. So there are a number of subspaces of the input space due to a number of fuzzy rules in the rule base. The premise is thus interpreted as a fuzzy hypervolume in the input space and hence the collection of the inferred fuzzy sets of all fuzzy rules in the rule base constitutes the so-called input fuzzy partition. The concept of the input fuzzy partition will be used to describe the mechanism of fuzzy inference employed in fuzzy systems for modeling and control.

DEFINITION 5. Consider a fuzzy system with the rule base in Eq. (2) or an equivalent rule base in Eq. (3). Let $\Omega \subset Z_+$ be an indexed set for the input fuzzy terms and let $\mu_K: \Xi_X \rightarrow [0, 1]$, $k \in \Omega$, be membership functions. Then $A^k = \{(x, \mu_k(x)) \mid x \in U \subset \Xi_X\}, k \in \Omega$, are fuzzy sets representing input fuzzy terms in the input domain $U \subset \Xi_X$ characterized by the membership function μ_k. Let Ψ be a collection of all subsets of $\{A^k, k \in \Omega\}$. The subset P_I of $\Phi = \{F \mid F = \bigcap \chi, \chi \in \Psi\}$ is called an input fuzzy partition of the input domain U if the union of the support of all elements of P_I is equal to the union of the support of A^k, $k \in \Omega$, and the support of any element of P_I is not a subset of the support of any other element of P_I. An element of an input fuzzy partition P_I is called an input fuzzy region. If all elements of P_I are the fuzzy sets A^k, $k \in \Omega$, P_I is called a direct input fuzzy partition, otherwise P_I is called an indirect input fuzzy partition.

EXAMPLE. Suppose there are fuzzy sets A^1, A^2, and A^3. Then

$$\Psi = \{\emptyset, \{A^1\}, \{A^2\}, \{A^3\}, \{A^1, A^2\}, \{A^1, A^3\}, \{A^2, A^3\}, \{A^1, A^2, A^3\}\},$$

and

$$\Phi = \left\{ F \mid F = \bigcap \chi, \chi \in \Psi \right\}$$
$$= \left\{\emptyset, A^1, A^2, A^3, A^1 \cap A^2, A^1 \cap A^3, A^2 \cap A^3, A^1 \cap A^2 \cap A^3 \right\}.$$

Thus a subset P_I of Φ is an input fuzzy partition if P_I satisfies the requirements in Definition 5. For example, see Fig. 2.

Remark 1. It is noted that each element of a direct input fuzzy partition is characterized by the membership function of an input fuzzy term, whereas each element of an indirect fuzzy partition is characterized by more than one membership function of input fuzzy terms. Nevertheless, the membership function of an input fuzzy region in an indirect fuzzy partition can be obtained by fuzzy set operations of input fuzzy terms and hence the indirect input fuzzy partition is

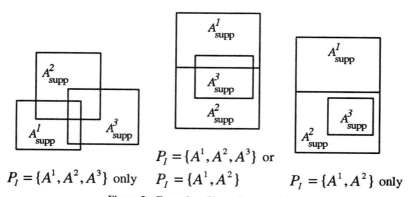

$$P_I = \{A^1, A^2, A^3\} \text{ only} \qquad P_I = \{A^1, A^2, A^3\} \text{ or} \\ P_I = \{A^1, A^2\} \qquad P_I = \{A^1, A^2\} \text{ only}$$

Figure 2 Examples of input fuzzy partition.

equivalent to the direct input fuzzy partition functionally. They are rooted from different partitioning methods (which will be discussed in the design examples) but their characteristics are the same—representing the premise of fuzzy rules.

DEFINITION 6.   Let $\Omega \subset Z_+$ be an index set for the rule base in Eq. (2) or Eq. (3). Let fuzzy set X^k with membership function μ_k be an input fuzzy region of an input partition P_I and X_{sup}^k be the support of X^k. If $\bigcup_{k\in\Omega} X_{\text{sup}}^k = U$ and $X_{\text{sup}}^k \neq \emptyset, \forall k \in \Omega$, and μ_k are normalized bell-shaped functions, then the input fuzzy partition P_I is said to be normal. If an input fuzzy partition P_I is normal, a set $\overline{P}_I = \{\{x \mid \mu_l(x) > \mu_i, i \in \Omega \setminus \{l\}\}, l \in \Omega\}$ can be defined and is called the input crisp partition. An element of the input crisp partition $\overline{P}_I$ is called the input crisp region.

Remark 2.   It is noted that the input fuzzy regions are fuzzy sets and the input crisp regions are crisp (classical) sets, whereas both the input fuzzy partition and the input crisp partition are classical sets of input fuzzy regions and input crisp regions, respectively.

G. *AND* MATRIX FOR INPUT FUZZY TERMS AND INPUT FUZZY REGIONS

The premise of a fuzzy rule is represented by input fuzzy regions which are obtained via *AND* fuzzy operation of input fuzzy terms. Thus, the linkage of input fuzzy terms to input fuzzy regions needs to be specified. The linkage of input fuzzy terms and input fuzzy regions can be many to many, which is expressed by the *AND* matrix M_{AND} with binary entries ("1" represents that an input fuzzy term links to an input fuzzy region and "0" represents no linkage) as shown in Fig. 3. The structure of the *AND* matrix depends on the input fuzzy partition selected. For instance, the *AND* matrix is an identity matrix for the case that input fuzzy partition is direct (Definition 5).

Input fuzzy terms

Input fuzzy regions

```
1 0 0 0 0 1 0
0 1 0 1 0 0 0
0 1 0 0 1 0 0
1 0 0 0 0 0 1
```

Figure 3   *AND* matrix.

H. OUTPUT FUZZY PARTITIONS

The formation of an output fuzzy partition in the output domain W is related to the setting up of the consequences of fuzzy rules. Same as the input fuzzy partition, the concept of the output fuzzy partition will be used to describe the mechanism of fuzzy inferences employed in fuzzy systems for modeling and control.

For conventional fuzzy systems, the consequent parts of fuzzy rules are fuzzy sets and the fuzzy partition of the output domain is clear. However, for Takagi–Sugeno fuzzy systems, the consequent parts are linear functions $y = f(x)$ without any fuzzy set. We can imagine that there is a fuzzy set B with ordered pairs of the output y and the membership function $\mu(y)$. Because the output range of the linear function $y = f(x)$ of a fuzzy rule with finite input domain U is a subset $\overline{B}$ of the output domain W, the membership function of the imagined fuzzy set can be defined as $\mu(y) = 1, \forall y \in \overline{B}$. The imagined fuzzy set is thus defined as $B = (\overline{B}, \mu(y) = 1)$ which is a crisp set, a special case of fuzzy sets.

DEFINITION 7. Consider a fuzzy system with the rule base in Eq. (2) or the equivalent rule base in Eq. (3). Let $\Omega \subset Z_+$ be an index set for the output terms and let $\mu_l: \Xi_Y \to [0, 1]$, $l \in \Omega$, be membership functions. Then $B^l = \{(y, \mu_l(y)) \mid y \in W \subset \Xi_Y\}$, $l \in \Omega$, are fuzzy sets representing output terms in the output domain $W \subset \Xi_Y$ characterized by the membership function $\mu_l(y)$. Let Ψ be the collection of all subsets of $\{B^l, l \in \Omega\}$. The subset P_O of $\Pi = \{G \mid G = \bigcap \chi, \chi \in \Psi\}$ is called an output fuzzy partition of the output domain W if the union of the support of all elements of P_O is equal to the union of the support of B^l, $l \in \Omega$, and the support of any element of P_O is not a subset of the support of any other element of P_O. An element of an output fuzzy partition P_O is called the output fuzzy region.

DEFINITION 8. Let $\Omega \subset Z_+$ be an index set for output terms. Let the fuzzy set G^l with a membership function μ_l be the output fuzzy region of the output fuzzy partition P_O and G^l_{sup} be the support of G^l. If $\bigcup_{l \in \Omega} G^l_{\text{sup}} = W$ and $G^l_{\text{sup}} \neq \emptyset$, $\forall l$, and μ_l is a normalized bell-shaped membership function, the output fuzzy partition P_O is said to be normal. If an output fuzzy partition P_O is normal, a set $\overline{P}_O = \{\{y \mid \mu_l(y) > \mu_i, i \in \Omega \setminus \{l\}\}, l \in \Omega\}$ is called the output crisp partition. An element of the output crisp partition $\overline{P}_O$ is called the output crisp region.

I. *OR* MATRIX FOR INPUT FUZZY REGIONS AND OUTPUT FUZZY REGIONS

The premise and consequence of a fuzzy rule are represented by input fuzzy regions and output fuzzy regions, respectively. To completely represent a fuzzy rule, the linkage of input fuzzy regions and output fuzzy regions needs to be spec-

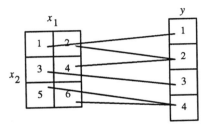

Figure 4   Linkage of input fuzzy regions into output fuzzy input regions.

ified. The linkage of input fuzzy regions and output fuzzy regions of conventional fuzzy systems is one to one, whereas in fuzzy systems with the rule base in Eq. (2) or the equivalent rule base in Eq. (3) the linkage of input fuzzy regions and output fuzzy regions is many to one (Fig. 4). This approach obviously reduces the number of consequences because a consequence can be shared by a number of premises of fuzzy rules.

The linkage of input fuzzy regions and output fuzzy regions can be expressed by the *OR* matrix M_{OR} with binary entries ("1" represents that an input fuzzy region links an output fuzzy region and "0" represents no linkage) as shown in Fig. 5. It should be noted that there is at least one input fuzzy region linked to each output fuzzy region. The structure of the *OR* matrix M_{OR} depends on the input fuzzy partition and the output fuzzy partition selected. For instance, the *OR* matrix can be an identity matrix for the case that an output fuzzy region links to one input fuzzy region only (the conventional fuzzy system).

Output fuzzy regions

Input
fuzzy
regions
$$
\begin{array}{cccc}
1 & 0 & 0 & 0 \\
0 & 1 & 0 & 0 \\
0 & 0 & 1 & 0 \\
0 & 1 & 0 & 0 \\
0 & 0 & 0 & 1 \\
0 & 0 & 0 & 1 \\
\end{array}
$$

Figure 5   *OR* matrix.

J. FUZZIFICATION

The task of fuzzification is to map a crisp input of the system to a fuzzy input.

DEFINITION 9.   Fuzzification is a mapping F of the crisp input domain U with x into the set $\mathfrak{I}(X)$ with fuzzified input X.

In fuzzy systems for modeling and control, singleton fuzzification is usually employed. The fuzzified input for the crisp input x is a fuzzy set (fuzzy singleton) of an ordered pair $(x, \mu(x) = 1)$ only.

K. INFERENCE ENGINE

The inference engine attempts to simulate human decision making based on fuzzy concepts. It aims to infer fuzzy outputs by employing fuzzy implication and rules of inference in fuzzy logic. For each "If A then B" rule, a fuzzy relation is defined based on fuzzy set operations. The rule can be expressed as a fuzzy implication denoted by $A \rightarrow B$ using the defined fuzzy relation. Thus, an individual fuzzy output can be inferred by a fuzzy rule in response to the input. The fuzzy output of the system inferred is the aggregated result derived from all individual fuzzy rules.

1. Fuzzy Relations

A fuzzy rule represents some linguistic relationship of input and output; the product of the input fuzzy region and the output fuzzy region (linked by M_{OR}) forms the fuzzy relation of the input and the output for a fuzzy rule. See Fig. 6 for illustration.

DEFINITION 10.   Let A be an input fuzzy region with element x and membership function μ_A and let B be an output fuzzy region with element y and membership function μ_B. The fuzzy relation on the fuzzy product $A \times B$ is a mapping such that $\mu_R: A \times B \rightarrow [0, 1]$ where $\mu_R(x, y) = \mu_A(x) \wedge \mu_B(y)$ and the fuzzy relation set is defined to be $R = \{((x, y), \mu_R(x, y)) \mid (x, y) \in A \times B\}$.

For a fuzzy rule in the rule base in Eq. (2), if there is more than one input fuzzy region in relation to an output fuzzy region, then the fuzzy relation is combined and defined as follows:

DEFINITION 11.   Let $A_j, i = 1, 2, \ldots, N$, be input fuzzy regions with element x and membership functions $\mu_{A_j}, j = 1, 2, \ldots, N$, and let B be an output

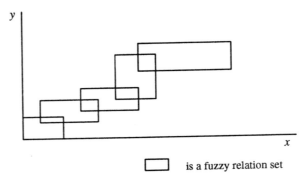

is a fuzzy relation set

Figure 6 Graphic representation of fuzzy relation on the product of input fuzzy regions and output fuzzy regions.

fuzzy region with element y and membership function μ_B. The combined fuzzy relation on the fuzzy product $\bigcup_{j=1}^{N} A_j \times B$ is a mapping such that

$$\mu_{R_C} : \bigcup_{j=1}^{N} A_j \times B \to [0, 1],$$

where

$$\mu_{R_C}(x, y) = \left(\bigvee_{j=1}^{N} \mu_{A_j}(x) \right) \wedge \mu_B(y)$$

and the combined fuzzy relation set is defined to be

$$R_C = \left\{ ((x, y), \mu_{R_C}(x, y)) \mid (x, y) \in \bigcup_{j=1}^{N} A_j \times B \right\}.$$

The fuzzy relation in a fuzzy system is depicted in Fig. 7.

Remark 3. It can be seen that $R_C = \bigcup_{j=1}^{N} R_j$ where R_j is the fuzzy relation set for rule j.

2. Fuzzy Implications

Using the defined fuzzy relation, a fuzzy rule can be expressed by fuzzy implication which means that each point in the input fuzzy region maps a point in the output fuzzy region.

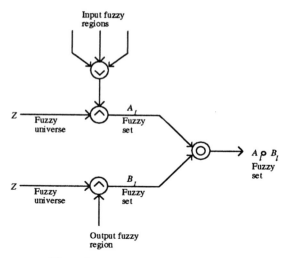

Figure 7 Fuzzy relation of fuzzy systems.

DEFINITION 12. Fuzzy implication is a mapping ϑ of an input fuzzy region A into an output fuzzy region B such that, to each ordered pair $(x, \mu_A(x))$ of A, an ordered pair $(y, \mu_B(y)) = \vartheta((x, \mu_a(x)))$ of B is assigned according to the defined fuzzy relation on $A \times B$.

3. Fuzzy Inference

Fuzzy logic inference is the mechanism to deduce an output y corresponding to an input x by operating fuzzy rules in the rule base.

In fuzzy systems for modeling and control, one level forward data-driven inference is employed for inference mechanism, that is, the fuzzy implication inference rule, which is the generalized modus ponens (GMP). It is of the form

> premise 1: x is X
> premise 2: if x is A, then y is B
> $- -$
> consequence: y is Y

where A, X, B, and Y are fuzzy predicates.

For a fuzzy rule expressed as a fuzzy implication using the defined fuzzy relation R, the linguistic value Y of consequence variable y induced from premise variable x with linguistic value represented by the fuzzy set X with membership function μ_X is given by the fuzzy set $Y = X \circ R$ which is characterized by the membership function $\mu_Y = \mu_X \wedge \mu_R$.

DEFINITION 13.   Fuzzy inference is a mapping φ of the set $\mathfrak{I}(X)$ with all fuzzified inputs X into the set $\mathfrak{I}(Y)$ with all fuzzified outputs Y such that, to each fuzzified input X, a fuzzified output $Y = \varphi(X)$ is assigned by the operation $Y = X \circ R$ according to the fuzzy relation R of a fuzzy rule.

The inference methods of different fuzzy systems are different because of the different structures of fuzzy systems although the fuzzy relations of the fuzzy systems can be expressed as general implication function as mentioned previously. Takagi–Sugeno fuzzy systems differ from conventional fuzzy systems in that local linear functions in the Takagi–Sugeno fuzzy systems are the consequences of fuzzy rules instead of output fuzzy sets. A local function delivers quantitative information to the consequence of a fuzzy rule in response to a quantitative input. In conventional fuzzy systems, the local functions are absent so no quantitative output information is available but qualitative output information (the output fuzzy sets) are given. Figures 8 and 9 illustrate the inference of different fuzzy systems.

We now consider the overall fuzzy inferences given by all fuzzy rules in a rule base. Let Y^k be the inferred output fuzzy set and R^k be the fuzzy relation set

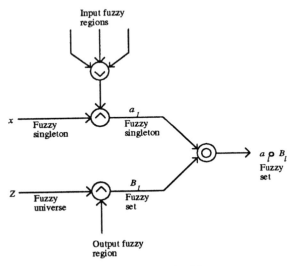

Figure 8   Inference of conventional fuzzy systems.

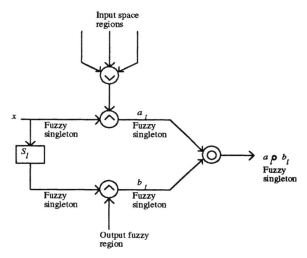

Input space
regions

x

Fuzzy
singleton

S_l

Fuzzy
singleton

a_l

Fuzzy
singleton

b_l

Fuzzy
singleton

$a \, \rho \, b_l$
Fuzzy
singleton

Output fuzzy
region

Figure 9 Inference of Takagi–Sugeno fuzzy systems.

corresponding to a fuzzy rule k in the rule base in Eq. (3) of K fuzzy rules. The overall inferred output fuzzy set is given by

$$Y = \bigcup_{k=1}^{K} Y^k = \bigcup_{k=1}^{K} X \circ R^k = X \circ \bigcup_{k=1}^{K} R^k.$$

Let Y_C^k be the combined inferred output fuzzy set and let R_C^k be the combined fuzzy relation set corresponding to a fuzzy rule l in the rule base in Eq. (2) of L fuzzy rules. The overall inferred output fuzzy set is given by

$$Y = \bigcup_{l=1}^{L} Y_C^l = \bigcup_{l=1}^{L} X \circ R_C^l = X \circ \bigcup_{l=1}^{L} R_C^l.$$

L. DEFUZZIFICATION

The task of defuzzification is to map a fuzzy output to a crisp output of the system.

DEFINITION 14. Defuzzification is a mapping D of the set $\mathfrak{I}(Y)$ with fuzzified outputs Y into the crisp output domain W with y.

A number of schemes have been proposed. A widely used method is the center-of-area method [35, 36]:

$$\frac{\sum_{k=1}^{K} \left(\int_{y \in W} \mu^k(x) \wedge \mu^k(y) y \, dy \right)}{\sum_{k=1}^{K} \left(\int_{y \in W} \mu^k(x) \wedge \mu^k(y) \, dy \right)},$$

where $\mu^k(x)$ and $\mu^k(y)$ are the membership functions of input fuzzy regions and output fuzzy regions, respectively, or

$$\frac{\sum_{l=1}^{L} \left(\int_{y \in W} \mu_C^l(x) \wedge \mu^l(y) y \, dy \right)}{\sum_{l=1}^{L} \left(\int_{y \in W} \mu_C^l(x) \wedge \mu^l(y) \, dy \right)},$$

where $\mu_C^l(x)$ and $\mu^l(y)$ are the membership functions of combined input fuzzy regions and output fuzzy regions, respectively.

Remark 4. For a Takagi–Sugeno system with fuzzy singleton input, the inferred output fuzzy regions are crisp sets and hence the center-of-area method can be reduced to the weighted average of fuzzy singletons:

$$\frac{\sum_{k=1}^{K} \mu^k(x) y_k}{\sum_{k=1}^{K} \mu^k(x)} \quad \text{or} \quad \frac{\sum_{l=1}^{L} \mu_C^l(x) y_l}{\sum_{l=1}^{L} \mu_C^l(x)}.$$

M. CONCLUDING REMARKS

This section has considered some basic concepts and the structure of fuzzy systems for modeling and control which are used for the illustration of fuzzy neural networks. In what follows, we will discuss two fuzzy neural network designs.

III. DESIGN 1: A FUZZY NEURAL NETWORK WITH AN ADDITIONAL *OR* LAYER

A. INTRODUCTION

This section reports the first attempt to solve the problem of neural network implementation of higher-order fuzzy systems with fewer hardware requirements and faster learning schemes. The fuzzy system used in this work is based on Takagi–Sugeno fuzzy systems modified with the introduction of an additional *OR* layer. A local linear system may be associated with more than one input fuzzy region. With this structure, the number of input fuzzy regions can be large, whereas the size of the matrix for local system parameters estimation remains small. Thus, the proposed system is suitable for higher-order complex system modeling and control. The proposed system also has the capability of rules generation.

B. INPUT DIMENSIONAL SPACE PARTITIONING

The partitioning method is discussed first because the structure of the fuzzy system and the architecture of the fuzzy neural network are largely affected by the partitioning method used. The input dimensional subspace partitioning method, as it is called in this chapter, is a conventional fuzzy partitioning method which has been adopted by Takagi [33] and Sugeno [37]. Each dimensional subspace of the input space is first partitioned into a number of fuzzy regions and the input fuzzy partition is then the product of all input dimensional subspace partitions. The partition of the input space depends on the shape of the membership functions.

The idea of this partitioning method can be illustrated by a two-dimensional input fuzzy system with input vector $x = [x_1 \quad x_2]^T$. It is assumed that the system has three fuzzy sets A_1, A_2, and A_3 with membership functions $\mu_{A_1}(x_1)$, $\mu_{A_2}(x_1)$, and $\mu_{A_3}(x_1)$, respectively, at dimension 1 and two fuzzy sets B_1 and B_2 with $\mu_{B_1}(x_2)$ and $\mu_{B_2}(x_2)$, respectively, at dimension 2, and the membership functions are of the form

$$\mu_j(x_i) = \exp\left(-\frac{(x_i - w_{ij})^2}{\sigma_{ij}^2}\right).$$

The premises of the fuzzy rules are

R_1: If x_1 is A_1 and x_2 is B_1, then ...   or   If x is $A_1 \times B_1$, then ...,

R_2: If x_1 is A_1 and x_2 is B_2, then ...   or   If x is $A_1 \times B_2$, then ...,

R_3: If x_1 is A_2 and x_2 is B_1, then ...   or   If x is $A_2 \times B_1$, then ...,

R_4: If x_1 is A_2 and x_2 is B_2, then ...   or   If x is $A_2 \times B_2$, then ...,

R_5: If x_1 is A_3 and x_2 is B_1, then ...   or   If x is $A_3 \times B_1$, then ...,

R_6: If x_1 is A_3 and x_2 is B_2, then ...   or   If x is $A_3 \times B_2$, then ...,

where $A_j \times B_k$, $j = 1, 2, 3$ and $k = 1, 2$, are two-dimensional fuzzy sets with membership function $\mu_{A_j}(x_1) \wedge \mu_{B_k}(x_2)$. If a numeric product is chosen for the t-norm $\wedge$, the membership function of $A_j \times B_k$ is

$$\mu_{A_j}(x_1) \cdot \mu_{B_k}(x_2) = \exp\left(-\frac{(x_1 - w_{1j})^2}{\sigma_{ij}^2} - \frac{(x_2 - w_{2j})^2}{\sigma_{ij}^2}\right).$$

Each two-dimensional fuzzy set $A_j \times B_k$ is represented by an input fuzzy region. The input dimensional subspace fuzzy partitions and the resulting input fuzzy partition of the two-dimensional fuzzy system are shown in Fig. 10.

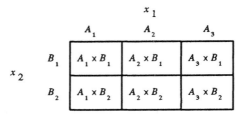

Figure 10   Input dimensional subspace partitioning.

C. STRUCTURE OF THE FUZZY SYSTEM

The proposed fuzzy system is evolved from Takagi–Sugeno fuzzy systems. The main idea of the structure of the Takagi–Sugeno fuzzy system is that in each input fuzzy region of the input domain a local linear function is formed. A membership function $\mu(x) \in [0, 1]$ of each region is a map indicating the degree of the output of the associated linear function belonging to the region. The output of the system is the "fuzzy" combination of the output of linear functions in all regions. The proposed fuzzy system, on the other hand, has added one more *OR* layer, which allows a local linear system to be associated with more than one input fuzzy region. The proposed fuzzy system also uses a singleton fuzzifier, product and sum inference, bell-shaped membership functions, and weighted average defuzzifier.

All input fuzzy terms in the premise part of the fuzzy rules of the proposed system are associated with a bell-shaped membership function $\mu(x)$ chosen to be

$$\mu(x) = \exp\left(-\left(\left(\frac{x - w}{\sigma}\right)^2\right)^b\right),$$

where w, σ, and b are its center, width, and shape, respectively, which are tuned premise parameters at learning.

The proposed fuzzy system has a rule base of L fuzzy rules of the form

$$R_l: \text{If } (x_1 \text{ is } A_1^{l_1} \text{ and } x_2 \text{ is } A_2^{l_1} \text{ and } \dots \text{ and } x_n \text{ is } A_n^{l_1})$$
$$\text{or } (x_1 \text{ is } A_1^{l_2} \text{ and } x_2 \text{ is } A_2^{l_2} \text{ and } \dots \text{ and } x_n \text{ is } A_n^{l_2})$$
$$\vdots$$
$$\text{or } (x_1 \text{ is } A_1^{l_{K_l}} \text{ and } x_2 \text{ is } A_2^{l_{K_l}} \text{ and } \dots \text{ and } x_n \text{ is } A_n^{l_{K_l}}),$$
$$\text{then } f_l = a_0^l + a_1^l x_1 + a_2^l x_2 + \dots + a_n^l x_n.$$

Equivalently, the fuzzy rules can be formulated as

$$R_{l_1}: \text{If } x_1 \text{ is } A_1^{l_1} \text{ and } x_2 \text{ is } A_2^{l_1} \text{ and } \dots \text{ and } x_n \text{ is } A_n^{l_1},$$
$$\text{then } f_l = a_0^l + a_1^l x_1 + a_2^l x_2 + \cdots + a_n^l x_n$$
$$R_{l_2}: \text{If } x_1 \text{ is } A_1^{l_2} \text{ and } x_2 \text{ is } A_2^{l_2} \text{ and } \dots \text{ and } x_n \text{ is } A_n^{l_2},$$
$$\text{then } f_l = a_0^l + a_1^l x_1 + a_2^l x_2 + \cdots + a_n^l x_n$$

$$\vdots$$

$$R_{l_{K_l}}: \text{If } x_1 \text{ is } A_1^{l_{K_l}} \text{ and } x_2 \text{ is } A_2^{l_{K_l}} \text{ and } \dots \text{ and } x_n \text{ is } A_n^{l_{K_l}},$$
$$\text{then } f_l = a_0^l + a_1^l x_1 + a_2^l x_2 + \cdots + a_n^l x_n.$$

The defuzzification is given by

$$\text{Output} = \frac{\sum_{l=1}^{L} q_l f_l}{\sum_{l=1}^{L} q_l},$$

where

$$q_l = \sum_{k_l=1}^{K_l} \prod_{i=1}^{n} \mu_{A_i^{l_{k_l}}}(x_i), \qquad l = 1, 2, \dots, L,$$

is the combined firing strength of a set of rules R_{k_l}, $k_l = 1, 2, \dots, K_l$, whose input fuzzy regions are linked to local system l. The combination of the outputs of the local linear systems is the output of the system under consideration.

It is not difficult to see that the proposed fuzzy system will have fewer consequence parameters than Takagi–Sugeno fuzzy systems because the number of local systems may usually be less than the number of input fuzzy regions for higher-order systems.

Consider an nth-order system with each input dimension partitioned into J input fuzzy regions. Takagi–Sugeno fuzzy systems require the number of premise and consequence parameters of all rules to be $3nJ$ and $(1 + n)J^n$, respectively, and hence the size of the matrix for least-squares estimation is $(1 + n)J^n \times (1 + n)J^n$. As for the proposed fuzzy system, with the input fuzzy regions combined and mapped to L local systems, the required number of premise parameters and consequence parameters of all rules are $3nJ$ and $(1+n)L$, respectively, and hence the size of the matrix for least-squares estimation is $(1 + n)L \times (1 + n)L$. For a higher-order system, this leads to a great reduction in matrix size, and hence the proposed fuzzy system is more suitable for higher-order systems.

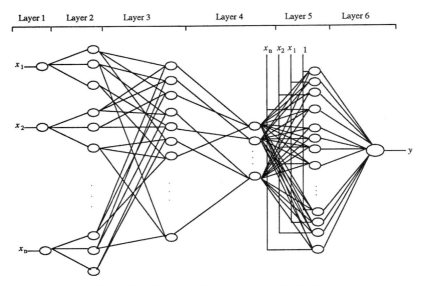

Figure 11   Structure of the proposed neural network.

D. ARCHITECTURE OF THE PROPOSED NEURAL NETWORK

Figure 11 shows the structure of the proposed fuzzy neural network. The network is composed of six layers which are made up of a number of neurons. All neurons in the same layer are identical in their functions, but neurons may have different functions in different layers. A typical neuron is depicted in Fig. 12.

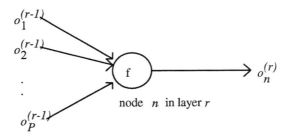

Figure 12   Neuron in layer r.

On the left of Fig. 12 are the multiple inputs $o_1^{(r-1)}, o_2^{(r-1)}, \ldots, o_P^{(r-1)}$ to the node, each arriving from another neuron in the preceding layer $r - 1$. The neuron performs a function f and delivers output to other neurons in the next layer $r + 1$:

$$o_n^{(r)} = f\left(o_1^{(r-1)}, o_2^{(r-1)}, \ldots, o_P^{(r-1)}\right),$$

where $o_n^{(r)}$ denotes the output of the nth neuron in layer r and $o_j^{(r-1)}$ denotes the jth output of the neuron in layer $r - 1$, $j = 1, 2, \ldots, P$.

We now consider the neurons in each layer.

Layer 1

This is an input layer whose neurons represent input variables. The neurons just transmit input values to the next layer directly because input fuzzy sets are fuzzy singletons:

$$o_i^{(1)} = x_i.$$

Layer 2

This is an input term layer whose neurons represent the membership functions associated with each linguistic term of input variables. Links at this layer are fully connected between input neurons and their corresponding terms. We choose the bell-shaped membership function as

$$o_j^{(2)} = \exp\left(-\left(\frac{(o_i^{(1)} - w_j)^2}{\sigma_j^2}\right)^{b_j}\right), \tag{4}$$

where w_j, σ_j, and b_j are the center, width, and shape of the membership function of the jth term of the input variable x_i.

Layer 3

Layer 3 is an input partition layer whose neurons represent the premise of fuzzy rules (input fuzzy regions). Links at this layer are formed in response to the *AND* preconditions of the rules (*AND* matrix). The neurons perform the fuzzy *AND* operation

$$o_k^{(3)} = \prod_j o_j^{(2)} \qquad \text{for some } j. \tag{5}$$

Layer 4

This is an output partition layer (combined input partition) whose neurons represent the output fuzzy regions. Links at this layer are formed in response to the *OR* preconditions of the rules (*OR* matrix). The neurons perform the fuzzy *OR* operation to integrate the fired rules

$$o_k^{(4)} = \sum_k o_k^{(3)} \qquad \text{for some } k. \tag{6}$$

Hence, layers 3 and 4 function as the premise of fuzzy rules.

Layer 5

This is a consequence layer whose neurons represent the weighted local linear systems. Links at this layer are fully connected.

$$o_l^{(5)} = \frac{f_l o_l^{(4)}}{\sum_m o_m^{(4)}}, \tag{7}$$

where

$$f_l = a_0^l + \sum_{i=1}^{n} a_i^l x_i$$

(n is the system order, i.e., the number of input variable x_i, $i = 1, \ldots, n$).

Layer 6

This is an output layer whose neurons represent the output variables. This architecture shows a single output only. It can be extended easily to a multiple-output system. Links at this layer are fully connected. The output of the network is

$$y = o^{(6)} = \sum_l o_l^{(5)}. \tag{8}$$

E. HYBRID LEARNING ALGORITHM

The learning algorithms aim at constructing the fuzzy system by locating the initial membership functions, generating the required fuzzy rules, tuning the membership functions, and finding the consequence parameters so that the performance is optimized through the whole set of training data pairs. However, before applying the learning algorithm, we need to choose the input regions for each input variable x_i and the output regions for output y. Because each neuron (input fuzzy region) in layer 3 is connected to one of the input term neurons (input fuzzy terms) in layer 2, the initial number of rules is equal to the product of the number of input regions of all input variables $\prod_i N(T(x_i))$, where $N(T(x_i))$ denotes the number of terms of input x_i.

All neurons are initially fully interconnected between layers 3 and 4. After rule generation (which will be discussed in phases 2 and 3 of the learning algorithm in this subsection), it is expected that the final number of rules will be reduced. Each neuron in layer 4 is only connected to one neuron in layer 3.

The learning scheme consists of four phases.

Phase 1: Finding the Initial Center and Width of the Membership Functions

In this phase, the centers w_j and widths σ_j of the membership functions of input fuzzy terms are determined. The centers of two membership functions are placed at the upper limits and lower limits of the input range at each dimension. The other centers of the remaining membership functions are located evenly over the input range. The width of the membership function can be simply determined by

$$\sigma_j = \tfrac{1}{2}|w_{j+1} - w_j| \quad \text{or} \quad \sigma_j = \tfrac{1}{2}|w_j - w_{j-1}|.$$

As for the output partition layer, the number of output regions needs to be chosen. It is expected that more accurate output can be obtained if the output layer is assigned more output regions. Each output region is associated with a local system.

Phase 2: Determining Fuzzy Rules by Competitive Learning

The purpose of this phase is to determine the relationship between input fuzzy regions and output fuzzy regions. Initially, the links between layers 3 and 4 are fully interconnected. The weight of the link connecting the kth neuron in layer 3 and the lth neuron in layer 4 is denoted by α_k^l and assigned a value of 0.5. A competitive learning algorithm is adopted. For the set of training data pairs (x, y), the weights are adjusted as follows:

$$\Delta\alpha_k^l = o_l^{(4)}\left(-\alpha_k^l + o_k^{(3)}\right),$$

where $o_k^{(3)}$ is the output of neurons (the output of the input fuzzy region) in layer 3 and $o_l^{(4)}$ is the output of neurons (the output of the combined input fuzzy region) in layer 4.

Hence, $o_l^{(4)}$ serves as a win–loss index of competition. After competitive learning, the weight α_k^l will approach either zero or some other value. The convergence proof of this law can be found in [20].

Phase 3: Generating Rules

In the previous phase, the links at layer 4 are fully interconnected; that is, a maximum number of rules are considered. However, not all the rules are vital to the fuzzy system. The purpose of this phase is to delete those unimportant rules and to retain the essential ones based on the result of competitive learning through the whole set of training data pairs. The weight of a link that connects a neuron in layer 3 (representing an input fuzzy region) and a neuron in layer 4 (representing an output region) indicates the strength of the rule affecting the output region. The weights of the links that connect the same neuron in layer 4 are compared. If the weight of the link is found to be small compared to the maximum one, the weight of the link is assigned a 0 (10% is chosen in our simulation example). The remaining weights are then assigned a 1. Hence, α_k^l will be either 1 or 0, which indicates the existence of the links connecting neuron l in layer 4 and neuron k in layer 3. If there is no link connecting a neuron in layer 3 and a neuron in layer 4, the neuron in layer 3 is regarded as deleted (see Tables I and II). The remaining $\{\alpha_k^l\}$ of the reduced network forms the *OR* matrix and represents the linkage of layers 3 and 4.

Phase 4: Optimizing the Parameters of Membership Functions by Error Back Propagation and Finding the Parameters of Local Systems by Recursive Least-Squares Estimation

After the first three phases, the structure of the whole network has been determined. In this phase, error back propagation is applied to tune the parameters of the membership functions and recursive least-squares estimation is applied to find the parameters of the local linear systems simultaneously. The network can be considered as a cascade of a nonlinear system and a linear system. Error back propagation is applied to the nonlinear part and recursive least-squares estimation the linear part. Figure 13 shows the block diagram for the learning scheme.

1. Error Back Propagation

For each training pair (x, z), the system output $y = o^{(6)}$ is obtained in forward pass after feeding x into the network. Thus, the purpose of this learning phase is that, for a given tth training data pair $(x(t), z(t))$, the parameters are adjusted so as to minimize the error function

$$E(t) = \tfrac{1}{2}\big(z(t) - y(t)\big)^2,$$

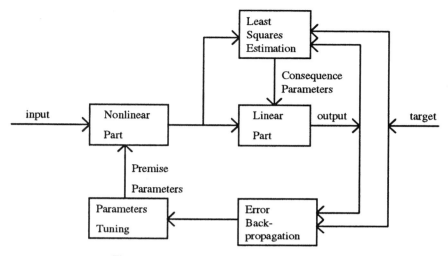

Figure 13 Block diagram for the learning scheme.

where $z(t)$ is the target output of the tth training data pair and $y(t)$ is the current output of the network.

The parameter update laws for w_j, σ_j, and b_j are found to be (see the Appendix)

$$\Delta w_j = -\eta \sum_k \frac{\partial E}{\partial o_j^{(2)}} \frac{2b_j}{o_i^{(1)} - w_j} D_{ij},$$

$$\Delta \sigma_j = -\eta \sum_k \frac{\partial E}{\partial o_j^{(2)}} \frac{2b_j}{\sigma_j} D_{ij},$$

and

$$\Delta b_j = \eta \sum_k \frac{\partial E}{\partial o_j^{(2)}} \log\left(\frac{(o_i^{(1)} - w_j)^2}{\sigma_j^2}\right) D_{ij},$$

where

$$D_{ij} = \left(\frac{(o_i^{(1)} - w_j)^2}{\sigma_j^2}\right)^{b_j} \exp\left(-\left(\frac{(o_i^{(1)} - w_j)^2}{\sigma_j^2}\right)^{b_j}\right).$$

2. Recursive Least-Squares Estimation

In addition to applying error back propagation for tuning the membership functions, recursive least-squares estimation is used to find the consequence parameters of the local linear systems. From Eqs. (7) and (8), we have

$$y = o^{(6)} = \frac{\sum_l f_l o_l^{(4)}}{\sum_l o_l^{(4)}},$$

where $f_l = a_0^l + \sum_{i=1}^n a_i^l x_i$ (n is the system order, i.e., the number of input variable x_i, $i = 1, \ldots, n$). a_i^l are the parameters needed to be estimated.

Let us define at the tth training data pair the cost function J:

$$J = \sum_t E(t) = \frac{1}{2} \sum_t e(t)^2,$$

where $e(t)$ is the estimation error given by

$$e(t) = z(t) - y(t) = z(t) - \phi(t)^T \hat{\theta}(t-1),$$

$\hat{\theta}(t) = \left[\hat{a}^1(t) \quad \hat{a}^2(t) \quad \cdots \quad \hat{a}^L(t)\right]^T$ with $\hat{a}^l(t) = \left[\hat{a}_0^l(t) \quad \hat{a}_1^l(t) \quad \hat{a}_2^l(t) \cdots \hat{a}_n^l(t)\right]$,

and

$$\phi(t) = \left[\frac{o_1^{(4)}(t)}{\sum_l o_l^{(4)}(t)} \quad \frac{o_2^{(4)}(t)}{\sum_l o_l^{(4)}(t)} \quad \cdots \quad \frac{o_L^{(4)}(t)}{\sum_l o_l^{(4)}(t)}\right]^T \otimes \left[1 \quad x_1 \quad x_2 \cdots x_n\right]^T.$$

$\otimes$ denotes the Kronecker matrix operator.

Then recursive least-squares estimation can readily be applied to find the parameters $\hat{\theta}$ such that the cost function J is minimized. The algorithm for updating the parameters is

$$\hat{\theta}(t) = \hat{\theta}(t-1) + \frac{P(t-1)\phi(t)}{1 + \phi(t)^T P(t-1)\phi(t)} e(t),$$

$$P(t) = P(t-1) - \frac{P(t-1)\phi(t)\phi(t)^T P(t-1)}{\phi(t)^T P(t-1)\phi(t)},$$

with $\hat{\theta}(0)$ given and $P(-1)$ a positive-definite matrix.

F. SIMULATION EXAMPLES

EXAMPLE 1.   The proposed neural network is trained to model a three-input nonlinear function $y = (1 + x_1^{0.5} + x_2^{-1} + x_3^{-1.5})^2$ which was also used by Takagi and Hayashi [38] and Sugeno and Kang [37] to verify their approaches. An input

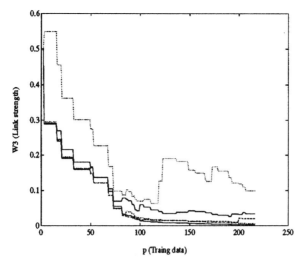

Figure 14 Convergence of some strengths of the links between input fuzzy regions and output fuzzy regions.

range $[1, 6]$ of each dimension is divided into three fuzzy regions, each of which is associated with a Gaussian membership function. The output range is divided into eight output fuzzy regions, each of which is associated with a local linear system. Twenty-seven rules are thus constructed initially, and 216 training data pairs, the input part of which is randomly generated within the input range, are used to train the network.

The initial shape of the membership functions after the first phase of hybrid learning is shown in Fig. 17 (later in this chapter). After the second phase of hybrid learning, the strengths of the links connecting input partition neurons and output partition neurons converge. Figure 14 shows the convergence of the strengths. After the third phase of hybrid learning, the rules are determined. Table I shows the rules before the rule reduction, and Table II shows the rules after the rule reduction. Nine rules are deleted. In the last phase, a step size $k = 0.001$ was selected for the proposed system. In addition, both the adaptive network fuzzy inference system (ANFI) [19] and the fuzzy radial basis function (FRBF) [21] are also simulated in order to evaluate the proposed system. The step sizes for them are also selected to be $k = 0.001$. For the sake of comparison, we use the same performance index adopted in [37]:

$$\text{Average percentage error (APE)} = \frac{1}{P} \sum_{p=1}^{P} \frac{|z(p) - y(p)|}{|z(p)|} \times 100\%,$$

Table I

Output partition

	1	0	0	0	0	0	0	0
	1	0	0	0	0	0	0	0
	1	0	0	0	0	0	0	0
	0	0	1	0	0	0	0	0
	0	0	1	0	0	0	0	0
	0	0	1	0	0	0	0	0
	0	0	1	0	0	0	0	0
R	0	1	0	0	0	0	0	0
u	0	0	0	0	0	0	1	0
l	1	0	0	0	0	0	0	0
e	1	0	0	0	0	0	0	0
	0	0	0	0	0	0	1	0
	0	0	0	0	0	0	0	1
n	0	0	0	0	1	0	0	0
o	0	0	0	0	0	0	1	0
d	0	0	0	0	0	0	0	1
e	0	1	0	0	0	0	0	0
	0	0	0	0	0	0	1	0
	0	0	0	1	0	0	0	0
	0	0	0	1	0	0	0	0
	0	0	0	1	0	0	0	0
	0	0	0	0	0	1	0	0
	0	0	0	0	0	1	0	0
	0	0	0	0	0	1	0	0
	0	0	0	0	0	0	0	1
	0	0	0	0	0	1	0	0
	0	0	0	0	0	0	1	0

Table II

Output partition

	0	0	0	0	0	0	0	0	deleted
	1	0	0	0	0	0	0	0	
	0	0	0	0	0	0	0	0	deleted
	0	0	1	0	0	0	0	0	
	0	0	1	0	0	0	0	0	
	0	0	1	0	0	0	0	0	
	0	0	0	0	0	0	0	0	deleted
R	0	1	0	0	0	0	0	0	
u	0	0	0	0	0	0	1	0	
l	1	0	0	0	0	0	0	0	
e	1	0	0	0	0	0	0	0	
	0	0	0	0	0	0	0	0	deleted
	0	0	0	0	0	0	0	1	
n	0	0	0	0	1	0	0	0	
o	0	0	0	0	0	0	1	0	
d	0	0	0	0	0	0	0	0	deleted
e	0	1	0	0	0	0	0	0	
	0	0	0	0	0	0	1	0	
	0	0	0	0	0	0	0	0	deleted
	0	0	0	1	0	0	0	0	
	0	0	0	1	0	0	0	0	
	0	0	0	0	0	0	0	0	deleted
	0	0	0	0	0	1	0	0	
	0	0	0	0	0	1	0	0	
	0	0	0	0	0	0	0	0	deleted
	0	0	0	0	0	0	0	0	deleted
	0	0	0	0	0	0	1	0	

where P is the number of data pairs and $z(p)$ and $y(p)$ are the pth desired output and network output, respectively.

It is noted that the number of consequence parameters in ANFI is 108 and in the proposed system it is 32. The results of the performance of the networks are shown in Figs. 15 and 16. The results indicate that the performance of the proposed system is close to that of ANFI and is much better than that of FRBF. Figures 15 and 16 show that the curve is "L" shaped. The average percentage error approaches its optimal value after two-epoch training. This is due to the fact that the parameters of the local systems have converged. This implies that the convergence of the parameters of local systems plays a dominate role for system estimation accuracy. The remaining time is just for fine tuning the parameters of the membership functions. Thus, the training required to achieve acceptable accuracy

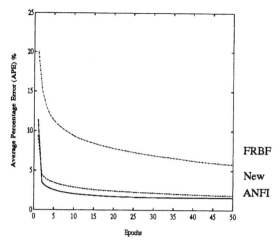

Figure 15 Performance comparison.

for the proposed network is expected to be fast. Figure 17 shows the membership functions before training; Figs. 18–20 show the membership functions after training. Figure 21 shows the convergence of the consequence parameters. Figure 22 shows the performance of the proposed neural network.

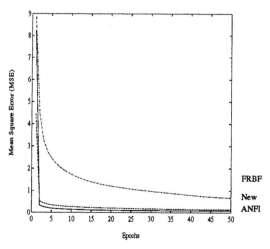

Figure 16 Performance comparison.

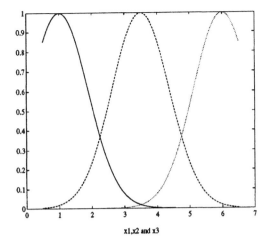

Figure 17 Initial shape of membership functions for x_1, x_2, and x_3.

EXAMPLE 2.   The proposed neural network is also trained to model an operator's control of a chemical plant [39]. In [39] the first three inputs were identified as being significant to the model. Thus, in this simulation, only the first three inputs are selected. Each input range is divided into two fuzzy regions. The output range is divided into four fuzzy regions. The simulation results after 30-epoch

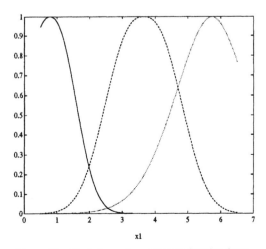

Figure 18 Final shape of membership function for x_1.

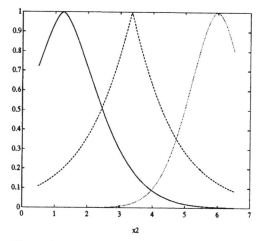

Figure 19 Final shape of membership function for x_2.

training are shown in Fig. 23. These results indicate good performance of the proposed neural network.

EXAMPLE 3. Finally, we deal with an example of the trend data of stock prices [39]. The data set consists of ten inputs and one output (a higher-order

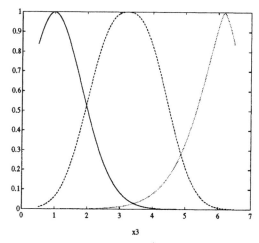

Figure 20 Final shape of membership function for x_3.

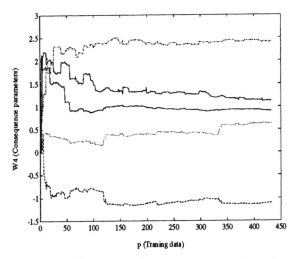

Figure 21   Convergence of some consequence parameters.

system). Each input range is divided into two fuzzy regions. The output range is divided into seven fuzzy regions. The simulation results after 100-epoch training are shown in Fig. 24. The simulation results demonstrate that the proposed neural network performs well.

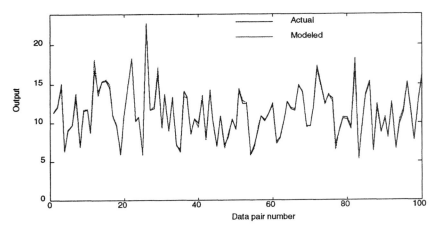

Figure 22   Output of the nonlinear function.

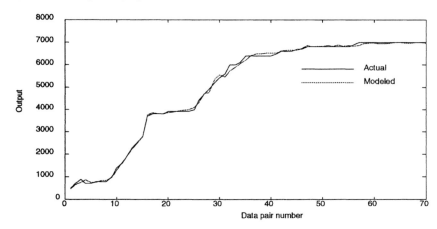

Figure 23   Output of plant operation model.

G. CONCLUDING REMARKS

A neural network implementation of a new fuzzy system has been proposed. Unlike the standard Takagi–Sugeno fuzzy system (in which the number of local linear systems is the same as the number of input fuzzy regions), the proposed system introduces an additional *OR* layer, which is a means of controlling the

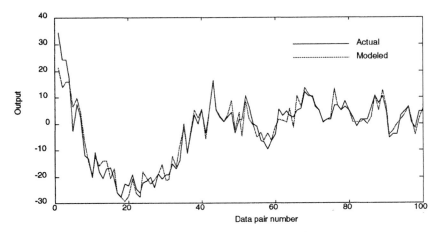

Figure 24   Output of stock price model.

growth of the number of local linear systems when the order of the system under consideration increases so that least-squares estimation can be applied without much performance degradation. The simulation results showed that even though the number of local linear systems is reduced, the performance of the proposed system is encouraging. It is expected that a better performance can be achieved if more local linear systems are allowed by dividing the output domain into more output fuzzy regions.

IV. DESIGN 2: A FUZZY NEURAL NETWORK BASED ON HIERARCHICAL SPACE PARTITIONING

A. INTRODUCTION

In this design, the Takagi–Sugeno fuzzy systems is implemented within the framework of a sigmoid function neural network, which is one of the most popular feedforward neural networks. The fuzzy neural network adopts the hierarchical space partitioning method for its structure selection. The partitioning method is based on the idea of recursively partitioning the regions of the worst performance. The performance of the system improves as this partitioning process continues until some performance criterion is satisfied. Thus, the number of input fuzzy regions (corresponding to fuzzy rules or neurons) is determined automatically in accordance with the prespecified error. The fuzzy neural network is suitable for higher-order fuzzy system implementation.

B. HIERARCHICAL INPUT SPACE PARTITIONING

The input fuzzy partition is formed by hierarchical partitioning of the input domain, that is, by recursive hyperplane cutting of the input domain. Figure 25 illustrates the idea of hierarchical input space partitioning. For each cutting, two input fuzzy terms A^{2j-1} and A^{2j} are formed by the cutting plane $g_j(x)$ shown in Fig. 26. The combination of the input fuzzy terms with fuzzy *AND* operations results in a set of input fuzzy regions which represent the premise of fuzzy rules. The relationship between the input fuzzy terms and the input fuzzy regions can be expressed by the *AND* matrix M_{AND}, and the matrix can be constructed systematically in accordance with the mechanism of hierarchical input space partitioning.

The mechanism of hierarchical input space partitioning can be illustrated as follows. Given a set of input and output data, a linear cutting plane $g_1(x) = 0$ is searched in the input space, which divides the input domain U into two input

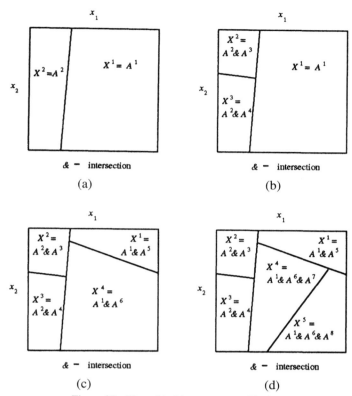

Figure 25   Hierarchical input space partitioning.

crisp regions, say, G^1 and G^2, to optimize some performance index. On the linear cutting plane $g_1(x) = 0$, there are two membership functions assigned which are complementary to each other. It can be seen that two fuzzy sets (input fuzzy terms), say, A^1 and A^2, corresponding to G^1 and G^2, respectively, can be formed as shown in Fig. 26. The two fuzzy sets are complementary to each other. The input fuzzy partition becomes $\{A^1, A^2\}$ as shown in Fig. 25a corresponding to the first step of input space partitioning. The *AND* matrix

$$M_{AND} = \begin{bmatrix} 1 & 0 \\ 0 & 1 \end{bmatrix}$$

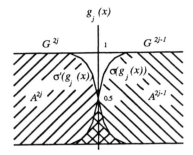

Figure 26 Input fuzzy terms and membership functions.

can be constructed. The generate rule base consisting of two fuzzy rules is

$$R_1: \text{If } x \text{ is } A^1, \text{then.} \ldots,$$
$$R_2: \text{If } x \text{ is } A^2, \text{then.} \ldots.$$

If the performance is not satisfactory, one of the input fuzzy regions with the worst performance, say, A^2, is selected to be partitioning again. Suppose another linear cutting plane $g_2(x) = 0$ is searched in the input domain U, which cuts U into two other input crisp regions, G^3 and G^4. On the linear cutting plane $g_2(x) = 0$, there are two other membership functions assigned which are complementary to each other. It can be seen that there exist another two fuzzy sets, say, A^3 and A^4, corresponding to G^3 and G^4. The four fuzzy sets (input fuzzy terms) A^1, A^2, A^3, and A^4 with *AND* fuzzy set operations can constitute a number of different input fuzzy partitions (Definition 5). Because A^2 is selected to be partitioned, the resulting input fuzzy partition is $\{A^1, A^2 \cap A^3, A^2 \cap A^4\}$ as shown in Fig. 25b corresponding to the second step of input space partitioning. The *AND* matrix M_{AND} becomes

$$M_{AND} = \begin{bmatrix} \begin{bmatrix} 1 & 0 \\ 0 & 1 \\ 0 & 1 \end{bmatrix} & \begin{matrix} 0 & 0 \\ 1 & 0 \\ 0 & 1 \end{matrix} \end{bmatrix}.$$

The generated rule base consisting of three rules is

$$R_1: \text{If } x \text{ is } A^1, \text{then.} \ldots,$$
$$R_2: \text{If } x \text{ is } A^2 \text{ and } x \text{ is } A^3, \text{then.} \ldots,$$
$$R_3: \text{If } x \text{ is } A^2 \text{ and } x \text{ is } A^4, \text{then.} \ldots.$$

This procedure is repeated again and again until some criterion is satisfied.

After four cuttings, for example, the hierarchical input space partitioning shown in Fig. 25d is completed. The generated fuzzy rule base consisting of five fuzzy rules is as follows:

$$R_1: \text{If } x \text{ is } A^1 \text{ and } x \text{ is } A^5, \text{then}\ldots,$$
$$R_2: \text{If } x \text{ is } A^2 \text{ and } x \text{ is } A^3, \text{then}\ldots,$$
$$R_3: \text{If } x \text{ is } A^2 \text{ and } x \text{ is } A^4, \text{then}\ldots,$$
$$R_4: \text{If } x \text{ is } A^1 \text{ and } x \text{ is } A^6 \text{ and } x \text{ is } A^7, \text{then}\ldots,$$
$$R_5: \text{If } x \text{ is } A^1 \text{ and } x \text{ is } A^6 \text{ and } x \text{ is } A^8, \text{then}\ldots,$$

where A^j are input fuzzy terms.

The corresponding *AND* matrix which describes the previous structure (the relationship of the input fuzzy terms and the input fuzzy regions) is updated recursively and can be represented as

$$M_{AND} = \begin{bmatrix} 1 & 0 & 0 & 0 & 1 & 0 & 0 & 0 \\ 0 & 1 & 1 & 0 & 0 & 0 & 0 & 0 \\ 0 & 1 & 0 & 1 & 0 & 0 & 0 & 0 \\ 1 & 0 & 0 & 0 & 0 & 1 & 1 & 0 \\ 1 & 0 & 0 & 0 & 0 & 1 & 0 & 1 \end{bmatrix}. \tag{9}$$

With the *AND* matrix, the preceding rule base can be expressed as

$$R_1: \text{If } x \text{ is } X^1, \text{then}\ldots,$$
$$R_2: \text{If } x \text{ is } X^2, \text{then}\ldots,$$
$$R_3: \text{If } x \text{ is } X^3, \text{then}\ldots,$$
$$R_4: \text{If } x \text{ is } X^4, \text{then}\ldots,$$
$$R_5: \text{If } x \text{ is } X^5, \text{then}\ldots,$$

where X^j are input fuzzy regions. The structure can also be represented by a binary tree structure shown in Fig. 27. The structure of the fuzzy neural network shown in Fig. 28 can be evolved because of the hierarchical input space partitioning which leads to the generation of fuzzy rules.

C. STRUCTURE OF THE FUZZY SYSTEM

The fuzzy system to be implemented is a Takagi–Sugeno fuzzy system with a singleton fuzzifier, product inference, sigmoid membership functions, and weighted average defuzzifier.

Let us consider an n-input and m-output Takagi–Sugeno fuzzy system with input vector $x = [x_1 \quad x_2 \quad \cdots \quad x_n]^T$ in input domain U and output vector $y =$

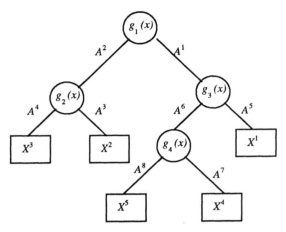

Figure 27 Binary tree structure representation of fuzzy rules.

$[y_1 \quad y_2 \quad \cdots \quad y_m]^T$ in output domain W. The hierarchical space partitioning method is adopted for the fuzzy system. The cutting plane $g_k(x) = 1 + w_k x = 0$ cuts the input domain U to form two input fuzzy terms A^{2k-1} and A^{2k}. Two complementary membership functions for the two input fuzzy terms at the cutting

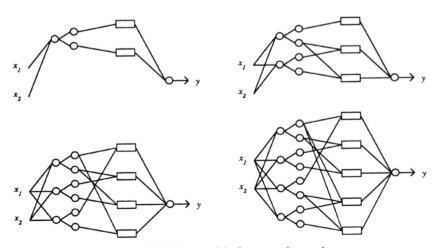

Figure 28 Evolution of the fuzzy neural network.

plane are chosen to be, respectively,

$$\mu_{A^{2k-1}}(x) = \sigma\left(\beta_k g_k(x)\right), \tag{10}$$

where $\sigma(\beta_k g_k(x)) = 0.5(1 + \tanh(-\beta_k g_k(x)))$ and

$$\mu_{A^{2k}}(x) = \sigma'\left(\beta_k g_k(x)\right), \tag{11}$$

where $\sigma'(\beta_k g_k(x)) = 0.5(1 - \tanh(-\beta_k g_k(x)))$.

With hierarchical space partitioning, an input fuzzy partition of the input domain is thus formed which is characterized by a structured *AND* matrix M_{AND}. The inferred membership function of the input fuzzy region X^j (corresponding to the rule R_j) is thus obtained by $\mu_{X^j}(x) = \prod_i \mu_{A^i}(x)$ for all input fuzzy terms A^i linking to the input fuzzy region X^j (for all entries x with value 1 in row j of the *AND* matrix). If the input fuzzy terms formed by cutting planes are complementary to one another, it can be shown that $\sum_j \mu_{X^j}(x) = 1$ for all j (inherent normalized membership functions). The fuzzy rule R_j is

$$R_j: \text{If } x \text{ is } X^j, \text{ then } y^j = a^j \bar{x},$$

where

$$a^j = \begin{bmatrix} a_{10}^j & a_{11}^j & a_{12}^j & \cdots & a_{1n}^j \\ a_{20}^j & a_{21}^j & a_{22}^j & \cdots & a_{2n}^j \\ a_{30}^j & a_{31}^j & a_{32}^j & \cdots & a_{3n}^j \\ \vdots & \vdots & \vdots & \ddots & \vdots \\ a_{m0}^j & a_{m1}^j & a_{m2}^j & \cdots & a_{mn}^j \end{bmatrix}, \qquad j = 1, 2, \ldots, L,$$

$x = [x_1 \quad x_2 \quad \cdots \quad x_n]^T$, $\bar{x} = [1 \quad x_1 \quad x_2 \quad \cdots \quad x_n]^T$, $y^j = [y_1^j \quad y_2^j \quad \cdots \quad y_m^j]^T$, and X^j is the input fuzzy region derived from input fuzzy terms A^i using the *AND* matrix M_{AND}.

The preceding equations form a fuzzy rule with multidimensional input variables and multidimensional membership functions. The fuzzy rule R_j is implemented by fuzzy implication $R_j : X^j \rightarrow Y^j$ and defined as follows:

$$\mu_{R_j} = \mu_{X^j} \wedge \mu_{Y^j} = \mu_{X^j},$$

where Y^j is a crisp set of local system output y^j.

The fuzzy inference engine is a decision-making logic which employs fuzzy rules from the fuzzy rule base to determine the weight output of each local linear system. The inference output Y^j of the rule R_j is $Y^j = X \circ R_j$ where Y^j is a fuzzy set characterized by membership function $\mu_{Y^j} = \mu_X(x) \wedge \mu_{R_j}$ with input fuzzy set X. Consider the fact that the input fuzzifier is a singleton. Then we have $y^j = x \circ R_j$ where x and y^j are fuzzy singletons.

The defuzzification is the weighted average of local linear systems output

$$y = \frac{\sum_{j=1}^{L} \mu_{X^j}(x) y^j(x)}{\sum_{j=1}^{L} \mu_{X^j}(x)} = \sum_{j=1}^{L} \mu_{X^j}(x) y^j(x) \qquad \text{because} \sum_{j=1}^{L} \mu_{X^j}(x) = 1.$$

D. ARCHITECTURE OF PROPOSED FUZZY NEURAL NETWORK

In this section, the architecture of the proposed fuzzy neural network is addressed. The proposed fuzzy neural network is constructed according to hierarchical input space partitioning discussed in Section IV.B and the fuzzy system structure discussed in Section IV.C. As shown in Fig. 29, the network is a six-layer sigmoid function neural network with a number of neurons in each layer. A typical neuron performs a function f and delivers its output to neurons in the next layer

$$o_n^{(r)} = f\left(o_1^{(r-1)}, o_2^{(r-1)}, \ldots, o_P^{(r-1)}\right),$$

where $o_n^{(r)}$ denotes the output of the nth neuron in layer r and $o_j^{(r-1)}$ denotes the output of the jth neuron in layer $r-1$, $j = 1, 2, \ldots, P$.

We now consider the neurons in each layer.

Layer 1

This is an input layer. Its neurons represent input variables. The neurons just transmit input values to the next layer directly because input fuzzy sets are fuzzy singletons:

$$o_i^{(1)} = u_i^{(1)} = x_i.$$

Layer 2

This is the cutting plane layer. The output of its neurons represents the output of cutting plane $g_j(x)$. The links to the neurons represent the coefficients of the cutting plane. The neuron function is

$$o_j^{(2)} = g_j(x) = 1 + w_j x,$$

where the w_j's are the coefficient vectors of cutting plane $g_j(x)$. The neuron performs the weighted summation.

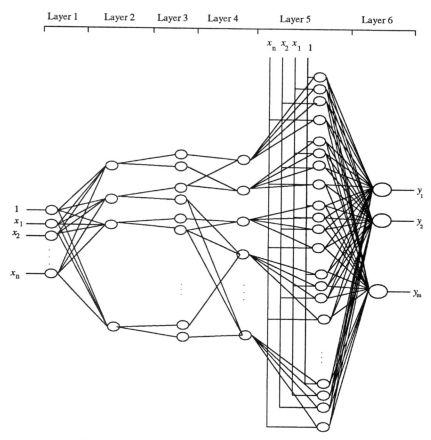

Figure 29 Architecture of the proposed fuzzy neural network.

Layer 3

This is an input term layer. Each of its neurons represents an input fuzzy term of a fuzzy rule. There are two input fuzzy terms associated with one cutting plane. The complementary membership functions are chosen to be a sigmoid function for the pair of input fuzzy terms A^{2j-1} and A^{2j}:

$$o^{(3)}_{2j-1} = \sigma\left(-\beta_j o^{(2)}_j\right) = 0.5\left(1 + \tanh\left(-\beta_j o^{(2)}_j\right)\right)$$

and

$$o^{(3)}_{2j} = \sigma'\left(-\beta_j o^{(2)}_j\right) = 0.5\left(1 - \tanh\left(-\beta_j o^{(2)}_j\right)\right).$$

The neuron performs the sigmoid function. The weights of the links have a value of 1.

Layer 4

This is an input fuzzy partition layer. Each of its neurons represents an input fuzzy region. The links represent the entries of the *AND* matrix M_{AND}. The neurons perform the *AND* function $o_l^{(4)} = \prod o_j^{(3)}$ for all input fuzzy terms A^j linking to input fuzzy regions X^l (for all entries x with value 1 in row l of the *AND* matrix).

Layer 5

This is a local system layer whose neurons represent local linear systems. The neurons perform multiplication and a local linear function. The weights of the input links have a value of 1. The neuron function is

$$o_{lp}^{(5)} = o_l^{(4)} f_{lp}, \qquad p = 1, \ldots, m \text{ and } l = 1, 2, \ldots, L, \qquad (12)$$

where $f_{lp} = a_{p0}^l + \sum_{i=1}^n a_{pi}^l x_i$.

Layer 6

This is an output layer whose neurons represent the output variables. The outputs of the network are

$$y_p = o_p^{(6)} = \sum_{l=1}^L o_{lp}^{(5)}, \qquad p = 1, 2, \ldots, m. \qquad (13)$$

The neurons perform summation functions. The weights of the links have a value of 1. The outputs of the neurons are the outputs of the neural network.

E. LEARNING ALGORITHM

A hybrid algorithm with the capability of structure selection and parameter tuning is developed for MIMO systems with input $x \in R^n$ and output $z \in R^m$.

The performance index is defined as the normalized root mean square error (NRMSE) ε which is given by

$$\varepsilon = \sqrt{\sum_{t=1}^N \sum_{p=1}^m \left(\frac{y_p(t) - z_p(t)}{\bar{z}_p(t)} \right)^2 \Big/ N}, \qquad (14)$$

where y_p is the pth output of the fuzzy neural network and z_p is the pth output of the modeled system, $\bar{z}_p = \max_t (z_p(t)) - \min_t (z_p(t))$, N is the number of training data pairs, and m is the number of output dimensions.

Suppose modeling error ε is required to be less than a prespecified error γ for the fuzzy neural network with the given $(x(t), z(t))$ data pairs for $t = 1, 2, \ldots, N$. For the convenience of applying the hybrid algorithm, x's are normalized within 1. The following hybrid algorithm is developed which consists of a number of steps:

1. Determine a linear system.
2. Create a cutting plane.
3. Expand the *AND* matrix.
4. Create the structure of local systems and initialize the local system parameters.
5. Calculate the output of the inferred membership function of each input fuzzy region.
6. Search for cutting plane parameters and local system parameters.
7. Find the derivatives of the output of the fuzzy neural network with respect to cutting plane parameters.
8. Find the derivatives of the output of the fuzzy neural network with respect to local system parameters.
9. Check stopping conditions and find the worst region.
10. Go to 2.

1. Determine a Linear System

When the training starts, there is no partition of input space and thus there exists only one linear system and one input fuzzy region which is the input domain U. It is a special case of the fuzzy neural network in which there is no rule because there is no input fuzzy partition. In this special case, the fuzzy neural network is equivalent to a linear system.

A pseudo-inverse technique can readily be applied to find the parameter a of the linear system by

$$a = \left(\overline{X}\,\overline{X}^T\right)^{-1}\overline{X}Z^T,$$

where

$$\overline{X} = \begin{bmatrix} 1 & 1 & \cdots & 1 \\ x(1) & x(2) & \cdots & x(N) \end{bmatrix} \quad \text{and} \quad Z = \begin{bmatrix} z(1) & z(2) & \cdots & z(N) \end{bmatrix}.$$

The modeled output is $Y = a^T \overline{X}$. If the modeling error ε defined in Eq. (14) is less than γ, the learning is completed. The structure of the fuzzy neural network is merely a linear function. Otherwise, the following learning steps are required. In this case, the parameter a of the linear system is denoted by a^1 for later use.

2. Create a Cutting Plane

There is only one region which can be selected to be cut if this step immediately follows step 1 because fuzzy partitioning has not started yet. Otherwise, the region required to be cut has been chosen at step 9 and the output of the inferred membership function of each input fuzzy region has been calculated at step 5. At this step, a cutting plane is created to cut the selected region. For generality, suppose that it is the kth cutting plane $g_k(x)$ and the input fuzzy region X^j is selected to be cut.

The output of the inferred membership function $\mu_{X^j}(x) = o_j^{(4)}$ of the selected region governs the importance of the data to the selected region. The weighted average of the input data x_o with respect to this region is used for setting the initial cutting plane, that is,

$$x_o = \frac{\sum_{t=1}^{N} \mu_{X^j}(x(t))x(t)}{\sum_{t=1}^{N} \mu_{X^j}(x(t))},$$

where $\mu_{X^j}(x(t)) = o_j^{(4)}(t)$. It should be noted that x_o cannot be 0.

The initial setting of the cutting plane is to find an equation $g_k(x) = 1 + w_k x = 0$ such that it passes through the point x_o. However, there are many cutting planes passing through x_o. One of the planes is randomly selected as the initial solution. The values of the entries of w_k are obtained by

$$w_k = -\frac{\overline{w}}{\overline{w}x_o},$$

where the values of the entries of $\overline{w}$ are randomly assigned between 0 and 1. If $\overline{w} = 0$, $\overline{w}$ will be generated again.

3. Expand the *AND* Matrix

With the cutting plane initially selected, the selected region can be cut into two splitting regions. This cutting plane actually cuts the whole input domain U into two parts as well as the selected region. Two membership functions $\mu_{A^{2k-1}}(x)$ and $\mu_{A^{2k}}(x)$ are assigned to form two input fuzzy terms A^{2k-1} and A^{2k}. The membership functions selected for the input fuzzy terms of the fuzzy neural network are, respectively,

$$o_{2k-1}^{(2)} = \mu_{A^{2k-1}}(x) = 0.5\big(1 + \tanh\big(-\beta_k g_k(x)\big)\big)$$

and

$$o_{2k}^{(2)} = \mu_{A^{2k}}(x) = 0.5\big(1 - \tanh\big(-\beta_k g_k(x)\big)\big).$$

The input fuzzy terms A^{2k-1} and A^{2k} with the selected membership functions are obviously complementary to one another.

Because the input fuzzy region X^j is selected to be split (one more rule is generated) by the cutting plane $g_k(x)$ (there are a total of $k+1$ regions formed), the *AND* matrix is expanded and a new matrix is obtained as follows:

$$M_{AND}^{new} = \begin{bmatrix} & & 0 & 0 \\ & & \vdots & \vdots \\ M_{AND}^{old} & & 1 & 0 \\ & & \vdots & \vdots \\ & & 0 & 0 \\ M_{AND}^{old}(j) & & 0 & 1 \end{bmatrix} \begin{matrix} \\ \\ \} \ j\text{th } row \\ \\ \\ \} \ (k+1)\text{th } row \end{matrix} ,$$

where $M_{AND}^{old}(j)$ denotes the jth row of M_{AND}^{old}.

4. Create the Structure of Local Systems and Initialize the Local System Parameters

The additional local system f^{k+1} is required to be added in response to input fuzzy region splitting. The matrix of the newly added local linear system f^{k+1} is initialized by $a^{k+1} = a^j$. It should be noted that the two newly added membership functions $\mu_{A^{2k-1}}(x)$ and $\mu_{A^{2k}}(x)$ will not affect other input fuzzy regions except for the region selected to be split. However, because the initial membership functions are complementary to each other and the parameters of the newly added local linear system a^{k+1} is the same as the one a^j associated with the region X^j selected to be cut, the performance of the splitting regions remains the same. That is, the performance of the fuzzy neural network is retained at the time of splitting. This property is very attractive for real-time learning processes.

5. Calculate the Output of the Inferred Membership Function of Each Input Fuzzy Region

The value of the output of membership functions of input fuzzy regions can be obtained by fuzzy set operation on the membership functions of input fuzzy terms. With respect to membership functions, the outputs of input fuzzy regions are the t-norm of the output of input fuzzy terms. In this design, the numerical product is selected as the t-norm for connective *AND*. Because the linkage of input fuzzy terms and input fuzzy regions is expressed by the structured *AND* matrix, the matrix provides a systematic way of performing the calculation. Therefore, the output of the input fuzzy regions is simply obtained using the *AND* matrix by performing the following steps:

1. Reversing the 0's of the *AND* matrix into 1's and the 1's into 0's.
2. Adding each row of the *AND* matrix with the output of input fuzzy terms $o^{(3)}$ (which are obtained by substituting x into the membership functions of input fuzzy terms because input is a fuzzy singleton).
3. Performing $\min(x, 1)$ for each entry x of the *AND* matrix.
4. Multiplying all entries at the same row of the *AND* matrix.

For example, consider the following *AND* matrix M_{AND} as in Eq. (9) which represents five input fuzzy regions (five fuzzy rules) and eight input fuzzy terms (four cutting planes). Suppose the output of input fuzzy terms is

$$o^{(3)} = [0.9998 \quad 0.0002 \quad 1.0000 \quad 0.0000 \quad 0.6154 \quad 0.3846 \quad 0.9977 \quad 0.0023].$$

Thus, the previously mentioned four steps lead to

1. $\begin{bmatrix} 0 & 1 & 1 & 1 & 0 & 1 & 1 & 1 \\ 1 & 0 & 0 & 1 & 1 & 1 & 1 & 1 \\ 1 & 0 & 1 & 0 & 1 & 1 & 1 & 1 \\ 0 & 1 & 1 & 1 & 1 & 0 & 0 & 1 \\ 0 & 1 & 1 & 1 & 1 & 0 & 1 & 0 \end{bmatrix}$,

2. $\begin{bmatrix} 0.9998 & 1.0002 & 2.0000 & 1.0000 & 0.6154 & 1.3846 & 1.9977 & 1.0023 \\ 1.9998 & 0.0002 & 1.0000 & 1.0000 & 1.6154 & 1.3846 & 1.9977 & 1.0023 \\ 1.9998 & 0.0002 & 2.0000 & 0.0000 & 1.6154 & 1.3846 & 1.9977 & 1.0023 \\ 0.9998 & 1.0002 & 2.0000 & 1.0000 & 1.6154 & 0.3846 & 0.9977 & 1.0023 \\ 0.9998 & 1.0002 & 2.0000 & 1.0000 & 1.6154 & 0.3846 & 1.9977 & 0.0023 \end{bmatrix}$,

3. $\begin{bmatrix} 0.9998 & 1.0000 & 1.0000 & 1.0000 & 0.6154 & 1.0000 & 1.0000 & 1.0000 \\ 1.0000 & 0.0002 & 1.0000 & 1.0000 & 1.0000 & 1.0000 & 1.0000 & 1.0000 \\ 1.0000 & 0.0002 & 1.0000 & 0.0000 & 1.0000 & 1.0000 & 1.0000 & 1.0000 \\ 0.9998 & 1.0000 & 1.0000 & 1.0000 & 1.0000 & 0.3846 & 0.9977 & 1.0000 \\ 0.9998 & 1.0000 & 1.0000 & 1.0000 & 1.0000 & 0.3846 & 1.0000 & 0.0023 \end{bmatrix}$,

4. $\begin{bmatrix} 0.6153 \\ 0.0002 \\ 0.0000 \\ 0.3836 \\ 0.0009 \end{bmatrix}$,

and $o^{(4)} = [0.6153 \quad 0.0002 \quad 0.0000 \quad 0.3836 \quad 0.0009].$

6. Search for Cutting Plane Parameters and Local System Parameters

Both the parameters w_k and β_k of the new cutting plane $g_k(x)$ and the parameters a_j and a_{k+1} of the local systems f^j and f^{k+1} of splitting regions X^j and X^{k+1}, respectively, are required to be searched so as to achieve locally optimal performance.

The extended Kalman filter algorithm [40–42] is adopted for the search. The update laws of the extended Kalman filter algorithm are

$$K_\theta(t) = \lambda^{-1} P_\theta(t-1) H_\theta(t) \left(I + \lambda^{-1} H_\theta^T(t) P_\theta(k-1) H_\theta(t) \right)^{-1},$$
$$\hat{\theta}(t) = \hat{\theta}(t-1) + K_\theta(t) \left(z(t) - y(t) \right),$$
$$P_\theta(t) = \lambda^{-1} P_\theta(t-1) - \lambda^{-1} K_\theta(t) H_\theta^T(t) P_\theta(t-1),$$

where K_θ is the Kalman gain and H_θ is the gradient matrix. The entries of the gradient matrix H_θ are the derivatives of the output of the fuzzy neural network with respect to the tunable parameters

$$\theta = \begin{bmatrix} w_k \\ \beta_k \\ a^j \\ a^{k+1} \end{bmatrix},$$

that is,

$$H_\theta(t) = \begin{bmatrix} \left. \dfrac{\partial y^T(w_k, x(t))}{\partial w_k} \right|_{w_k = \hat{w}_k(t-1)} \\ \left. \dfrac{\partial y^T(\beta_k, x(t))}{\partial \beta_k} \right|_{\beta_k = \hat{\beta}_k(t-1)} \\ \left. \dfrac{\partial y^T(a^j, x(t))}{\partial a^j} \right|_{a^j = \hat{a}^j(t-1)} \\ \left. \dfrac{\partial y^T(a^{k+1}, x(t))}{\partial a^{k+1}} \right|_{a^{k+1} = \hat{a}^{k+1}(t-1)} \end{bmatrix}.$$

The algorithm is initialized with $P_\theta(0) = I$ and $\hat{w}_k(0)$ and $\hat{a}^{k+1}(0)$ to the values found at steps 2 and 4, respectively.

7. Find the Derivatives of the Output of the Fuzzy Neural Network with Respect to Cutting Plane Parameters

The derivatives of the output of the fuzzy neural network with respect to cutting plane parameters can be illustrated by an example. Consider a two-input m-output fuzzy neural network having four cutting planes with parameter matrix w of size 4×2 and five local linear systems with five parameter matrices a^j,

$j = 1, 2, \ldots, 5$, each of size $m \times 3$. The *AND* matrix of the fuzzy neural network is as in Eq. (9).

From Eqs. (12) and (13), the system function is obtained

$$
\begin{aligned}
y &= \sum_{k=1}^{5} o_k^{(4)} f_k \\
&= \sigma\big(\beta_1(1 + w_1 x)\big)\sigma\big(\beta_3(1 + w_3 x)\big) f_1 + \sigma'\big(\beta_1(1 + w_1 x)\big)\sigma\big(\beta_2(1 + w_2 x)\big) f_2 \\
&\quad + \sigma'\big(\beta_1(1 + w_1 x)\big)\sigma'\big(\beta_2(1 + w_2 x)\big) f_3 \\
&\quad + \sigma\big(\beta_1(1 + w_1 x)\big)\sigma'\big(\beta_3(1 + w_3 x)\big)\sigma\big(\beta_4(1 + w_4 x)\big) f_4 \\
&\quad + \sigma\big(\beta_1(1 + w_1 x)\big)\sigma'\big(\beta_3(1 + w_3 x)\big)\sigma'\big(\beta_4(1 + w_4 x)\big) f_5,
\end{aligned}
$$

where $y \in R^m$ is the output of the neural network and $f_k \in R^m$, $k = 1, 2, \ldots, 5$, are the outputs of the local linear systems.

For implementing the extended Kalman filter algorithm, the derivatives of y with respect to all w_j, $j = 1, 2, \ldots, 5$, are required. $\partial y^T / \partial w_3$ are selected for illustration:

$$
\frac{\partial y^T}{\partial w_3} = \frac{\partial \sigma(\beta_3(1 + w_3 x))}{\partial w_3} N_3^T + \frac{\partial \sigma'(\beta_3(1 + w_3 x))}{\partial w_3} N_3'^T,
$$

where

$$
\begin{aligned}
N_3 &= \sigma\big(\beta_1(1 + w_1 x)\big) f_1, \\
N_3' &= \sigma\big(\beta_1(1 + w_1 x)\big)\sigma\big(\beta_4(1 + w_4 x)\big) f_4 \\
&\quad + \sigma\big(\beta_1(1 + w_1 x)\big)\sigma'\big(\beta_4(1 + w_4 x)\big) f_5.
\end{aligned}
$$

Then

$$
\begin{aligned}
\frac{\partial y^T}{\partial w_3} &= \frac{\partial}{\partial w_3}\big(0.5\big(1 + \tanh\big(\beta_3(1 + w_3 x)\big)\big)\big) N_3^T \\
&\quad + \frac{\partial}{\partial w_3}\big(0.5\big(1 - \tanh\big(\beta_3(1 + w_3 x)\big)\big)\big) N_3'^T \\
&= 0.5\beta_3 x \big(1 - \tanh^2\big(\beta_3(1 + w_3 x)\big)\big) N_3^T \\
&\quad - 0.5\beta_3 x \big(1 - \tanh^2\big(\beta_3(1 + w_3 x)\big)\big) N_3'^T \\
&= 0.5\beta_3 x \big(1 - \tanh^2\big(\beta_3(1 + w_3 x)\big)\big)\big(N_3^T - N_3'^T\big).
\end{aligned}
$$

Similarly, the derivatives of y with respect to β_3 are

$$
\begin{aligned}
\frac{\partial y^T}{\partial \beta_3} &= \frac{\partial}{\partial \beta_3}\big(0.5\big(1 + \tanh\big(\beta_3(1 + w_3 x)\big)\big)\big) N_3^T \\
&\quad + \frac{\partial}{\partial \beta_3}\big(0.5\big(1 - \tanh\big(\beta_3(1 + w_3 x)\big)\big)\big) N_3'^T
\end{aligned}
$$

$$= 0.5(1 + w_3 x)\left(1 - \tanh^2\left(\beta_3(1 + w_3 x)\right)\right) N_3^T$$
$$- 0.5(1 + w_3 x)\left(1 - \tanh^2\left(\beta_3(1 + w_3 x)\right)\right) N_3'^T$$
$$= 0.5(1 + w_3 x)\left(1 - \tanh^2\left(\beta_3(1 + w_3 x)\right)\right)\left(N_3^T - N_3'^T\right).$$

The expressions of N_3 and N_3' represent outputs of two subnetworks. Thus, the evaluation of N_3 and N_3' can be achieved by using reduced M_{AND}, that is,

$$M_{AND}^{N_3} = [1 \quad 0 \quad 0 \quad 0 \quad 1 \quad 0 \quad 0 \quad 0]$$

and

$$M_{AND}^{N_3'} = \begin{bmatrix} 1 & 0 & 0 & 0 & 0 & 1 & 1 & 0 \\ 1 & 0 & 0 & 0 & 0 & 1 & 0 & 1 \end{bmatrix},$$

which are obtained by deleting the rows of M_{AND},

$$
\begin{array}{cccc}
w_1 & w_2 & w_3 & w_4 \\
\left[\begin{bmatrix} 1 & 0 \\ 0 & 1 \\ 0 & 1 \\ 1 & 0 \\ 1 & 0 \end{bmatrix}\right.
&
\begin{bmatrix} 0 & 0 \\ 1 & 0 \\ 0 & 1 \\ 0 & 0 \\ 0 & 0 \end{bmatrix}
&
\begin{bmatrix} 1 & 0 \\ 0^* & 0 \\ 0^* & 0 \\ 0^* & 1 \\ 0^* & 1 \end{bmatrix}
&
\left.\begin{bmatrix} 0 & 0 \\ 0 & 0 \\ 0 & 0 \\ 1 & 0 \\ 0 & 1 \end{bmatrix}\right]
&
\begin{array}{l} \\ \textit{delete} \\ \textit{delete} \\ \textit{delete} \\ \textit{delete} \end{array}
\end{array}
$$

and

$$
\begin{array}{cccc}
w_1 & w_2 & w_3 & w_4 \\
\left[\begin{bmatrix} 1 & 0 \\ 0 & 1 \\ 0 & 1 \\ 1 & 0 \\ 1 & 0 \end{bmatrix}\right.
&
\begin{bmatrix} 0 & 0 \\ 1 & 0 \\ 0 & 1 \\ 0 & 0 \\ 0 & 0 \end{bmatrix}
&
\begin{bmatrix} 1 & 0^* \\ 0 & 0^* \\ 0 & 0^* \\ 0 & 1 \\ 0 & 1 \end{bmatrix}
&
\left.\begin{bmatrix} 0 & 0 \\ 0 & 0 \\ 0 & 0 \\ 1 & 0 \\ 0 & 1 \end{bmatrix}\right]
&
\begin{array}{l} \textit{delete} \\ \textit{delete} \\ \textit{delete} \\ \\ \end{array}
\end{array}
$$

corresponding to the 0's (which are marked by *) of the first column and the second column of w_3, respectively.

The procedure for finding N_3 and N_3' is similar to the procedure described at step 5 with the modification that the entries of $o^{(3)}$ corresponding to w_3 are replaced by 1's so the output of input fuzzy terms for $M_{AND}^{N_3}$ and $M_{AND}^{N_3'}$ is modified to be

$$[0.9998 \quad 0.0002 \quad 1.0000 \quad 0.0000 \quad 1.0000 \quad 1.0000 \quad 0.9977 \quad 0.0023].$$

Thus, $N_3 = 0.9998\, f_1$ and $N_3' = 0.9975\, f_4 + 0.0023\, f_5$.

In general,

$$\frac{\partial y^T}{\partial w_j} = 0.5\beta_j x \left(1 - \tanh^2\left(\beta_j(1 + w_j x)\right)\right)\left(N_j^T - N_j'^T\right).$$

The evaluation of N_j and N_j' can be achieved by using reduced M_{AND}, that is, $M_{AND}^{N_j}$ and $M_{AND}^{N_j'}$ which are obtained by deleting the rows of M_{AND} corresponding to the 0's of the first column and the second column of w_j, respectively. Using the same procedure as before, all $\partial y^T/\partial w_j$ can be evaluated. Similarly,

$$\frac{\partial y^T}{\partial \beta_j} = 0.5\left(1 + w_j x\right)\left(1 - \tanh^2\left(\beta_j(1 + w_j x)\right)\right)\left(N_j^T - N_j'^T\right)$$

can also be found by the same procedure.

8. Find the Derivatives of the Output of the Fuzzy Neural Network with Respect to Local System Parameters

The derivatives of the output of the fuzzy neural network with respect to local system parameters a^j are

$$\left.\frac{\partial y^T(a^j, x(t))}{\partial a^j}\right|_{a^j = \hat{a}^j(t-1)}.$$

From Eqs. (12) and (13),

$$y = \sum_{l=1}^{k+1} o_l^{(4)} f_l,$$

where $o^{(4)}$ has been calculated at step 5. Then

$$\frac{\partial y^T}{\partial a^j} = o_j^{(4)} \frac{\partial f_j^T}{\partial a^j} = o_j^{(4)} x^T.$$

9. Check Stopping Conditions and Find the Worst Region

After the selected region is split so as to add a new local system and the required parameters are updated, the performance of the system is checked. If it satisfies the criterion, that is, if it is within the prespecified modeling error γ, the learning stops and the trained fuzzy neural network is ready for application. Otherwise, the fuzzy neural network needs further structure evolution (partitioning). The performances of all existing regions are inspected to see which one is the worst. The performance of the input fuzzy region X^j (modeling error ε_j) is defined as follows:

$$\varepsilon_j = \sqrt{\frac{\sum_{t=1}^{N} \mu_{X^j}(x(t)) \sum_{p=1}^{m}(z_p(t) - y_p(t))^2}{\sum_{t=1}^{N} \mu_{X^j}(x(t))}},$$

where $\mu_{X^j}(x(t)) = o(t)_j^{(4)}$.

The region with the worst performance is selected to be divided.

F. SIMULATION EXAMPLES

EXAMPLE 1 (Approximating a Second-Order Highly Nonlinear Function). The algorithm is illustrated by a simulation example of modeling a two-input one-output nonlinear function $z = 0.5(1 + \tanh(5(\sqrt{x_1} + x_2^2 - 1)))$ with input vector $x = [x_1 \quad x_2]^T$. $N = 1000$ data pairs are generated randomly for the learning. The error threshold γ is chosen to be 0.03 and membership functions are chosen to be $\mu_{A^{2j-1}}(x) = 0.5(1 + \tanh(-\beta_j g_j(x)))$ and $\mu_{A^{2j}}(x) = 0.5(1 - \tanh(-\beta_j g_j(x)))$ with initial $\beta_j = 10$. The whole set of data is presented one time for each step of the partitioning.

The progress of learning is listed in Table III.

Figure 28 shows the evolution of the fuzzy neural network during the learning period. After learning, the fuzzy neural network has five rules (with modeling

Table III

Progress of Learning

Step of partitioning	1	2	3	4	5
Number of fuzzy rules generated	1	2	3	4	5
Modeling error before tuning	0.1410	0.1410	0.0470	0.0396	0.0324
Modeling error after tuning	0.1410	0.0470	0.0396	0.0324	0.0281
Modeling error of each region	0.1410	0.0417	0.0419	0.0181	0.0172
		0.0484	0.0305	0.0312	0.0304
			0.0259	0.0259	0.0259
				0.0314	0.0197
					0.0151
Selected region to be divided	1	2	1	4	Stop

error 0.0281). The structure of the fuzzy neural network is described by the *AND* matrix in Eq. (9). The matrices for the local systems are

$$a_1 = [-0.0654 \quad 0.2453 \quad 0.1253],$$
$$a_2 = [0.1949 \quad 0.2973 \quad 0.7599],$$
$$a_3 = [0.4234 \quad 0.2856 \quad 0.3882],$$
$$a_4 = [-0.3616 \quad 0.7633 \quad 0.3296],$$

and

$$a_5 = [-0.2263 \quad 0.4798 \quad 0.4247],$$

and the parameters for the cutting planes are

$$w = \begin{bmatrix} -0.8236 & -1.1106 \\ -0.7217 & -0.6269 \\ -1.9574 & -1.0580 \\ -0.5615 & -1.9895 \end{bmatrix} \quad \text{and} \quad \beta = \begin{bmatrix} 9.9517 \\ 10.0018 \\ 10.0001 \\ 10.0002 \end{bmatrix}.$$

The performance is then verified by $21 \times 21 = 441$ checking data pairs. The checking modeling error is found to be 0.0430. The shapes of the nonlinear function and the modeling function are shown in Figs. 30 and 31, respectively. The shape of the input fuzzy regions (partition) is depicted in Fig. 32. The performance is shown in Fig. 33.

EXAMPLE 2 (Modeling a Bioreactor [42]). The aim of this simulation is to show that the proposed fuzzy neural network can model a multioutput dynamical

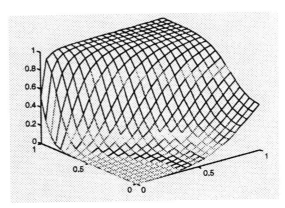

Figure 30 Shape of the nonlinear function.

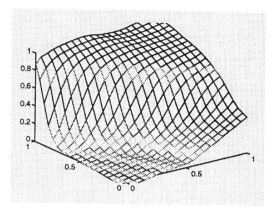

Figure 31 Shape of the modeling function.

system efficiently with a few number of rules. In this simulation, the model of the bioreactor describing a process involving a continuous-flow stirred tank reactor is given by

$$\dot{x}_1 = -x_1 u + x_1(1 - x_2)\exp\left(\frac{x_2}{0.48}\right),$$

$$\dot{x}_2 = -x_2 u + x_1(1 - x_2)\exp\left(\frac{x_2}{0.48}\right)\frac{1.02}{1.02 - x_2},$$

$$y_1 = x_1, \qquad y_2 = x_2,$$

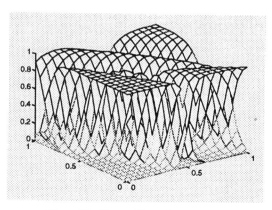

Figure 32 Shape of input fuzzy regions.

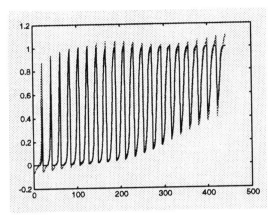

Figure 33   Outputs of the fuzzy neural network and the function.

where x_1 is the cell mass in dimensionless form, x_2 is the amount of nutrients in a constant-volume tank, bounded between zero and unity, and the control u is the flow rate of nutrients into the tank (the same rate at which contents are removed from the tank).

The dynamical equations are computed with MATLAB ode23. We define $\Delta T = 0.5$ s as the sampling period which defines the intervals for sampling the system states. The control $u(k\Delta T)$ to the system was assigned to be a sequence of random numbers between 0 and 1 at the first 100 samples but was assigned to be 0 and 1 at the last second 50 samples and the last 43 samples, respectively. Thus, 193 data pairs $(z_1(k\Delta T), z_2(k\Delta T), u(k\Delta T))$ were obtained. The first 30 data pairs and the last 163 data pairs are used for checking and training, respectively. The structure of the linear local systems j is chosen to be

$$y_1^j(t) = a_{10}^j + a_{11}^j y_1(t-1) + a_{12}^j y_2(t-1) + a_{13}^j y_1(t-2)$$
$$+ a_{14}^j y_2(t-2) + a_{15}^j u(t-1) + a_{16}^j u(t-2),$$
$$y_2^j(t) = a_{20}^j + a_{21}^j y_1(t-1) + a_{22}^j y_2(t-1) + a_{23}^j y_1(t-2)$$
$$+ a_{24}^j y_2(t-2) + a_{25}^j u(t-1) + a_{26}^j u(t-2).$$

The error threshold γ is chosen to be 0.05 and membership functions are chosen to be $\mu_{A^{2j-1}}(x) = 0.5(1 + \tanh(-\beta_j g_j(x)))$ and $\mu_{A^{2j}}(x) = 0.5(1 - \tanh(-\beta_j g_j(x)))$ with initial $\beta_j = 10$. The whole set of data is presented five times for each step of the partitioning because the amount of training data is small. The progress of learning is listed in Table IV.

Table IV

Progress of Learning

Step of partitioning	1	2	3	4
Number of fuzzy rules generated	1	2	3	4
Modeling error before tuning	0.1005	0.1005	0.0603	0.0526
Modeling error after tuning	0.1005	0.0603	0.0526	0.0467
Modeling error of each region	0.1005	0.0597	0.0526	0.0557
	0	0.0535	0.0525	0.0518
	0	0	0.0342	0.0341
	0	0	0	0.0212
Selected region to be divided	1	1	1	Stop

After learning, the fuzzy neural network will have four rules (with modeling error 0.0467). The structure of the fuzzy neural network is described by the *AND* matrix

$$M_{AND} = \begin{bmatrix} 1 & 0 & 1 & 0 & 1 & 0 \\ 0 & 1 & 0 & 0 & 0 & 0 \\ 1 & 0 & 0 & 1 & 0 & 0 \\ 1 & 0 & 1 & 0 & 0 & 1 \end{bmatrix},$$

the matrices for the local systems

$$a^1 = \begin{bmatrix} 0.4299 & -0.0798 & -0.5016 & 0.0155 & 0.1956 & 1.3347 & -0.4054 \\ 0.1991 & -0.0019 & -0.6027 & -0.1158 & -0.0488 & 1.0303 & 1.2425 \end{bmatrix},$$

$$a^2 = \begin{bmatrix} 0.3709 & 0.0685 & -0.2123 & -0.1683 & 0.1306 & 0.9048 & -0.4515 \\ 0.2612 & 0.1988 & -0.3438 & -0.1151 & 0.0553 & 0.3603 & 0.6290 \end{bmatrix},$$

$$a^3 = \begin{bmatrix} 0.3497 & -0.0119 & -0.3407 & -0.0549 & 0.2623 & 1.1379 & -0.4120 \\ 0.2216 & 0.2618 & -0.7121 & -0.0778 & -0.2333 & 0.8263 & 1.0669 \end{bmatrix},$$

and

$$a^4 = \begin{bmatrix} 0.4023 & -0.1009 & -0.2157 & -0.1350 & 0.1578 & 1.1694 & -0.4998 \\ 0.1643 & 0.0952 & -0.5536 & -0.1412 & -0.0972 & 0.9290 & 1.0771 \end{bmatrix},$$

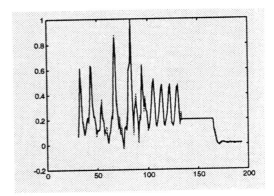

Figure 34 Performance of y_1.

and the parameters for the cutting planes

$$w = \begin{bmatrix} 0.0145 & 0.1322 & -0.3542 & -0.5017 & -0.6951 & -0.4850 \\ -0.3188 & -0.7451 & -0.0080 & 0.0267 & -0.4675 & -0.5322 \\ -0.0144 & -0.4121 & -0.0557 & -0.5296 & -0.7257 & -0.7301 \end{bmatrix}$$

and

$$\beta = \begin{bmatrix} 9.9938 \\ 10.0000 \\ 10.0017 \end{bmatrix}.$$

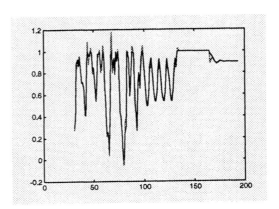

Figure 35 Performance of y_2.

The checking modeling error is found to be 0.0583. The performance is shown in Figs. 34 and 35.

G. CONCLUDING REMARKS

An adaptive neural network implementation of fuzzy systems has been proposed. The proposed fuzzy neural network has the capabilities of self-organization and adaptation through the proposed hybrid learning algorithm which can determine the structure of the fuzzy neural network and tune the parameters of the linear local systems and the membership functions. By applying *hierarchical input space partitioning*, the proposed fuzzy neural network can determine the number of rules or neurons automatically to achieve a prespecified modeling error. In addition, the fuzzy neural network has the following attractive properties:

1. Sigmoid neural network structure (i.e., perception neural network).
2. Inherent normalized membership functions.
3. Simple learning algorithm for implementation.
4. The performance of the fuzzy neural network is retained at the time of region splitting (evolution).

The simulation result showed that the proposed neural network has the capability of self-organizing and adaptive learning.

V. CONCLUSION

The aim of this chapter has been to investigate the techniques of implementing fuzzy systems within the framework of neural networks for modeling and control. Two fuzzy neural network designs have been developed in this chapter.

The structure of the fuzzy neural network in the first design has been introduced with an additional *OR* layer based on standard Takagi–Sugeno fuzzy systems. This makes it useful for the implementation of higher-order fuzzy systems; the proposed fuzzy neural network provides a means of controlling the growth of the number of local linear systems when the order of the system under consideration increases so that least-squares estimation can be applied without much performance degradation. In the second design, an attempt has been made to develop an adaptive fuzzy neural network by using hierarchical space partitioning. It has the capability of determining the structure of the network and the number of neurons automatically. Together with the extended Kalman filter algorithm proposed in the fuzzy neural network which requires no tunable parameter, the performance of the fuzzy neural network can be optimized in one training process.

It can be seen that the approach of the fuzzy neural network is a possible solution to fuzzy systems design. The learning capability of fuzzy neural networks makes the design procedures for fuzzy systems more systematic.

APPENDIX

Recall

$$\frac{\partial E}{\partial o^{(6)}} = -\left(z(t) - o^{(6)}\right).$$

From Eqs. (6)–(8), we have

$$o^{(6)} = \sum_l f_l o_l^{(4)} \Big/ \sum_l o_l^{(4)},$$

where $f_l = a_0^l + \sum_{i=1}^n a_i^l x_i$ and n is the system order. Thus,

$$\frac{\partial E}{\partial o_l^{(4)}} = \frac{\partial E}{\partial o^{(6)}} \frac{\partial o^{(6)}}{\partial o_l^{(4)}}$$

$$= -\left(z(t) - o^{(6)}\right)\left(f_l \sum_l o_l^{(4)} - \sum_l f_l o_l^{(4)}\right) \Big/ \left(\sum_l o_l^{(4)}\right)^2.$$

From Eq. (6), we have $o_l^{(4)} = \sum_k o_k^{(3)}$. Thus,

$$\frac{\partial E}{\partial o_k^{(3)}} = \frac{\partial E}{\partial o_l^{(4)}} \frac{\partial o^{(4)}}{\partial o_k^{(3)}} = \frac{\partial E}{\partial o_l^{(4)}}.$$

(*Note:* $\partial E/\partial o_k^{(3)}$ of the kth neuron in layer 3 is the same as $\partial E/\partial o_l^{(4)}$ of the lth neuron in layer 4 if the two neurons are linked.)

From Eq. (5), we have

$$o_k^{(3)} = \prod_i o_j^{(2)},$$

$$\frac{\partial E}{\partial o_j^{(2)}} = \sum_k \frac{\partial E}{\partial o_k^{(3)}} \frac{\partial o_k^{(3)}}{\partial o_j^{(2)}} = \sum_k \left(\frac{\partial E}{\partial o_k^{(3)}} \prod_{q,q\neq j} o_q^{(2)}\right).$$

We derive the parameter update laws:

$$\frac{\partial E}{\partial w_j} = \frac{\partial E}{\partial o_j^{(2)}} \frac{\partial o_j^{(2)}}{\partial w_j}, \qquad \frac{\partial E}{\partial b_j} = \frac{\partial E}{\partial o_j^{(2)}} \frac{\partial o_j^{(2)}}{\partial b_j}, \qquad \frac{\partial E}{\partial \sigma_j} = \frac{\partial E}{\partial o_j^{(2)}} \frac{\partial o_j^{(2)}}{\partial \sigma_j},$$

where w_j, b_j, and σ_j are the tuning parameters. From Eq. (4), we have

$$\frac{\partial o_j^{(2)}}{\partial w_j} = 2b_j \frac{(o_i^{(1)} - w_j)}{\sigma_j^2} \left(\frac{(o_i^{(1)} - w_j)^2}{\sigma_j^2}\right)^{b_j-1} \exp\left(-\left(\frac{(o_i^{(1)} - w_j)^2}{\sigma_j^2}\right)^{b_j}\right).$$

Thus,

$$\Delta w_j^{(2)} = -\eta \sum_k \frac{\partial E}{\partial o_j^{(2)}} 2b_j \frac{(o_i^{(1)} - w_j)}{\sigma_j^2} \left(\frac{(o_i^{(1)} - w_j)^2}{\sigma_j^2}\right)^{b_j-1}$$
$$\times \exp\left(-\left(\frac{(o_i^{(1)} - w_j)^2}{\sigma_j^2}\right)^{b_j}\right).$$

From Eq. (4), we also have

$$\frac{\partial o_j^{(2)}}{\partial \sigma_j} = 2b_j \frac{(o_i^{(1)} - w_j)^2}{\sigma_j^3} \left(\frac{(o_i^{(1)} - w_j)^2}{\sigma_j^2}\right)^{b_j-1} \exp\left(-\left(\frac{(o_i^{(1)} - w_j)^2}{\sigma_j^2}\right)^{b_j}\right).$$

Thus,

$$\Delta \sigma_j = -\eta \sum_k \frac{\partial E}{\partial o_j^{(2)}} 2b_j \frac{(o_i^{(1)} - w_j)^2}{\sigma_j^3} \left(\frac{(o_i^{(1)} - w_j)^2}{\sigma_j^2}\right)^{b_j-1}$$
$$\times \exp\left(-\left(\frac{(o_i^{(1)} - w_j)^2}{\sigma_j^2}\right)^{b_j}\right).$$

From Eq. (4), we also have

$$\frac{\partial o_j^{(2)}}{\partial b_j} = -\log\left(\frac{(o_i^{(1)} - w_j)^2}{\sigma_j^2}\right)\left(\frac{(o_i^{(1)} - w_j)^2}{\sigma_j^2}\right)^{b_j}\right)$$
$$\times \exp\left(-\left(\frac{(o_i^{(1)} - w_j)^2}{\sigma_j^2}\right)^{b_j}\right).$$

Thus,

$$\Delta b_j = \eta \sum_k \frac{\partial E}{\partial o_j^{(2)}} \log\left(\frac{(o_i^{(1)} - w_j)^2}{\sigma_j^2}\right)\left(\frac{(o_i^{(1)} - w_j)^2}{\sigma_j^2}\right)^{b_j}\right)$$
$$\times \exp\left(-\left(\frac{(o_i^{(1)} - w_j)^2}{\sigma_j^2}\right)^{b_j}\right).$$

REFERENCES

[1] L. A. Zadeh. *Inform. Control* 8:338–353, 1965.

[2] L. A. Zadeh. *Inform. Control* 12:94–102, 1968.

[3] L. A. Zadeh. *IEEE Trans. Systems Man Cybernet.* 3:28–44, 1973.

[4] E. H. Mamdani. *Proc. IEE Control Sci.* 121:1585–1588, 1974.

[5] J. J. Ostergaard. In *Fuzzy Automation and Decision Process* (M. M. Gupta, G. N. Saridis, and B. R. Gaines, Eds.), pp. 285–320. North-Holland, New York, 1977.

[6] E. H. Mamdani and S. Assilian. *Internat. J. Man Machine Studies* 7:1–13, 1975.

[7] P. J. King and E. H. Mamdani. *Automatica* 13:235–242, 1977.

[8] E. H. Mamdani and C. P. Pappis. *IEEE Trans. Systems Man Cybernet.* 7:707–717, 1977.

[9] L. P. Holmblad and J. J. Ostergaard. In *Fuzzy Information and Decision Processes* (M. M. Gupta and E. Sanchez, Eds.), pp. 389–399. North-Holland, New York, 1982.

[10] M. Sugeno and K. Murakami. In *Industrial Application of Fuzzy Control* (M. Sugeno, Ed.). North-Holland, Amsterdam, 1985.

[11] S. Murakami. In *Fuzzy Information, Knowledge Representation and Decision Analysis* (E. Sanchez, Ed.), pp. 43–48. Pergamon, Oxford, 1984.

[12] D. Lakov. *Fuzzy Sets Systems* 17:1–8, 1985.

[13] M. Uragami, M. Mizumoto, and K. Tanaka. *J. Cybernet.* 6:39–64, 1976.

[14] L. I. Larkin. In *Industrial Application of Fuzzy Control* (M. Sugeno, Ed.), pp. 87–103. North-Holland, Amsterdam, 1985.

[15] H. Takahashi, K. Ikeura, and T. Yamamori. In *Proceedings of the International Fuzzy Engineering Symposium '91 (IFES'91)*, pp. 1136–1137, 1991.

[16] H. Takagi and M. Sugeno. In *Proceedings of the IFAC Symposium on Fuzzy Information, Knowledge Representation and Decision Analysis*, pp. 55–60, 1983.

[17] C. C. Lee. *Internat. Intell. Systems* 6:71–92, 1991.

[18] L. X. Wang and J. M. Mendel. *IEEE Trans. Systems Man Cybernet.* 22:1414–1427, 1992.

[19] J.-S. Jang. *IEEE Trans. Systems Man Cybernet.* 23:665–685, 1993.

[20] C.-T. Lin and C. S. G. Lee. *IEEE Trans. Comput.* 40:1320–1336, 1991.

[21] L.-X. Wang and J. M. Mendel. *IEEE Trans. Neural Networks* 3:807–814, 1992.

[22] S. Horikawa, T. Furuhashi, and Y. Uchikawa. *IEEE Trans. Neural Networks* 3:801–806, 1992.

[23] J. M. Keller, R. R. Yager, and H. Tahani. *Fuzzy Sets Systems* 45:1–12, 1992.

[24] R. R. Yager. *Fuzzy Sets Systems* 48:53–64, 1992.

[25] W. Pedrycz. *Fuzzy Sets Systems* 56:1–28, 1993.

[26] M. M. Gupta and D. H. Rao. *Fuzzy Sets Systems* 61:1–18, 1994.

[27] H. Ishibuchi and H. Tanaka. *Fuzzy Sets Systems* 50:257–265, 1992.

[28] H. Ishibuchi, R. Fujioka, and H. Tanaka. *IEEE Trans. Fuzzy Systems* 1:85–97, 1993.

[29] Y. Jin, J. Jiang, and J. Zhu. *IEEE Trans. Systems Man Cybernet.* 25:990–997, 1995.

[30] C.-T. Lin and C. S. G. Lee. *IEEE Proc. 0-7803-0236-2* 1283–1291, 1992.

[31] C.-T. Lin and C. S. G. Lee. *IEEE Trans. Fuzzy Systems* 2:46–63, 1994.

[32] J.-S. Jang and C.-T. Sun. Functional equivalent between radial basis function networks and fuzzy inference systems. Department of Electrical Engineering and Computer Science, University of California, Berkeley, 1992.

[33] H. Takagi and M. Sugeno. *IEEE Trans. Systems Man Cybernet.* 15:116–132, 1985.

[34] W. Pedrycz. *Fuzzy Control and Fuzzy Systems.* Research Studies Press, 1989.

[35] C. C. Lee. *IEEE Trans. Systems Man Cybernet.* 20:404–415, 1990.

[36] C. C. Lee. *IEEE Trans. Systems Man Cybernet.* 20:419–435, 1990.

[37] M. Sugeno and G. T. Kang. *Fuzzy Sets Systems* 28:15–23, 1988.

[38] H. Takagi and I. Hayashi. *Internat. J. Approximate Reasoning* 5:191–212, 1991.

[39] M. Sugeno and T. Yasukawa. *IEEE Trans. Fuzzy Systems* 1:7–31, 1993.

[40] S. Singhal and L. Wu. In *Proceedings of the IEEE International Conference on Acoustics Speech and Signal Processing*, Glasgow, pp. 1187–1190. IEEE Press, New York, 1989.

[41] G. V. Puskorius and L. A. Feldkamp. In *Proceedings of the International Joint Conference on Neural Networks*, Seattle, IEEE Press, New York, 1991.

[42] G. V. Puskorius and L. A. Feldkamp. *IEEE Trans. Neural Networks* 5:279–297, 1994.

Neural Networks and Rule-Based Systems

Aldo Aiello
Istituto di Cibernetica C.N.R.
I-80072 Arco Felice, Italy

Ernesto Burattini
Istituto di Cibernetica C.N.R.
I-80072 Arco Felice, Italy

Guglielmo Tamburrini
Istituto di Cibernetica C.N.R.
I-80072 Arco Felice, Italy

I. INTRODUCTION

This chapter presents an approach to simulating, and in several cases efficiently so, a wide variety of rule-based reasoning processes by means of networks formed by nonlinear thresholded neural units. In particular, the following networks are examined:

1. networks representing knowledge bases formed by propositional production rules and performing forward chaining on them;
2. a network monitoring the elaboration of the forward chaining system and learning new production rules by an elementary chunking process;
3. networks performing qualitative forms of uncertain reasoning, such as hypothetical reasoning in two-level causal networks and the application of preconditions in default reasoning;
4. networks simulating elementary forms of quantitative uncertain reasoning.

The possible uses of these techniques are partially exemplified by the overall structure and implementation features of a purely neural, rule-based expert system for a diagnostic task. Here, the expression "purely neural" indicates that in addition to knowledge representation and processing proper also the control and synchronization functions that are needed to schedule the given diagnostic task are achieved by means of neural networks.

Fuzzy Logic and Expert Systems Applications
123

The neural representation of rules is based on a localist, rather than distributed, semantic interpretation: each propositional literal appearing in a rule is represented by means of an individual neuron. Moreover, even when rules are learned, they are not acquired by standard neural learning techniques. Finally, their applicability is governed by essentially rigid conditions, even when rules are used to simulate forms of uncertain reasoning. In the latter case, rule firing is tailored to reflect rigorous models of reasoning under incomplete or uncertain knowledge, so that the uncertainty attached to their conclusions can be evaluated in robust theoretical settings.

In view of these qualifications, one has to state explicitly what is the interest of a neural architecture of this sort for rule-based systems, because the learning and adaptivity typical of neural nets play a secondary role in the present approach: systems that are designed to fulfill these constraints remain brittle on the whole, much in the way that traditional symbolic systems are. Clearly, their interest lies elsewhere:

(i) Neural net implementations of production systems naturally lend themselves to parallel execution. Given appropriate hardware or software support, these networks can be used to build applications in domains where real-time responses are a crucial demand.

(ii) The absence of semantically opaque, hidden layers of neurons governed by learning algorithms guarantees the possibility of providing an informative justification for the conclusions obtained by stepwise inferential processes.

(iii) The neural simulation of various sorts of rule-based reasoning makes available a wide repertoire of technical tools for *unified* approaches to neurosymbolic integration where, in contrast with *hybrid* approaches, symbolic processing is carried out by a neural network, too.

(iv) Revisable reasoning is very naturally modeled by means of neural settings that include negative weights, as the neuron outputs of these networks are not intrinsically monotonic functions of their inputs.

These points will be taken up again both in the concluding remarks and in comments accompanying the presentation of more technical material in the main body of the chapter.

II. NONLINEAR THRESHOLDED ARTIFICIAL NEURONS

The artificial neurons used throughout this chapter are weighted-sum, nonlinear thresholded elements which may keep memory of past activity by means of a memory decay function. These artificial neurons are obtained by a modification of Caianiello's classical neural equations (see Caianiello [1]). The state equation for one of these neurons, say, h, is

$$u_h(t+1) = 1 \left[\sum_{j=1}^{n} \sum_{i=0}^{t} a_{j,h} \cdot u_j(i) \cdot \delta_h(t-i) - s_h \right], \qquad (1)$$

where $u_h(i)$ is the state (1 or 0) of the neuron h at time i; $a_{j,h}$ is the weight (or coupling coefficient) between neurons j and h; $\delta_h(i)$ is a monotone, nonincreasing function of the discrete time i for neuron h regulating a time variable memory of the excitation received by h from its neighbors (this memory "decay law" plays a crucial role in modeling various forms of uncertain reasoning by thresholded neural elements, as it allows one to encode numerical values by sequences of neuron firings); s_h is the threshold of h; and

$$1[x] = \begin{cases} 1, & \text{if } x > 0, \\ 0, & \text{if } x \leq 0, \end{cases}$$

is the step function determining the state of each neural unit.

The specific settings for each neural element h can be described by means of a "characteristic triple":

$$N_h \equiv \{A_h, \delta_h, s_h\},$$

where A_h is a set of pairs $\{\langle j, x \rangle\}$, where x is the value of the weight between neurons j and h; δ_h is the memory decay law for h; and s_h is the threshold of h.

The "characteristic triple" notation will be omitted whenever a detailed description of particular (types of) neural elements is not needed.

III. PRODUCTION RULES

A production rule is a pair consisting of a condition part and an action part (see, e.g., Genesereth and Nilsson [2, pp. 274–280] or Grzymala-Busse [3, pp. 17–28]). The particular production rules that we shall be concerned with can be cast in the form of conditional expressions of the form

$$p_1 \wedge \cdots \wedge p_k \rightarrow h, \qquad (R)$$

with the restriction that both the k elements of the condition part on the left-hand side of the arrow and the only element of the action part on the right-hand side are propositional literals (where a literal is a propositional letter or the negation of a propositional letter in sentential logic).

Because no specific inference scheme for handling negation will be introduced in this chapter, no greater inferential power is achieved by permitting literals, rather than just propositional letters, to appear in rules. Nonetheless, allowing for literals in the context of the present approach is useful in at least two respects. First, it furnishes external sources of information (typically, human users of rule-based systems) with greater expressive power during query processes (the external source may directly inform a system handling literals that the negation of a certain proposition holds). Second, it enables one to introduce a control mechanism for *explicit* contradictions, which detects whether contradictory pairs of literals occur in a database formed by the set of asserted literals expressing known facts (such as the literals asserted by an external source) and literals obtained by an inference engine working on production rules and asserted facts.

By adding to a production system formed by a finite set of rules of form (R) a *database* containing literals expressing *facts* and a *rule interpreter*, one can implement search processes on facts and rules. The *facts database* is a record of assertions, whether inferred by applying the rules or asserted by other means (for instance, by an external source of information). The *rule interpreter* works iteratively in recognize-and-act cycles, which can be used to implement various kinds of searches. *Forward* and *backward chaining* are basic search strategies for production systems, which may be suitably amalgamated to obtain mixed strategies.

The inferential strategy we shall be mainly concerned with in this chapter is forward chaining. In forward chaining, one checks whether the condition parts of production rules are satisfied and, if so, performs the corresponding action parts. This process is iterated until no rule with a satisfied condition part can be found.

The main building block of a neural inference engine for *parallel* forward chaining is the neural representation of individual rules. In view of the fact that the state function of the neurons described in the previous section can assume only values of 1 and 0, these thresholded elements can provide, under a localist semantic representation, Boolean-valued information about the literals to which they are associated.

Under such localist semantic interpretation, each rule of the form (R) can be represented as a net having k neurons $p_1, \ldots, p_k$ connected to a neuron h (see Fig. 1) with the following settings:

$$a_{j,h} = 1, \qquad 1 \le j \le k,$$

$$s_h = k - \varepsilon, \qquad 0 < \varepsilon < 1,$$

$$\delta_h(i) = \delta^0(i), \qquad \text{where } \delta^0 \text{ is } \begin{cases} 1, & \text{if } i = 0 \\ 0, & \text{if } i \ne 0 \end{cases} \qquad \text{(i.e., there is no memory).}$$

By (1) and the previous settings, one has

$$u_h(1) = \mathbf{1}\left[\sum_{j=1}^{k} u_{p_j}(0) - (k - \varepsilon) \right].$$

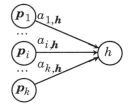

Figure 1 Neural rule model.

Thus,

$$u_h(1) = 1 \qquad \text{iff } \forall j \ u_{p_j}(0) = 1,$$

that is, the neuron h is active (its state is 1) at time $t = 1$ if and only if all $p_1, \ldots, p_k$ are simultaneously active at time $t = 0$. More in general,

$$u_h(t+1) = 1 \qquad \text{iff } \forall j \ u_{p_j}(t) = 1.$$

Thus, the behavior of the net formed by the neurons $p_1, \ldots, p_k$ and h reflects faithfully the behavior of a rule interpreter used in forward chaining when applied to a rule of the form (R): whenever the condition part of such a rule is satisfied, its action part is executed (in the case of an inferential process, this execution amounts to adding literal h to the database).

IV. FORWARD CHAINING

Using this representation of rules as basic building blocks, one can design neural networks representing production systems formed by rules of the form (R) and capable of performing a parallel process of forward chaining on them. Because the process is parallel, all rules whose condition part is satisfied can be simultaneously applied, and therefore no particular scheduling for rule firing is needed. However, there are some crucial problems that have to be addressed:

- *Correctness.* How to ensure that rules will fire only when their condition part is fully satisfied;
- *Control.* How to verify that the process has come to an end, namely, that no more literals can be inferred on the basis of the available information;
- *Output.* How the results of forward chaining are to be read off from the network when the inference process has been completed.

Let us examine in some detail how these constraints can be fulfilled in the specific case of the following system of rules:

$$b \rightarrow d,$$
$$e \wedge d \rightarrow a,$$
$$\neg d \wedge c \rightarrow a,$$
$$d \wedge a \rightarrow b.$$

A neural inferential engine, capable of carrying out forward search on this system of rules, starting from an initial set of asserted facts, is formed by five distinct layers of neurons (see Fig. 2). The first layer (*IN*) accepts external *inputs* to the net. It is formed by as many neurons IN_j as different propositional literals appear in the rules (in this particular case, $1 \leq j \leq 6$).

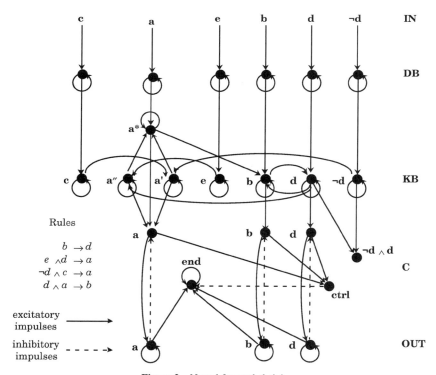

Figure 2   Neural forward chaining.

Each neuron IN_j is connected to the neuron DB_j representing the same literal in the second layer DB. The characteristic triple of the generic neuron IN_i in this layer is

$$N_{IN_i} \equiv \{EXT_i, \delta^0, \varepsilon\}.$$

Its first element represents the action of an external source, EXT_i; the second element is the decay law, while the third element is the threshold of IN_i, with $0 < \varepsilon < 1$. The symbol "ε" will assume, unless otherwise stated, an arbitrary positive value less than 1.

The second layer (DB) is a partial *database* formed by as many neurons as in the layer IN. These elements store the premises introduced in the IN layer: the neuron DB_j becomes active at time $t + 1$ whenever the neuron IN_j is active at time t and preserves this information by self-excitation. The characteristic triple for the generic neuron DB_i in this layer is

$$N_{DB_i} \equiv \{\{\langle IN_i, 1\rangle, \langle DB_i, 1\rangle\}, \delta^0, \varepsilon\}.$$

The third layer (KB) codifies the entire *knowledge base*. Here, each rule is represented as in the previous section, with the additional condition that if a literal p occurs as the right-hand side (or, as we shall also say, as the conclusion) of z rules, then z distinct neurons—each one representing an occurrence of p in the conclusion part of those rules—have to be introduced in this layer. This additional condition is crucial to ensure that rules are correctly activated. It is needed to avoid that a neuron representing p can be activated from a combination of premises belonging to different rules having p as a conclusion. These z neurons are connected to a neuron p^* which represents all occurrences of p as the left-hand sides of production rules and fires on the neural representative of the right-hand sides of those rules.

Because the elements represented by neurons are propositional *literals*, one may obtain an inconsistency in this layer if both an atomic proposition and its negation become simultaneously active. Even supposing that the knowledge base itself is consistent, inconsistencies may still be introduced if one allows external sources of information to assert new facts. The system is capable of signaling such explicit contradictions because each pair of neurons representing contradictory literals is connected to a neuron belonging to the control layer. The latter becomes active when both elements of the pair are active. (See, e.g., the pair formed by the neural representatives of d and $\neg d$ in Fig. 2.)

The characteristic triples of the neurons KB_i vary according to their role in the system of production rules. For neurons representing literals appearing only in condition parts of production rules, we have

$$N_{KB_i} \equiv \{\{\langle KB_i, 1\rangle, \langle DB_i, 1\rangle\}, \delta^0, \varepsilon\}.$$

The characteristic triple for a generic neuron KB_i representing a literal p_i occurring only once as conclusion in the system of rules, say in a rule r with q literals in its condition part, is

$$\mathbf{N}_{KB_i} \equiv \{\{\langle KB_i, q\rangle, \langle DB_i, q\rangle, \langle KB_j^1, 1\rangle, \ldots, \langle KB_j^q, 1\rangle\}, \delta^0, q - \varepsilon\},$$

where $KB_j^1, \ldots, KB_j^q$ are the q neurons in KB representing the literals in the condition part of rule r.

When a literal p_i occurs as a conclusion in z different rules (with $z > 1$), the characteristic triple for each of the neurons $KB_i^1, \ldots, KB_i^z$ representing these z occurrences is

$$\mathbf{N}_{KB_i^l} \equiv \{\{\langle KB_i^l, q_l\rangle, \langle KB_{jl}^1, 1\rangle, \ldots, \langle KB_{ji}^{ql}, 1\rangle\}, \delta^0, q_l - \varepsilon\}, \qquad l = 1, \ldots, z.$$

These z neurons fire on a neuron KB_i^* which represents all occurrences of p in the condition parts of the production rules. The characteristic triple for such KB_i^* is

$$\mathbf{N}_{KB_i^*} \equiv \{\{\langle KB_i^*, 1\rangle, \langle DB_i, 1\rangle, \langle KB_i^1, 1\rangle, \ldots, \langle KB_i^z, 1\rangle\}, \delta^0, \varepsilon\}.$$

The inferential process is triggered by exciting neurons in *IN* and its results are codified by active neurons in *KB*. One can easily verify that no scheduling is necessary for carrying out this process, because all rules whose condition part is satisfied in the *KB* layer are simultaneously applied.

The *control problem* can be solved by introducing a distinct layer C of m neurons, where m is the number of possible different conclusions in the system of production rules. Each neuron C_i in C represents a literal appearing as a conclusion; it receives impulses from all z neurons, with $z \geq 1$, that represent the same conclusion in *KB*, and activates the corresponding neuron OUT_i in the layer *OUT*. The latter, once excited, sends back an inhibition, equal to $z + 1$, to the neuron C_i in the layer C. Moreover, each neuron C_i can fire on the special neuron ***ctrl*** which is active as long as new conclusions are reached. Another special control neuron ***end*** is inhibited by ***ctrl***, with strength $m+1$, and is excited by each neuron OUT_i. Thus, ***end*** is inactive until ***ctrl*** is active, that is, until new conclusions are reached. When nothing else can be inferred, ***ctrl*** becomes inactive. As a result, ***end*** is no longer inhibited and becomes active, thus signaling that the forward process on the input data is terminated.

The characteristic triples for each neuron C_i vary according to the role it plays in the system of production rules. For a neuron C_i representing a literal appearing as a conclusion in just one rule, we have

$$\mathbf{N}_{C_i} \equiv \{\{\langle KB_i, 1\rangle, \langle OUT_i, -1\rangle\}, \delta^0, \varepsilon\}.$$

For a neuron C_i representing a literal appearing as a conclusion in z production rules

$$\mathbf{N}_{C_i} \equiv \left\{\left\{\langle KB_i^*, 1\rangle, \langle KB_i^1, 1\rangle, \ldots, \langle KB_i^z, 1\rangle, \langle OUT_i, -(z+1)\rangle\right\}, \delta^0, \varepsilon\right\}.$$

And the characteristic triples for the special control neurons are

$$\mathbf{N}_{ctrl} \equiv \left\{\bigcup_{C_i \in C}\{\langle C_i, 1\rangle\}, \delta^0, \varepsilon\right\},$$

$$\mathbf{N}_{end} \equiv \left\{\bigcup_{OUT_i \in OUT}\{\langle OUT_i, 1\rangle\} \cup \{\langle end, 1\rangle\} \cup \{\langle ctrl, -(m+1)\rangle\}, \delta^0, \varepsilon\right\}.$$

If *KB* contains pairs of neurons x and $\neg x$ representing contradictory literals, then for each pair a distinguished neuron $(\neg xx)$ is created, which becomes active if and only if both elements of the pair are simultaneously active in *KB*.

$$\mathbf{N}_{\neg xx} \equiv \left\{\{\langle\neg x, 1\rangle, \langle x, 1\rangle\}, \delta^0, 2 - \varepsilon\right\}.$$

This mechanism enables the system to signal that a contradiction has been derived.

The *output problem* is solved through a layer *OUT* of m neurons, where m is again the number of possible different conclusions in the system of production rules. Each neuron in *OUT* is excited by the corresponding neuron in C and is self-excited in order to store this information. The characteristic triple for the generic neuron in *OUT* is

$$\mathbf{N}_{OUT_i} \equiv \left\{\{\langle OUT_i, 1\rangle, \langle C_i, 1\rangle\}, \delta^0, \varepsilon\right\}, \qquad i = 1, \ldots, m.$$

When the forward process terminates, the **end** neuron becomes active and signals that the process has been completed; the active neurons in the layer *DB* store the initial input; other active neurons in *KB* indicate both asserted and inferred facts; the active neurons in the layer *OUT* represent the conclusions of production rules which have been reached by forward chaining under the initial assumptions stored in *DB*.

This neural architecture can be used for implementing neural forward chaining mechanisms for arbitrary systems of production rules of form (R). In particular, one can specify an algorithm which, given in input a system of propositional production rules of form (R) presented in a certain canonical form, outputs a neural network for executing forward chaining on that system of rules (see Burattini *et al.* [4]).

Any such system of rules is allowed to contain cycles (unlike, e.g., the KBANN neural production systems of Towell and Shavlik [5]): a literal appearing as a consequent in one rule can appear in the antecedent of another rule. Moreover, several rules may share the same consequent. As we pointed out previously, this latter possibility requires, to preserve the correctness of forward inferencing, that each occurrence of a literal appearing in the consequent part of the rules be represented by a distinct neuron.

Finally, it is worth emphasizing that the localist semantic interpretation of neurons in terms of literals enables one to provide an informative justification for the conclusions reached in each run of this forward chaining mechanism. In Burattini *et al.* [4, pp. 97–99], we have described how to organize a neural network monitoring the activity of a forward chaining net and exhibiting a trace of the shortest inferential paths from the initial premises to each one of the conclusions obtained by forward chaining.

V. CHUNKING

A. CHUNKING AND PRODUCTION SYSTEMS

In his *Unified Theories of Cognition*, Newell [6, p. 185] gives the following description of chunking in a rule-based problem solving system:

> Chunking is learning from experience. It is a way of converting goal-based problem-solving into accessible long-term memory (productions). Whenever problem-solving has provided some result, a new production will be created, whose actions are these just obtained results and whose conditions are the working-memory elements that existed *before* the problem-solving started that were used to produce the results. This newly minted production will be added to the long-term memory, and will henceforth be available to add its knowledge to the working memory in any future elaboration phase where its conditions are satisfied.

In the setting of the forward chaining system described in the previous section, the condition part of a new chunk that may be added to a long-term knowledge base is to be identified with the conjunction of the facts (propositional literals) from which a run of the forward chaining process starts; and the action part of the same chunk is the conjunction of the literals inferred in the same run of forward chaining.

More formally, a chunk may be viewed as an ordered pair $\langle I, C \rangle$, where, for $k, m \geq 1$, $I = \{p_1, \ldots, p_k\}$ is the set of initial data provided to the system in a given run of forward chaining and $C = \{q_1, \ldots, q_m\}$ is the set of literals derived in that run of forward chaining starting from I. These chunks may be cast in the form

$$p_1 \wedge \cdots \wedge p_k \rightarrow q_1 \wedge \cdots \wedge q_m$$

and thus may differ in their right-hand sides from the rules of form (R) we have considered so far, where $m = 1$.

Even independently of any consideration about their significance in cognitive modeling, chunking mechanisms in rule-based systems may play a significant role in artificial intelligence (AI) applications, both for the automatic acquisition of knowledge bases and for the design of more efficient problem-solving strategies. In this section, we are concerned with the use of chunking mechanisms for addressing the latter problem. In particular, we describe a chunking mechanism generating rules which codify associations between initial data and final outcomes of a forward chaining process. Once these rules are stored, these outcomes can be immediately recalled upon presentation of the same initial data, without having to repeat the forward chaining process. Chunking mechanisms generally give rise to what, following Tambe *et al.* [7], may be called *cognitive* and *computational* effects. The cognitive effect is the reduction of the number of (inferential) steps needed to carry out a given task. The computational effect is the increase in the amount of time needed to carry out each individual step. Thus, what is gained in efficiency by reducing the number of steps is often lost by an increase in execution time for each step. Clearly, when chunks take the form of production rules, the time required for executing the matching process between data in working memory and the conditions of production rules may increase. In fact, more rules have to be scanned, and the newly introduced rules may contain more complicated conditions than those present in the original system of rules. Another related phenomenon may be called the *memory saturation* effect: given preassigned finite memory capacities, a system endowed with a chunking mechanism, which cannot "forget" some of the previously stored chunks, will be eventually unable to make room for newly acquired and possibly more useful chunks. In view of the computational and memory saturation effects, an efficient use of chunking mechanisms requires a computational agent capable of

(a) leaving unaltered the access time to knowledge when new chunks are added, and

(b) attenuating the incidence of the memory saturation phenomenon.

The neural system for chunking on systems of production rules described here—extracting associations between initial data and final outcomes of a forward chaining process—does satisfy condition (a). Condition (b) is satisfied as well, if the relative frequency of the use of chunks is regarded as a satisfactory criterion for deciding which chunks have to be "unlearned" by the system. The chunks stored by the system enable it to reduce processing time: whenever a set of input data coincides with or strictly contains the literals in the first element I of a stored chunk, at the next step the system outputs the literals in its second element C.

Constructing a chunking mechanism of this sort, which can efficiently cope with the computational and memory saturation problems, requires solving the following problems:

(i) recognizing an input pattern previously presented to the system in order to recall the chunks with conditions matching the input pattern (or else storing an input pattern presented for the first time to the system);

(ii) keeping track of how often the stored chunks are used during the system operation, in order to discard less often used chunks when new chunks have to be acquired;

(iii) executing operations (i) and (ii) in a preassigned time, independently of the size of the input patterns and the number of stored chunks.

B. NEURAL MODULE FOR CHUNKING

The neural chunking module (**CM** for short) and the algorithm *Recorder* that we have implemented and described in previous papers affords a sensible solution to these problems. (See Burattini *et al.* [4] for a discussion of the difficulties which may arise in attempting to solve simultaneously problems (i)–(iii) by means of multilayer connectionist networks, Hopfield nets, or Grossberg's ART networks.)

At the end of each run of the forward chaining system, the algorithm *Recorder* isolates a chunk $\langle I, C \rangle$, where, for $k, m \geq 1$, $I = \{p_1, \ldots, p_k\}$ is the set of initial data provided to the system in that run and $C = \{q_1, \ldots, q_m\}$ is the set of literals derived by forward chaining starting from I. If the chunk $\langle I, C \rangle$ was already stored in **CM** as a result of previous runs of the system, this fact is signaled by **CM**, and the algorithm *Recorder* merely modifies the weights of the neural units in **CM** that are devoted to storing data about the relative frequency of use of chunk $\langle I, C \rangle$. Otherwise, the chunk $\langle I, C \rangle$ is stored in **CM**. In the latter case, the new chunk $\langle I, C \rangle$ may replace one of the previously stored chunks that have been less frequently used during the system operation.

Because the example of a forward chaining network examined in the previous section was representative of the fine structure of our networks, our descriptions shall be comparatively more sketchy in this section. The overall structure of the module **CM** is outlined in Fig. 3. The various kinds of information provided by the neural units belonging to its submodular components are briefly described hereinafter.

Let x be a variable ranging over literals and let I be the set of initial data currently provided to the forward chaining system. Then

- at the start, if x is an element of the set I, then the neural representative of x in *PATTERN_ON* and the neural representative of x in *PATTERN_OFF* become active;
- at the end, if the neural representative of x in *PATTERN_OUT* is active, this means that **CM** retrieves x from recorded chunks: x is a conclusion already inferred by a previously performed forward chaining from the set I.

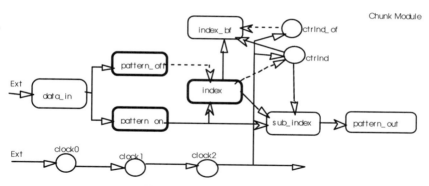

Figure 3   Chunk module (**CM**).

Let j be a variable ranging over indexes which can be associated by the system to a chunk. Then

- if the neural representative of j in *INDEX_BF* is active, then j is already associated to a stored chunk[1]; the system is not going to select j as the index for a new chunk;
- if the neural representative of j in *INDEX* becomes active, then the set I matches the first part of a chunk that was stored and associated to the index j;
- if the neural representative of j in *SUB_INDEX* becomes active, then the set I contains a subset of literals which matches the first part of a chunk previously stored and associated to the index j.

The algorithm *Recorder* stores each chunk $\langle I, C \rangle$ by constructing two different links. The first link $(L1)$ connects every literal (actually, their neural representatives) in I with a given index j, whereas the second link $(L2)$ connects index j with the neural representatives of all the literals in C. This mechanism enables one to eliminate interferences between nonorthogonal patterns affecting traditional neural associative memories, such as the multilayer perceptron (see Rumelhart and McClelland [8]) and the linear associators (see Kohonen [9, 10]): links $(L1)$ and $(L2)$ establish a one–one correspondence between the first and second element of each chunk.

Let us now describe how **CM** operates.

Upon presentation to the system of an input set of literals I, three different cases may occur. In the first case, for every index j the first element of the chunk $\langle I_j, C_j \rangle$ is different from I and the third case described in the following discussion does not hold. In this case, **CM** does not recall any stored chunk. However,

---

[1] See Section V.C for a description of the selection procedure and the role of the *INDEX_BF* layer.

when forward chaining on I is completed and outputs a nonempty set of literals C, the algorithm *Recorder* selects a new index j' and stores a new chunk $\langle I_{j'}, C_{j'} \rangle$, with $I_{j'} = I$ and $C_{j'} = C$.

In the second case, there is an index j such that $I = I_j$ for some previously acquired chunk $\langle I_j, C_j \rangle$. **CM** recognizes this situation, and provides the elements of C_j as outputs of the forward chaining process.

In the third case, for every index j the first element of the chunk $\langle I_j, C_j \rangle$ is different from I, but there are indexes $j_1, \ldots, j_k$ such that $I_{j_1}, \ldots, I_{j_k}$ are strictly included in I. Then **CM** provides the elements of $S = \cup \{C_{j_1}, \ldots, C_{j_k}\}$ as outputs. When forward chaining on I is completed and outputs a nonempty set of literals C, the algorithm *Recorder* selects a new index j' and stores a new chunk $\langle I_{j'}, C_{j'} \rangle$, with $I_{j'} = I$ and $C_{j'} = C$.

Let us now describe less schematically how the network behaves in each of these cases.

Case 1 $(\forall j\ I \neq I_j)$. When I is given as input to **CM** at time $t = 0$ (and the neural units representing its elements in *PATTERN_ON* and *PATTERN_OFF* become active at $t = 1$), the relation $\forall j\ I \neq I_j$ is recognized to hold at time $t = 2$ because, for every j, the neuron $\textbf{\textit{Index}}_j$, representing the index j in the layer *INDEX*, remains inactive at that time. At time $t = 3$, the impulse from control neuron $\textbf{\textit{clock2}}$ activates the other control neuron $\textbf{\textit{ctrind}}$, whose activity is necessary and, together with impulses from *PATTERN_ON* (which are absent in this case), sufficient to activating subpatterns in *SUB_INDEX*. Thus, at time $t = 4$ (resp. at time $t = 5$), all neurons of layers *SUB_INDEX* (resp. of *PATTERN_OUT*) are inactive.

As a consequence, the neural activity triggered by the input set I in the **CM** module does not recall the second element of any chunk represented by the links that bring to the layer *PATTERN_OUT*. So, at time $t = 6$, only the input set I is transferred to the forward chaining system. Finally, when forward chaining on I is completed and outputs a nonempty set of literals C, the algorithm *Recorder* selects a new index j' and stores a new chunk $\langle I_{j'}, C_{j'} \rangle$, with $I = I_{j'}$ and $C = C_{j'}$.

The recording of a new chunk consists of the change of the weights of the following connections: (1) the connections between the neural representatives of the elements of I in *PATTERN_ON* and *PATTERN_OFF* on the one hand, and the neural representatives of the selected index j' in *INDEX* and *SUB_INDEX* on the other hand; and (2) the connections between the neuron $\textbf{\textit{SubIndex}}_{j'}$ of *SUB_INDEX* and the neural representatives of the elements of C in *PATTERN_OUT*.

The weights of the connections between *PATTERN_ON* and *INDEX* are initialized with the following values:

$$w_{\textit{\textbf{Pon}}_i \textit{Index}_j} = 0, \qquad j = 1, 2, \ldots, M, \ i = 1, 2, \ldots, N,$$

where M is the maximum number of patterns which can be stored by the system (i.e., the maximum number of index neurons available in *INDEX*), N is the dimension (or number of components) of input patterns (i.e., the number of neural units in each layer of *PATTERN_ON* and *PATTERN_OFF*), and $Index_j$ and Pon_i are neurons of the *INDEX* and *PATTERN_ON* layers, respectively.

The weights of the connections between *PATTERN_ON* and $Index_{j'}$ are updated according to the rule

$$w'_{Pon_i\,Index_{j'}} = \begin{cases} 1/|I|, & \text{if } Pon_i \text{ active,} \\ 0, & \text{otherwise.} \end{cases} \tag{2}$$

The weights of the connections between *PATTERN_OFF* and *INDEX* are initialized with the following values:

$$w_{Poff_i\,Index_j} = -1, \qquad j = 1, 2, \ldots, M, \ i = 1, 2, \ldots, N.$$

The weights of the connections between *PATTERN_OFF* and $Index_{j'}$ are updated according to the rule

$$w'_{Poff_i\,Index_{j'}} = \begin{cases} 0, & \text{if } Poff_i \text{ active,} \\ -1, & \text{otherwise.} \end{cases} \tag{3}$$

Rules (2) and (3) ensure that exactly one neuron in *INDEX* becomes active when Case 2 occurs. (See Pasconcino [11] for a detailed justification of this claim.)

The algorithm *Recorder* initializes the weights of the connections from units of *SUB_INDEX* to the units of *PATTERN_OUT* in the following way:

$$w_{SubIndex_j\,Pout_i} = 0, \qquad j = 1, \ldots, M, \ i = 1, \ldots, Nc,$$

where Nc is the number of "actions" in the system of rules. The weights of the connections from the active unit $SubIndex_{j'}$ in *SUB_INDEX* to the neural representatives of the elements of $C_{j'}$ in *PATTERN_OUT* are updated by the following rule:

$$w'_{SubIndex_{j'}\,Pout_i} = \begin{cases} 1, & \text{if } q_i \in C_{j'}, \\ 0, & \text{otherwise.} \end{cases} \tag{4}$$

Rules (2), (3), and (4) ensure that when the input set $I_{j'}$ is presented to the network the neurons representing the elements of $C_{j'}$ in *PATTERN_OUT* are activated in the manner described in Case 2.

Case 2 $(\exists j\ I = I_j)$. As usual, I is given as input to **CM** at time $t = 0$ and the neural units representing its elements in *PATTERN_ON* and *PATTERN_OFF* become active at $t = 1$. Now, at time $t = 2$, the neuron representing index j in the layer *INDEX* becomes active, as determined by rules (2) and (3): the system has identified I with I_j.

When this identification is successfully completed, **CM** has to retrieve from index j the second element of the chunk $\langle I_j, C_j \rangle$. Because, at time $t = 2$, the index neuron $\textbf{\textit{Index}}_j$ is active, then, at time $t = 3$, the neural unit $\textbf{\textit{SubIndex}}_j$, representing the same index j in SUB_INDEX, will become active as well. And, at time $t = 4$, the neurons in PATTERN_OUT representing the elements of C_j, as associated to the input pattern $I = I_j$ by rule (4), will become active.

Case 3 ($\forall j$ $I \neq I_j$, but $\exists j_1, \ldots, j_k$: $I_{j_1} \subset I \wedge \cdots \wedge I_{j_k} \subset I$). I contains subpatterns $I_{j_1}, \ldots, I_{j_k}$ which consist of the literals in the first element of already stored chunks. Then the input set I activates in the PATTERN_OUT layer of **CM** the neural representatives of the elements of $S = \cup\{C_{j_1}, \ldots, C_{j_k}\}$.

Let us describe in more detail how this result is achieved. As in Case 1, all neurons of INDEX are inactive at time $t = 2$ when I is presented to **CM** and, at time $t = 3$, the impulse of control neuron $\textbf{\textit{clock2}}$ (see Fig. 5) activates the control neurons $\textbf{\textit{ctrindoff}}$ and $\textbf{\textit{ctrind}}$. At time $t = 4$, the impulse from $\textbf{\textit{ctrind}}$, combined with the impulses from the neural representatives of the elements of the sets $I_{j_1}, \ldots, I_{j_k}$ in PATTERN_ON, activates the neurons $\textbf{\textit{SubIndex}}_{j_n}$ in SUB_INDEX with $n = 1, \ldots, k$. Thus, at time $t = 5$, all neural representatives of the elements of S in PATTERN_OUT are activated.

The activation of the right neurons $\textbf{\textit{SubIndex}}_{j_n}$ upon presentation of input pattern I is determined by the weight values of connections from the PATTERN_ON layer to the SUB_INDEX layer. These weights are initialized with the following values:

$$w_{Pon_i SubIndex_j} = 0, \qquad j = 1, \ldots, M, \ i = 1, 2, \ldots, N.$$

The updating rule for these weights is analogous to (2):

$$w'_{Pon_i SubIndex_{j'}} = \begin{cases} 1/|I|, & \text{if } \textbf{\textit{Pon}}_i \text{ active,} \\ 0, & \text{otherwise.} \end{cases} \tag{5}$$

One can easily show that rule (5) guarantees that, if I_j is a subpattern of the new input set I, then, at time $t = 4$, neuron $\textbf{\textit{SubIndex}}_j$ is active.

C. SELECTING INDEXES

Let us now describe the role of the layer INDEX_BF and the selection criterion of indexes used by *Recorder* to store a new chunk. The algorithm *Recorder* modifies the weights of connections from neurons $\textbf{\textit{ctrind}}$ and $\textbf{\textit{ctrindoff}}$ to the neurons of INDEX_BF in order to codify, for every chunk $\langle I_j, C_j \rangle$, its frequency of recall, relative to the total number of network runs. In particular, the weight of the connection from the $\textbf{\textit{ctrind}}$ neuron to neuron $\textbf{\textit{IndexBf}}_j$ in INDEX_BF is increased by one[2] if and only if the presentation of set I activates the neuron **Index Bf_j**, that is to say, when the network recognizes the situation $I = I_j$, such as described in Case 2. At the same time, the weight of the connection from the **ctrindoff** neuron to each neuron of *INDEX_BF* is decreased with a real value (let us call it α). The parameter α can be interpreted as a frequency threshold suitably chosen by the user. Thus, if the difference between the weighted impulses from **ctrind** and **ctrindoff** to the neuron **Index Bf_j** is positive, then neuron **Index Bf_j** is activated. This signals that the frequency of recall of chunk $\langle I_j, C_j \rangle$ is greater than the α threshold (see Pasconcino [11] for more details).

Thus, when the **ctrind** and **ctrindoff** neurons are activated (at time $t = 3$), at the next instant of time only those index neurons which have frequency of recall greater than α are active in *INDEX_BF*. Every other neuron in *INDEX_BF* is inactive either because no chunk is associated to it or because the associated chunk has recall frequency below the α threshold. To store a new chunk in **CM**, *Recorder* randomly selects an index j' among the inactive neurons of *INDEX_BF*. It may be the case that such a "drawing" procedure selects an index j' which is associated to a chunk $\langle A_{j'}, B_{j'} \rangle$ below the α threshold, and therefore the new chunk $\langle I_{j'}, C_{j'} \rangle$ replaces an old one that scores a low frequency of use.

This simple criterion based on recall frequency provides a mechanism for managing the preassigned limited resources of the **CM** module. In other words, this drawing procedure provides a simple version of a *garbage collector* which, as is well known, does not eliminate the phenomenon of memory saturation, but simply reduces its incidence.

In concluding this section, we wish to emphasize that only one presentation of an input set I and the corresponding output set C suffices for the system to acquire a new chunk. Rules (2) and (3), which enable the system to acquire new chunks, eliminate the interference between the first elements of recorded chunks, as they change exclusively the weights of the input connections to the selected index neurons in the *INDEX* and *SUB_INDEX* layers. Furthermore, one can easily show that the computational complexity of the procedure *Recorder* is linear with respect to the size of the sets I and C.

A system endowed with this chunking mechanism can improve its performances on the basis of the previous activity. Reduction of the processing time is due both to the parallelism inherent in the general neural architecture of the system and to specific features of the **CM** module; the latter guarantee that the computational cost of access to stored chunks is independent of the number of chunks that are in memory. Clearly, the possibility of actually achieving a reduction in processing time is contingent on the availability of a computational agent capable of modifying neural weights and executing the parallel computations allowed by this neural model.

---

[2]Except for the first updating of this weight which increases the initial value by one plus the total number of system runs plus the total number of indexes.

VI. NEURAL TOOLS FOR UNCERTAIN REASONING: TOWARD HYBRID EXTENSIONS

A naive neural model of propositional reasoning from incomplete or uncertain information can be obtained, starting from the rules described in Section III, by modifying the threshold value for a neuron h representing the literal on the right-hand side of a rule of type (R) with k literals on its left-hand side, as follows:

$$s_h = k - (\varepsilon + \eta), \qquad 0 < \varepsilon < 1, \ \eta \geq 1.$$

Of course, rules implemented in this way can "fire" even if the input data match only partially their left-hand sides. However, the firing of rules under a partial match, like that achieved by means of the previous setting, may be a desirable property when rules are used for simulating similarity-based, commonsense reasoning, where loose contextual associations play a central role (for discussion, see, e.g., Sun [12]). Of course, these considerations must be cautiously generalized to the inherently more brittle modes of reasoning used in expert systems: principled restrictions on rule firing are required when the correctness or plausibility of diagnoses or classifications is at stake. The best one can do, in our view, to fulfill this desideratum is to have rule firing reflect rigorous models of reasoning under incomplete or uncertain knowledge, so that the uncertainty attached to the conclusions reached by an expert system can be evaluated within relatively robust conceptual frameworks.

To neurally implement various rigorous models of quantitative uncertain reasoning, the neural elements introduced in Section II must be shown to be capable of coding and operating on arbitrary integer values, in addition to the values 0 and 1 which correspond, respectively, to the active and quiescent state of an individual neuron. The basic idea pursued in this subsection is that of exploiting both excitation values and sequences of neuron firings as codes for positive integers. Various arithmetical functions can be implemented following this strategy, without modifying the simple neuron model adopted in Section II.

Unbroken sequences of unitary impulses traveling from neuron to neuron can be regarded as messages representing positive integers (the number of impulses contained in each such sequence represents an integer). Conversely, positive integers given in input as unbroken sequences of firings can be stored, for immediate or later elaboration, under the form of excitation values. Two examples are given in which excitation is transformed, respectively, into a proportional number of consecutive outgoing impulses and a number of consecutive incoming impulses is stored as a proportional excitation. Finally, it is shown that ordinary multiplication between two positive integers can be neurally implemented following this approach.

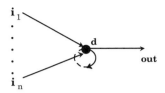

Figure 4 Dropper neuron.

A. TRANSFORMING EXCITATION VALUES INTO SEQUENCES OF FIRINGS

Transformation of excitation into impulses is the natural activity of neurons. Thus, to translate a certain value E of excitation into a number m of consecutive impulses, where m is proportional to E (say, $m \cdot c = E$), one needs just a single "dropper" neuron d (Fig. 4) defined by the following characteristic triple:

$$\mathbf{N}_d \equiv \left\{ \{ \langle i_1, a_1 \rangle, \ldots, \langle i_n, a_n \rangle, \langle d, -c \rangle \}, \delta^1, \eta \right\}$$

in which $i_1, \ldots, i_n$ are input neurons generating the excitation of d, $\langle d, -c \rangle$ is a negative feedback, $\delta^1 = 1$ is the constant decay function $\delta(i)$ ensuring permanent memory, and $\eta < c$ is a threshold allowing the neuron d to drop all the excitation collected from its inputs under the form of consecutive unitary output impulses.

Whenever d emits an impulse, the negative feedback connection determines a constant value c to be subtracted from the residual excitation value. Thus, d keeps on firing until all the excitation E stored in it is dropped away in the form of $m = E/c$ consecutive impulses.

B. TRANSFORMING SEQUENCES OF FIRINGS INTO EXCITATION VALUES

The converse problem is that of counting the number of components in a sequence of impulses and transforming this number into an excitation value equal to the sum of the incoming impulses. A solution to this problem is illustrated with an example for integer values in the interval $[0, 10]$, uniformly modifiable to deal with different numerical ranges (see Fig. 5).

The device behaves as a spring: unbroken sequences of impulses, coming from the input neuron a, charge the spring layer $a_1', \ldots, a_{10}'$. As soon as the input sequence ends, all the excitation stored up in the spring layer is projected at once toward the output neuron b through the gate layer $a_1'', \ldots, a_{10}''$.

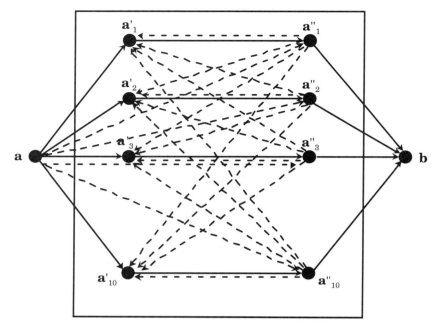

Figure 5   Spring net converting a train of impulses into an excitation value.

The spring layer is described by the following characteristic triple:

$$N_{a'_i} \equiv \{\{\langle a, 1 \rangle, \langle a''_1, -1 \rangle, \dots \langle a''_{10}, -1 \rangle\}, \delta^1, i - \varepsilon\}, \qquad \forall i = 1, \dots, 10.$$

Each neuron a'_i has permanent memory that enables it to cumulate the excitations due to impulses coming from the input neuron a. Threshold values are assigned such that neuron a'_1 starts firing after the first impulse, neuron a'_2 starts firing after the second impulse, and so on. Neurons in the gate layer receive excitatory impulses from the corresponding neurons in the spring layer and inhibitory impulses from the input neuron a. In this way, although their threshold is 0, they cannot transmit the incoming impulses to the output neuron b before the input sequence ends. The characteristic triples for neurons in the gate layer are

$$N_{a''_i} \equiv \{\{\langle a'_i, 1 \rangle, \langle a, -1 \rangle\}, \delta^0, \varepsilon\}, \qquad \forall i = 1, \dots, 10.$$

When at a certain time t the input sequence ends, the neurons in the gate layer are no longer inhibited by a. Therefore, the neurons in this layer that receive an impulse from the corresponding neurons in the spring layer fire on b at time $t + 1$. One can easily see that all neurons in the spring layer eventually reach excitation equal to k if the input sequence was made of k consecutive impulses. However, only neurons $a'_1, \ldots, a'_k$ in the spring layer have threshold lower than k (as one can easily verify from the characteristic triple for this sort of neuron) and fire on the corresponding $a''_1, \ldots, a''_k$ in the gate layer. The latter, in turn, fire on b. Eventually, b receives an excitation proportional (or equal) to the number of impulses in the input sequence. The negative connections from each neuron a''_i toward all neurons a'_i are needed to reset to 0 the excitation of all neurons in the spring layer at once.

In principle, networks performing this transformation can be designed for arbitrary ranges of integer values of excitation. However, this approach is rather impractical for large integer intervals, and the boxed subnet of Fig. 5 can be replaced by another type of processor computing the same function: the neural system making use of this processor becomes hybrid, but the overall parallel implementation afforded by the equivalent purely neural system including the boxed subnet is preserved in the simplified hybrid version.

C. PRODUCT OF POSITIVE INTEGERS

Ordinary multiplication between two positive integers requires a more complicated network. The basic idea is that of outputting the product $a \cdot b$ under the form of a unbroken sequences of b impulses. The value of the integer a is represented by a sequence of a impulses. A controlling subnet enables a distinguished neuron to output b impulses for each of the a input impulses. The network in Fig. 6 performs the product between two positive integers a and b codified as sequences of impulses, sent to the two input neurons a and b (a is not shown).

The structure shown in Fig. 6 includes a spring network quite similar to that shown in Fig. 5. In the new structure, the terminal neuron b' is a dropper neuron which outputs sequences of impulses toward the intermediate neuron c_4. The latter (whose threshold is $2 - \varepsilon$) transfers these sequences to other neurons if and only if it receives simultaneously impulses from the neuron a'. Also, b' sends its output sequence back to b. Neuron a' is the terminal node of a network that receives as input the factor a through the input neuron a. We omit this subnet but it is structurally equal to the one that codifies the trains of impulses coming from b into excitation values in b'. Thus, the neuron a', which has unbounded memory, receives an excitation equal to the input sequence a. Because its threshold is ε, it fires until its excitation value is close to 0. The input neuron b has threshold $2 - \varepsilon$ and transmits forward impulses coming back from b' until a' is active.

The first sequence of b impulses output by b at time t reaches b' at $t + 3$, to be transmitted to neuron c_4 which has been receiving from a' an unbounded sequence of impulses. The neuron c_4 will fire throughout the time it receives impulses from both b' and a'. Moreover, c_4 activates a subnet (c_5, c_8) which sends an inhibition

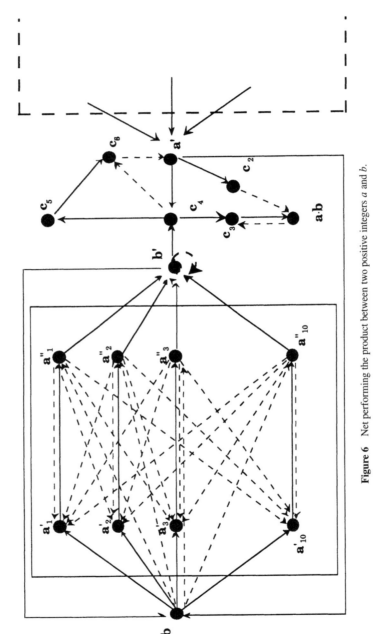

Figure 6 Net performing the product between two positive integers a and b.

to a' whenever c_4 stops firing. This inhibition decreases by one unit the current excitation of a'. Because b' sends back its output to b, a new sequence of impulses will again reach b' with a delay of three instants of time. This process is iterated as many times as the initial excitation value of a', because every sequence of impulses from b' has the effect of decreasing by one unit this excitation value.

The output of c_4 charges the neuron c_3, which acts like a dropper neuron (it has unbounded memory and threshold $2 - \varepsilon$) and fires on the output neuron $a \cdot b$. The latter has threshold ε and is inhibited by c_2 which, in turn, is activated by a'. Thus, $a \cdot b$ will start firing two instants after a' becomes inactive. The neuron $a \cdot b$ sends inhibitory impulses back to c_3 and, consequently, discharges it completely after having fired $a \cdot b$ impulses.

VII. QUALITATIVE AND QUANTITATIVE UNCERTAIN REASONING

A. PRECONDITIONS IN NONMONOTONIC INFERENCE

In many domains of interest for artificial intelligence, but also in everyday life, reasonings are often just plausible or approximately correct. The conclusions obtained by means of uncertain reasonings may have to be withdrawn if some of their premises are no longer verified or some additional piece of information modifies the inference pattern.

Neural networks seem particularly well suited for designing systems that perform "revisable" reasoning. Indeed, such systems should contain "restrictive" rules of the form "p is a theorem if $q_1, \ldots, q_n$ are not theorems" [13]. Hereafter, we show that similar restrictions can be easily implemented by means of neural networks set with appropriate inhibitory connections.

There are two main features rendering neural networks suitable for formalizing revisable reasoning:

(a) *Idleness does not encode negation.* If the nonlinear thresholded neural element n_A, representing sentence A, is not firing, this does not necessarily mean that A is (asserted to be) false. Indeed, in a neural representation of knowledge, the falsehood of A is declared through the activity of neuron $n_{\neg A}$ representing $\neg A$. The inactivity of both n_A and $n_{\neg A}$ is allowed for and may be interpreted as the absolute lack of knowledge about A. The neural network which includes the inactive elements n_A and $n_{\neg A}$ is intrinsically capable of carrying out processes that involve neither of them, that is, performing inferences from information which is unrelated to A. (For a similar view, and its relation to the so-called closed world assumption, see Valiant [14, pp. 172–177].)

(b) *Neuron outputs are not intrinsically monotonic functions of the inputs.* This is due to the fact that neural connections can be assigned either positive or negative weights. Indeed, let us consider a neuron n that receives both excitatory inputs (positive couplings) and inhibitory inputs (negative couplings). The following situation is likely to occur: n fires because the sum of positive and negative inputs exceeds its threshold, while certain neurons, connected to n by means of negative couplings, are left idle. If, at a later time, some of those idle neurons become active, the excitation of n decreases; n may even stop firing if the new inputs are such as to bring the excitation below its threshold value. The situation can change again and again in time, as long as there are further positive and negative inputs being left idle. And clearly, if active neural inputs become idle at a later time, the state of neuron n can change, too. Accordingly, inferences that are performed starting from knowledge of certain facts (active neurons) and ignorance of certain other facts (idle neurons) can be withdrawn if new facts, as well as new uncertainties, are added to the database and this new information changes the state of knowledge concerning the premises of those inferences (the set of active neurons and the set of idle neurons involved in the inferences).

In the following, we call "nonmonotonic neural networks" those neural networks in which both negative and positive weights are implemented. A nonmonotonic inference system contains rules whose application can dynamically be blocked. Some of these rules are specified together with applicability conditions or preconditions, whose verification can dynamically change as the set of available premises changes over time.

A well-known rule of this sort was introduced by Sandewall (see Sandewall [15] and Kramosil [16]):

$$\text{UNLESS}(q) \mid\sim p, \qquad\qquad (\text{nmR1})$$

where the symbol $\mid\sim$ denotes nonmonotonic inference and the argument of the UNLESS operator is the precondition of the inference. In the context of the rule-based systems we have been concerned with, UNLESS can be naively defined as follows:

- UNLESS(q) is true for a given propositional formula q if and only if q cannot be inferred from the set of facts and rules encoded into the knowledge base.

The nonmonotonic inference rule (nmR1) states that p can be inferred under the precondition that q cannot be inferred. In general, this precondition is not equivalent to requiring that $\neg q$ can be inferred. We shall focus on the following generalization of rule (nmR1) in which the conclusion of the rule depends on preconditions $q_1, \ldots, q_k$ that are signaled by the operator UNLESS, and finitely many ordinary premises $a_1, \ldots, a_m$:

$$\{a_1, \ldots, a_m, \text{UNLESS}(q_1, \ldots, q_k)\} \mid\sim p.$$

Using the metavariables A for $a_1 \wedge \cdots \wedge a_m$ and Q for $q_1 \vee \cdots \vee q_k$, this rule can be expressed under the more compact form:

$$A \wedge \text{UNLESS}(Q) \mid\sim p. \tag{nmR2}$$

Because verifying precondition $\text{UNLESS}(Q)$ may be computationally intractable or even impossible in some formal settings, in the context of actually implementable inference systems this rule has been usually (and more aptly) interpreted in the following way: p can be inferred from A if Q has not been inferred so far.

The neural implementation of this kind of nonmonotonic inference rule is quite straightforward. If A, Q, and p are represented by neurons n_A, n_Q, and n_p, respectively, then the neural subnet of Fig. 7 encodes the rule expressed by (nmR2). Here, the threshold of n_p is set equal to some constant T (throughout this section, we assume that thresholds are all given the same value T), the coupling from n_A to n_p has a positive weight $w_{A,p}$ greater than T (e.g., $w_{A,p} = T + \varepsilon$), and the inhibitory connection from n_Q to n_p can nullify the possible excitation coming from A (e.g., with a negative weight $w_{Q,p}$ equal to -2ε). Clearly, neuron n_p is active if and only if n_A is active and n_Q is idle, just as it must be if the neural system has to encode the rule given by (nmR2).

In a nonmonotonic setting, the verification of the premises, that is, preconditions and ordinary conditions, can dynamically change with the set of formulas already inferred by the system. Our neural implementation reflects that dynamic behavior. Indeed, if at any instant t either n_A stops firing or n_Q starts firing (i.e., either A no longer holds or Q becomes inferable), then the excitation of n_p decreases to a value below the threshold and n_p becomes idle (signaling absolute ignorance about the status of p).

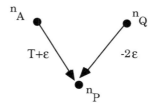

Figure 7 UNLESS operator subnet.

Setting conditions, preconditions, and any sort of inference scheme on the same ground can be a source of instability in the inference system (Reiter and Criscuolo [17]). Indeed, assume p was inferred nonmonotonically applying a certain rule (nmRA) on the occurrence of certain facts A and the nonoccurrence of certain other facts Q. It may well happen that by inferring p the system can apply another rule (nmRB) to produce the nonmonotonic inference of Q. Thus, the preconditions of (nmRA) no longer hold and p is to be withdrawn. In turn, the withdrawing of p might trigger the withdrawing of Q, and hence a return to the starting point, with the net ready to repeat the cycle: apply (nmRA) to infer p, then apply (nmRB) to infer Q, then withdraw p because of Q and then withdraw Q because of p.

Now, let us examine how neural implementations can cope with such classical problems of swinging decisions in nonmonotonic reasoning. The situation described previously can be reproduced in a knowledge base that contains the following nonmonotonic rules:

$$A \wedge \text{UNLESS}(Q) \mid\sim P, \qquad\qquad \text{(nmRA)}$$

$$A \wedge \text{UNLESS}(P) \mid\sim Q. \qquad\qquad \text{(nmRB)}$$

Furthermore, it is assumed that, initially, A holds and that nothing is known about P and Q or their negations. This knowledge base can be encoded in a neural network in which A is represented by an active neuron n_A, whereas P and Q are represented by two idle neurons n_P and n_Q. The connections can be set as in the scheme of Fig. 8.

In this neural implementation, the two rules are applied in parallel. Assume that neuron n_A starts firing at instant t and keeps on firing indefinitely. Then neurons n_P and n_Q, excited by n_A, will both fire at instant $t + 1$. However, because they

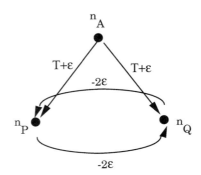

Figure 8   Swinging subnet.

send to each other inhibitory impulses, they both stay idle at instant $t + 2$. The situation evolves with a swinging behavior in which n_P and n_Q oscillate between the firing and the idle state, indefinitely.

To find a way out, let us concentrate on just one of these nonmonotonic rules, say, (nmRA), analyzing in more detail the sequence of facts and inferences involving that rule:

1. A holds.
2. Nothing is known about Q.
3. UNLESS(Q) is justified on the basis of step 2.
4. P can be nonmonotonically inferred from steps 1 and 3, by (nmRA).
5. Q is later on recognized to hold (no matter how).
6. UNLESS(Q) must be withdrawn, because of step 5.
7. P, being no longer inferable by (nmRA), must be withdrawn.
8. Q is withdrawn (may be because of step 7).
9. Go to step 3.

The source of the inferring–withdrawing oscillation is the complete and unlimited freedom in retracting and reintroducing the same preconditions. This unrestrained use is perhaps pragmatically justifiable on the ground that preconditions are meant as signals of untypical situations, rather than normally occurring events. However, when the need arises for avoiding or at least attenuating the incidence of loops, solutions have to be searched for in a different model of revisable reasoning. For example, one may devise a system in which the withdrawing of a conclusion, previously inferred by application of a certain nonmonotonic rule, should produce a suitable transformation of that rule. More specifically, convergent nonmonotonic inference procedures should be developed, by introducing mechanisms that make nonmonotonicity an expendable property of nonmonotonic rules.

Neural networks are suitable for modeling various degrees of expendable nonmonotonicity. For example, let us consider the following neural implementation and interpretation of the nonmonotonic rule (nmRA), in which the rule itself is practically withdrawn when the precondition is retracted (Fig. 9). Here, UNLESS(Q) is treated as an individual proposition, represented by an appropriate neuron $n_{U(Q)}$. Initially, $n_{U(Q)}$ is set in the firing state: nothing is known about Q, except that it represents a rare event. Because it is endowed with a self-excitatory connection, $n_{U(Q)}$ keeps on firing until neuron n_Q begins to fire and inhibits it.

Let us emphasize that neuron $n_{U(Q)}$ cannot be switched on again, after being switched off the first time. In other words, this network implements a nonmonotonic inference scheme in which inferred statements can be withdrawn and, in addition, nonmonotonic rules can be suspended indefinitely. Let us now try to introduce a nonmonotonic inference rule which may be viewed as governing the behavior of this neural net. A straightforward modification of rule (nmR2),

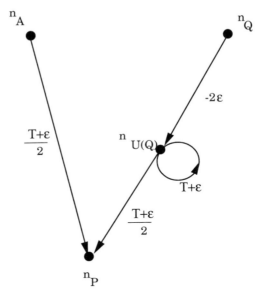

Figure 9 UNTIL operator subnet.

obtained by substituting the operator UNTIL for UNLESS, may serve this purpose:

$$A \wedge \text{UNTIL}(Q) \mid\sim p. \tag{nmtR}$$

The new operator may be defined in the following way:

- UNTIL(q) holds for a given propositional formula q, at a certain step of the inference process, if and only if q has "never" been inferred before that step.

The reactivating of suspended rules can be modeled by neural networks, too. Conditions of reactivating must be carefully chosen so as to avoid new sources of loops. One may allow for the switching on of neuron $n_{U(Q)}$ upon the occurrence of some special event Z (such as, for example, the reinitialization of the system, the inference of $\neg Q$, etc.). Of course, reactivating UNTIL(Q) on the withdrawing of Q, that is, putting Z equivalent to UNLESS(Q), boils down to modeling the classical nonmonotonic rule (nmR2). Figure 10 shows the neural network suspending (on the occurrence of Q) and reactivating (on the occurrence of Z) the inference of p from A.

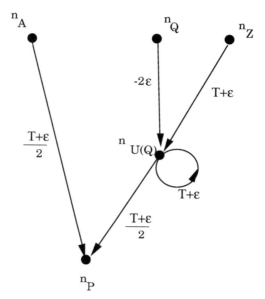

Figure 10 BETWEEN operator subnet.

After being reactivated by just one firing of neuron n_Z, the neuron $n_{B(Z,Q)}$ remains active until Q, even if Z is withdrawn at a later time. This means that neuron $n_{B(Z,Q)}$ retains memory of which neuron, out of n_Q and n_Z, fired last: it is active in the interval between the firing of n_Z and the firing (if any) of n_Q; it is idle between the firing of n_Q and the firing (if any) of n_Z. A straightforward formalization of this behavior can be given by means of the following rule:

$$A \wedge \text{BETWEEN}(Z, Q) \mid\sim p, \qquad\qquad \text{(nmbR)}$$

where the definition of BETWEEN is

- BETWEEN(Z, Q) holds for the given propositional formulas Z (representing the resetting event) and Q (representing the suspending event), at a certain stage of the inference process, if and only if one of the following situations occurs: (i) Q has never been inferred; or (ii) Z has been inferred (or asserted by other means) after the last inference of Q.

Let us consider in some detail the behavior of the neural network implementing rule (nmBR). The initial conditions are encoded into the setting of the net at time $t = 0$:

$t = 0$:

n_A	starts firing	(A holds)
$n_{B(Z,Q)}$	is firing	(initially nothing is known about Q)

$t = 1$:

n_P	starts firing	(P is nonmonotonically inferred)

...   things keep on unchanged until some time later:

$t = q$:

n_Q	fires	(Q is asserted to hold, no matter how)

$t = q + 1$:

$n_{B(Z,Q)}$	stops firing	(BETWEEN(Z, Q) must be withdrawn)

$t = q + 2$:

n_P	stops firing	(P is no longer inferable and is with-

...   drawn) the rule remains suspended and no new inference
...   can be performed even if, some time later:

$t = s$:

n_Q	stops firing	(possibly, because of P)

$t \geq s + 1$

...   the rule remains suspended and no new inference can be
...   performed until, at a later time, if any, n_Z fires.
                       For example, the process may develop as follows:

$t = x$:

n_Z	fires	(Z is asserted to hold, no matter how)

$t = x + 1$:

$n_{B(Z,Q)}$	fires	(the nonmonotonic rule is reactivated)
	elaboration starts	again as from $t = 1$

...
...

$t = y$:

n_Z	stops firing	(but the nonmonotonic rule remains ac-
		tive until Q is again asserted to hold).

B. QUALITATIVE HYPOTHESIS SELECTION IN TWO-LEVEL CAUSAL NETWORKS

In this subsection, we consider a qualitative form of causal reasoning which is commonly used in diagnostic tasks. A basic inferential strategy of diagnostic problem solving is that of considering the abnormal observed manifestations *OBS* relative to the system under examination and isolating hypotheses that *may* explain their occurrence on the basis of known causal relationships between manifestations and hypotheses (see Josephson and Josephson [18] and Peng and Reggia [19]). This strategy is grounded on abductive inference schemes, such as:

$$\frac{m_1, \ldots, m_n \in OBS \quad h_i \text{ can cause } m_1, \ldots, m_n}{h_i} \qquad (AB1)$$

Parsimonious set covering models of diagnosis (see Peng and Reggia [19] and Reggia *et al.* [20]) are based, for $n = 1$, on $(AB1)$. A connectionist approach to parsimonious set covering diagnoses can be found in Ahuja *et al.* [21]. And clearly, each particular instance of $(AB1)$ can also be represented as a neural production rule within a localist semantic approach.

Inferences based on $(AB1)$ do not use any information about the degree of support that manifestations lend to given hypotheses, even though this sort of information may prove crucial to converge on the more plausible explanations. Consider the following schematic example (Fig. 11): the observed manifestations are m_1, m_2, and m_3, connected via causal relations to hypotheses h_1, h_2, and h_3. The only manifestation supporting h_1 is m_1—but strongly so, because it is a highly specific manifestation for h_1. Manifestations m_1 and m_2 weakly support h_2, whereas h_3 is supported by m_3.

If one restricts admissible explanations E for this diagnostic problem to minimal cardinality covers of the observed manifestations, one has the counterintuitive result that $\{h_2, h_3\}$ is a solution to this diagnostic problem, whereas h_1 is discarded. Reiter [22] replaces the minimal cardinality restriction with the weaker condition that an explanation E must be an irredundant cover of the observed manifestations (where E is an irredundant cover of the observed manifestations iff no proper subset of E is a cover of the observed manifestations). However, even in this framework one has that no irredundant cover E includes hypothesis h_1.

The inferential scheme $(AB1)$ can be generalized so as to take into account information enabling one to decide whether the observations provide significant support for, and thus have to be covered by, candidate explanations (see Console and Torasso [23]). A reasonable solution to this problem is that of evaluating the total degree of support for h_i provided by the observed data that are causally re-

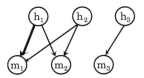

Figure 11 Causal system.

lated to h_i, and comparing this value with a threshold a_i, expressing the minimum degree of support needed to advance h_i as a candidate hypothesis: for each hypothesis h_i and the set $\mathrm{Man}(h_i) = \{m_j: h_i \text{ can cause } m_j\}$, one has to define a suitable function $f_i: X \subseteq \mathrm{Man}(h_i) \rightarrow f_i(X) \in \mathbb{R}$:

$$
\begin{array}{c}
m_1, \ldots, m_n \in OBS \\
h_i \text{ can cause } m_1, \ldots, m_n \\
\underline{f_i(m_1, \ldots, m_n) \geq a_i} \\
h_i
\end{array}
\qquad (AB2)
$$

Scheme $(AB1)$ is a particular case of $(AB2)$. Within this more general framework, new problems may arise; for example, f_i may be computationally intractable or the data needed to compute it may be lacking. Nevertheless, appropriate solutions can be found in significant cases: many diagnostic expert systems make use of various methods to compute specific f_i's that enable the system to perform reasonably well. And, in turn, some of these methods lend themselves to neural representation and processing. We describe here a neural implementation, presented in Burattini and Tamburrini [24], of an instance of $(AB2)$, developed within the framework of an expert system for a medical domain [25] and based on a qualitative approach to uncertain reasoning [26].

Human experts are often reluctant to set precise numerical weights for manifestation–hypothesis relationships, and more confidently advance qualitative judgments on the support that observable manifestations lend to a hypothesis. For example, a physician may consider a given manifestation as "moderately" or "very" suggestive of a certain disease. One can reasonably assume that there is a finite bound h on such discrimination power of human experts, and, namely, that the number of different qualifying labels available to human experts is at most h. Thus, when qualitative judgments of this sort are expressed by experts for every pair (h_i, m_j) such that h_i can cause m_j (e.g., "m_1 strongly suggests h_2"), the elements in each set $\mathrm{Man}(h_i) = \{m_j: h_i \text{ can cause } m_j\}$ can be partitioned into h disjoint classes, each class containing all manifestations with a given degree of relevance with respect to h_i. As a consequence, these classes can themselves be ordered according to the degree of relevance of their elements. One can make the following additional assumption:

Assumption 1. The manifestations belonging to a lower-ranked class, even when taken as a whole, cannot be more relevant with respect to h_i than any manifestation belonging to a higher-ranked class.

Then this qualitative information can be readily represented and processed in a neural system.

Let $\mathcal{H}_i$ be the neuron representing hypothesis h_i, and let $M_i = \{f_{i_1}, \ldots, f_{i_n}\}$ be the set of neurons representing the elements of $\mathrm{Man}(h_i)$. Given the qualitative information provided by human experts, M_i can be partitioned into a totally ordered series of h disjoint classes $M_{i_1}, \ldots, M_{i_h}$. Let their cardinality be

$K_i, \ldots, K_h$, respectively. For each M_{i_j} and each $f_{i_m} \in M_{i_j}$, the weight between f_{i_m} and $\mathcal{H}_i$ must assume the value

$$\frac{j}{(j+1) \cdot K_j \cdot K_{j+1} \cdot \cdots \cdot K_h} \cdot$$

A rough idea of the causal net which reflects the previous settings is given in Fig. 12. Given these settings, for any $j < k \leq h$, one has that the sum of excitations sent to $\mathcal{H}_i$ by all elements belonging to M_{i_j} is less than the minimum excitation sent by any element belonging to M_{i_k}. $\mathcal{H}_i$ receives these excitations with weight equal to 1, and their sum expresses the degree of confidence reached by hypothesis h_i which may or may not exceed $\mathcal{H}_i$'s threshold. Such a threshold is set by the human experts. Furthermore, this value can be compared with the degree of confidence reached by other competing hypotheses (see Burattini and Tamburrini [24, p. 543]) by means of another neural module.

Assumption 1 was adopted in the design of an expert system applied to a particular medical domain (see Section VIII), but is inappropriate in many other diagnostic domains. This inferential strategy, however, can be modified to a certain extent without having to relinquish neural representation and processing. For example, one may wish to model situations in which aggregations of manifestations from lower-ranked classes provide more significant evidence with respect to h_i than manifestations from higher-ranked classes. Aggregations forming so-called "typical patterns" for h_i are a case in point. These can be dealt with in the framework of neural representation and processing by introducing, between the layers of neurons representing individual manifestations and hypotheses, an intermediate layer of neurons representing such aggregations of manifestations.

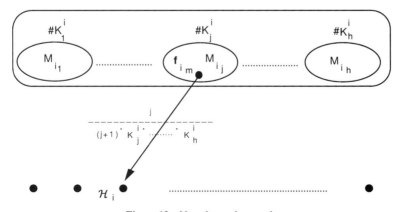

Figure 12   Neural causal network.

C. QUERY PROCESSES AND THE PROBABILISTIC CAUSAL METHOD

Once a set of candidate diagnostic hypotheses has been advanced by applying, for example, production rules enabling one to perform the qualitative form of causal reasoning considered in the previous subsection, a deeper probing of the selected hypotheses may be needed for the system to settle on a final diagnosis. For example, the system may have to choose between two competing hypotheses or to look for additional evidence corroborating the candidate hypotheses.

At this stage of the diagnostic process, the system has to use the causal knowledge that the candidate hypotheses (e.g., a disease) may give rise to a number of manifestations (e.g., symptoms), in order to identify additional observations that enable it to turn a prediagnosis into a final diagnosis. Thus, supposing that h_1 was selected as a candidate hypothesis on the basis of observation m_1, the system may follow the strategy of testing this hypothesis by verifying whether other manifestations in $\text{Man}(h_1)$ are actually present. In many situations, the system will have to choose (in view of, e.g., priority criteria set by the experts) which subset of $\text{Man}(h_1)$ is to be investigated first. An appropriate ranking of the manifestations $\text{Man}(h_1)$, given that the hypothesis h_1 holds, may reflect these selection criteria.

A method for obtaining this ranking in several diagnostic domains is encompassed by the probabilistic causal model (see Peng and Reggia [19] and Reggia *et al.* [20]).

Let the expression $h_i \rightarrow m_j$ denote the event that h_i actually causes m_j, and let $P(h_i \rightarrow m_j|h_i)$ be the conditional probability that h_i causes m_j given that h_i is present. Under certain assumptions (see Peng and Reggia [19]) that seem reasonable for many types of cause–effect relationships, one can prove that

$$P\left(h_i \rightarrow m_j|h_i\right) = P\left(m_j|[h_i]\right),$$

where $[h_i]$ stands for the event that h_i is present and all other possible causes of m_j are definitely not present. This result indicates that the value of $P(h_i \rightarrow m_j|h_i)$ can be obtained from the statistical analysis of the population of individuals that have h_i without being affected by any other possible cause of m_j.

Given the values of the $P(h_i \rightarrow m_j|h_i)$ for all h_i such that h_i can cause m_j, an algorithm for finding the probability of m_j given the presence of a subset of its causes is made available by the following theorem, because the conditional probabilities of m_j are equal to 1 minus the corresponding conditional probabilities of m_j:

THEOREM (Peng and Reggia [19]).   *Let $c_{ij} = P(h_i \rightarrow m_j|h_i)$ and let D be a subset of* $\text{Causes}(m_j) = \{h_i: h_i \text{ can cause } m_j\}$. *Then*

$$P\left(\neg m_j|\{h_i:\ h_i \in D\}\&\left\{\neg h_i:\ h_i \in \text{Causes}(m_j) - D\right\}\right) = \prod_{h_i \in D} (1 - c_{ij}).$$

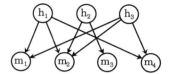

Figure 13 Causal system.

A network enabling one to compute these conditional probabilities can be easily constructed by means of the tools described in previous sections. An example is given in the following discussion for the causal system of Fig. 13. The network for this causal system, in which probability values are transmitted under the form of unbroken sequences of impulses, is formed by four layers (see Fig. 14).

The neurons of the first layer (*HYP*) represent hypotheses h_i and their negations $\neg h_i$. A neuron representing a hypothesis h_i has, consistent with the schema in Fig. 14, as many outgoing arcs as the number of manifestations $m_{i_1}, \ldots, m_{i_n}$ which are possibly caused by h_i. The coupling coefficients of these connections are $(1 - c_{i_1 j}), \ldots, (1 - c_{i_n j})$, respectively. Thus, h_1 has three arcs, and their coupling coefficients are $(1 - c_{11})$, $(1 - c_{41})$, and $(1 - c_{22})$. Similarly, the element representing the negation $\neg h_i$ of some h_i has i_n outgoing arcs, but their coupling

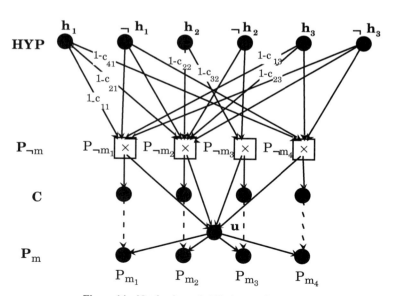

Figure 14 Net for the probabilistic causal system.

coefficients are equal to 1 (the latter are not displayed in Fig. 14). The second layer ($P_{\neg m}$) is formed by as many small subnets $\mathbf{P}_{\neg m_j}$ as the number of possible manifestations. In our example, $1 \leq j \leq 4$. These subnets perform the product of the incoming input values from the layer HYP, outputting conditional probability values of the various $\neg m_j$, in accordance with the preceding theorem.

Because the conditional probabilities of m_j are equal to 1 minus the corresponding conditional probabilities of $\neg m_j$ the remaining part of the network is devoted to computing this simple operation. The third synchronizing layer (C) is formed by as many neurons as the number of possible manifestations. The layer C permits, through a distinguished neuron $\boldsymbol{u}$ and some synchronizing neurons, evaluation of the conditional probabilities of m_j. The impulses fired by $\mathbf{P}_{\neg m_j}$ through a negative coupling coefficient are subtracted from the impulses fired by $\boldsymbol{u}$ and representing the value 1 (suitably normalized to 10, or to 100, etc.). The output of the neurons belonging to the layer $\mathbf{P}_m$ is a sequence of impulses equal to the chosen normalization of the value 1 minus the conditional probability of $\neg m_j$ given the presence or absence of the hypotheses from a given set.

The parallel execution afforded by this neural implementation may not be beneficial in terms of computation time because of the numerical computations that have to be performed by the subnets in layer $P_{\neg m}$. These subnets can be profitably replaced by other types of processors computing the same function. Again, the system thus obtained becomes hybrid, but the parallel architecture introduced by means of the purely neural system including layer $P_{\neg m}$ is on the whole preserved in the simplified hybrid version.

VIII. PURELY NEURAL, RULE-BASED DIAGNOSTIC SYSTEM

A. ABDUCTION–PREDICTION CYCLE

The task of a diagnostic expert system can be roughly described as that of isolating a set of explanatory hypotheses for the insurgence of anomalies observed in objects belonging to its domain of application. Thus, a diagnostic expert system can be viewed as a particular type of problem solver. The statement of the problem is a description of an abnormal state, and a solution is given by an explanation for the occurrence of this abnormal state.

To produce a diagnosis, an expert system makes use of a knowledge base which must include relationships between observable anomalies and their possible explanations (a simple example being "symptom x is a likely manifestation of disease y"). However, knowledge bases of diagnostic systems may encompass relationships between observable facts as well as between possible explanatory hypotheses (e.g., incompatibility relations between pairs of observables or causes, groups of anomalies characterizing pathological patterns, etc.), and sometimes even simplified models of correct or abnormal behaviors of the objects in their domains.

Designing a particular diagnostic expert system involves, as a crucial preliminary step, analyzing human expert knowledge and reasoning applied to the given diagnostic domain. This analysis, which amounts to extracting from human experts reports on the data, theories, and inferential processes used in their problem-solving activity, is the starting point for the knowledge engineer engaged in the task of specifying the knowledge base and inferential schemes of a particular diagnostic expert system.

However, there are some stages of diagnostic problem solving that remain invariant across particular applications:

(i) *Data entry*. Abnormality observations to be accounted for have to be recorded and possibly refined.

(ii) *Prediagnosis and diagnosis*. The data have to be evaluated with the aim of focusing on, refining groups of (possibly incompatible) diagnostic hypotheses, and advancing a final diagnosis.

(iii) *Hypothesis-driven query*. New data may have to be collected, between prediagnosis and diagnosis proper, to test the hypotheses selected at the prediagnostic stage.

(iv) *Justification*. One must be capable of providing, upon request, an informative justification for the conclusions that have been reached.

Similarly, there are inferential schemes that play a significant role in most cases of diagnostic hypothesis formation and testing. Abductive inferences [such as rule $(AB1)$ examined in Section VII.B] enable one to select possible explanatory hypotheses for observed facts, and predictive inferences enable one to isolate possible observable manifestations of the explanatory hypotheses selected by abductive inferences. Unlike deductive rules, abductive rules of inference such as $(AB1)$ may fail to satisfy the correctness requirement: even when the premises of an abductive rule are verified to hold, its conclusion might be shown to be false in the light of new evidence, and has to be withdrawn. However, because abductive rules enable one to generate *possible* explanatory hypotheses, they play a key role in hypothesis formation processes. As already emphasized in Section VII.B, the question whether a possible explanation is also a *plausible* one requires additional considerations, transcending observations and relationships involved in simple abductive inferences such as $(AB1)$. In particular, one may need information about how much the various observable facts are suggestive of or support explanatory hypotheses, in order to assign a plausibility degree to a hypothesis, and thus to induce a ranking between possible explanations. Various methods for handling this information (e.g., probability theory, certainty measures, qualitative nonnumerical orderings) have been used in the setting of diagnostic problem solving. The

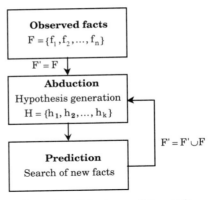

Figure 15   Abduction–prediction cycle.

application described in the following section makes use of the particular non-numerical method for ranking hypotheses described in Section VII.B, whereas in other situations numerical methods may be more appropriately adopted and neurally implemented by means of tools presented in Sections VI and VII.C.

Predictive inferences are also involved in the process of assessing the plausibility of explanatory hypotheses selected by means of abductive rules of inference. From the knowledge that Φ is a possible cause for the occurrence of Ψ and that Φ is an explanatory hypothesis which is being entertained, one can infer Ψ as a possible manifestation of this hypothesis. This kind of inference can be applied in diagnostic reasoning to test explanatory hypotheses. In fact, suppose one has both $\Phi \rightarrow \Psi$ and $\Phi \rightarrow \chi$ in a knowledge base, where Φ is an explanatory hypothesis, Ψ and χ are observable facts. After observing Ψ one can assume, by applying, for example, inference scheme $(AB1)$, Φ as a possible explanation for the presence of Ψ. Which additional facts could be detected if Φ were the right hypothesis? Using predictive inferences, one produces the set of possible observable manifestations of Φ (in our case just $\{\chi\}$). This information can be used to test hypothesis Φ by determining which of the observable manifestations of Φ are actually present, and possibly to advance new explanatory hypotheses. The abduction–prediction cycle is schematized in Fig. 15.

B. Diagnoses in Pediatric Gastroenterology

The abduction–prediction cycle, supplemented by a qualitative method for hypothesis ranking, is at the heart of the inference engine for the diagnosis of pediatric gastroenterological diseases [25], which is schematically described in this subsection: from the initial observations, a set of diagnostic hypotheses is focused on by abductive inferences; then a hypothesis-driven query process enables one to isolate, by predictive inferences, additional evidence that is subsequently used to differentiate between the selected hypotheses. A purely neural implementation of this system is described, and its advantages in terms of computational time are emphasized. Physicians use various procedures for gathering information about their patients: history taking (i.e., the patients' subjective accounts of their medical problems), direct physical examination, and diagnostic tests. Each procedure gives rise to a specific class of evidence: histories, symptoms detected as a result of physical examination, and test outcomes. Signs belonging to each of these classes are further classified as *generic* signs (those signs that are possible manifestations of more than one disease) and *specific* signs (that are possible manifestations of one disease only). Accordingly, signs can be divided into six, mutually exclusive classes: generic histories, symptoms, and tests; specific histories, symptoms, and tests.

In the specific domain of pediatric gastroenterology, medical experts proposed a "clinical relevance" hierarchy between these types of evidence, with a clinical relevance gradient from generic histories to specific tests: for example, the information that a patient manifests a specific symptom s_i, which is associated to disease d only, provides more supporting evidence for d than a generic symptom s_j which is a possible manifestation of d and other diseases as well. Accordingly, if one denotes by "$x >_d y$" the relationship "x provides more support for d than y," one has that $[s_i >_d s_j]$. More in general, one has that

$$[\text{any specific test} >_d \text{any specific symptom} >_d$$
$$\text{any specific history} >_d \text{any generic test} >_d$$
$$\text{any generic symptom} >_d \text{any generic history}].$$

This hierarchical organization can be occasionally overridden when evidences belonging to a lower-ranked class constitutes pathological patterns that are highly suggestive of a certain diagnostic hypothesis.

Evidence–disease causal relationships are used by experts to focus on explanatory hypotheses, in the sense that each evidence e and each relationship of the form "Disease d is a possible explanation for the insurgence of e" can serve as a premise for an abductive inference to disease d as a possible diagnosis. Diagnostic hypotheses supported by each type of evidence are more confidently advanced by physicians, whereas a collection of evidence of one type only is often regarded as insufficient to achieve a diagnosis. When one or more types of evidence are lacking, physicians generally attempt to gather further information, using predictive inferences to isolate additional possible manifestations of the diagnostic hypotheses they are already entertaining, and focusing on those manifestations that are more significant for assessing the hypotheses. The plausibility of a diagnostic hypothesis d focused on by abductive inferences depends on both the significance and approximation to "completeness" of the supporting evidence. By "compleness" of supporting evidence, in this context, we mean diagnostic hypotheses supported by evidence from each of the previously mentioned six classes.

The knowledge base of the expert system designed on the basis of this analysis is chiefly formed by causal relationships between evidence and disease. The set of such relationships can be visualized as a bipartite graph (see Fig. 16). The upper and lower sets of nodes represent evidence and disease, respectively. Edges connect the evidence with their possible explanations.

The set of evidence nodes is partitioned into six classes. Splitting the set of evidence nodes into classes allows one to introduce, in accordance with the comparative relevance judgments expressed by medical experts, a nonnumerical, qualitative evaluation of the support that each evidence lends to a candidate hypothesis. For example, for the evidence present in Fig. 16, one can assert that $s_7 >_{d_4} s_4 >_{d_4} s_2$.

Organizing the knowledge base as a bipartite graph facilitates the implementation of an inference engine which applies an abductive–predictive inferential cycle. Each edge of the graph is interpreted by the system as a causal relationship between one evidence and one disease. For example, in Fig. 16 the causal relationships represented by the edges of the graph are: $d_1 \rightarrow s_1, d_1 \rightarrow s_4, \ldots, d_4 \rightarrow s_8$.

Let $O_s = \{s_2, s_3, s_8\}$ be the set of initially observed evidence relative to a given patient. To explain observations O_s, the system goes from evidence nodes to disease nodes following the arcs of the graph, so as to obtain a set of possible diseases explaining the observations O_s. The result of this abductive inference is the set of diseases $D_s = \{d_2, d_3, d_4\}$. Then the system makes use of predictive inferences to test and differentiate between these hypotheses.

The arcs of the graph are followed starting from the disease nodes corresponding to d_2, d_3, d_4 so as to isolate, among the evidence nodes, the possible manifestations OD_s of the diseases in D_s that are not elements of O_s. This inferential

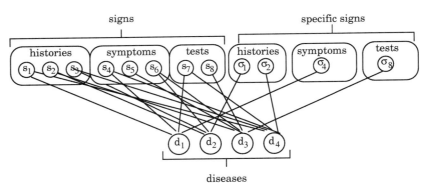

Figure 16   Structure of the knowledge base.

step can be viewed as an attempt to answer the following question: "Given D_s, which evidence could be detected in addition to the elements of O_s?" Thus, by applying this abductive–predictive cycle, the system is capable of generating a set of diagnostic hypotheses D_s for explaining the initial data O_s and isolating the possible observable consequences OD_s of these hypotheses.

Subsequently, the system starts an interaction with the user concerning the elements of OD_s, with the goal of collecting information useful for testing and differentiating between the diagnostic hypotheses D_s. More specifically, (i) the system focuses on the manifestations in OD_s which, if observed in the patient, would prove more useful for assessing the hypotheses in D_s, and (ii) the system initiates a query process asking whether, by applying some medical procedure, these manifestations can be actually detected in the patient.

The selection of the elements in OD_s to be investigated must reflect the basic heuristic strategies adopted by physicians. After collecting information about the patient's anamnesis, physicians look for other signs deriving from a physical examination of the patient, and possibly prescribe tests and/or therapies. This heuristic strategy captures the assumption, discussed previously, that the degree of support for a disease depends, at least partially, on the diversity or "completeness" of the observed evidence (i.e., a diagnostic hypothesis which explains evidence belonging to different classes is more credible than a hypothesis which explains only evidence belonging to the same class).

C. NEURAL IMPLEMENTATION

The main components of a purely neural system for this diagnostic task are schematically represented in Fig. 17. The overall system is organized into five distinct subnets: *evidence, abductive, hypothesis, predictive,* and *justification.*

The global network of neurons can also be viewed as formed by two interacting parts. The first part codifies the declarative and procedural knowledge relative to each specific domain of application (facts, hypotheses, and their relations). The second part is the invariant structure of the shell, which embodies computational utilities supporting and synchronizing the activity of the whole shell; it is not modifiable by experts and users. In the following, we give a brief description of the internal organization of each subnet.

1. Evidence Subnet

This is an input subnet and accepts the information that a set of facts has been detected. A network of neurons stores these data and checks their internal coherence by controlling whether input neurons representing incompatible facts (e.g., f and $\neg f$) have both been erroneously activated. If incoherences are not detected,

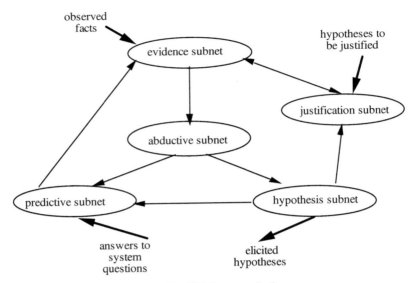

observed
facts

evidence subnet

hypotheses to
be justified

justification subnet

abductive subnet

predictive subnet

hypothesis subnet

answers to
system
questions

elicited
hypotheses

Figure 17   Global net organization.

the data are given as input to the abductive subnet; otherwise, the incoherences are declared to the user.

In this subnet, there are four main layers of neurons (*entry, memory, fact, restart*), each layer with n neurons, where n is the number of observable facts in the given diagnostic domain. Furthermore, there is a layer of k neurons, whose role is that of signaling incompatibility between pairs of facts, and k is the number of such incompatible pairs.

Let us consider first the n *entry neurons* (e neurons) (Fig. 18). From now on, we will associate to each type of neural element its graphical representation and characteristic triple. Each e neuron e_i receives information from (i) the neuron ext_i that fires on e_i, through a coupling coefficient equal to 1, to signal that fact f_i

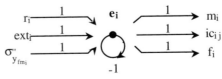

r_i ——1——▶ e_i ——1——▶ m_i

ext_i ——1——▶ ——1——▶ ic_{ij}

$\sigma''_{y f_{mi}}$ ——1——▶ ——1——▶ f_i

-1

Figure 18 $e_i \equiv \{\{\langle r_i, 1\rangle, \langle ext_i, 1\rangle, \langle \sigma''_{y f_{mi}}, 1\rangle, \langle e_i, -1\rangle\}, \delta^0, 1 - \varepsilon\}.$

Figure 19 $m_i \equiv \{\{\langle e_i, 1 \rangle, \langle m_i, 1 \rangle\}, \delta^0, 1 - \varepsilon\}$.

is given as input; (ii) the restart neuron r_i firing on e_i when the net is restarted (i.e., r_i guarantees, when the net is restarted on an augmented input, that the presence of fact f_i is not forgotten); and (iii) the synchronizing neuron σ''_{yfm_i} firing on e_i when the question whether fact f_i has been detected obtains a positive answer in the predictive subnet (for details, see Burattini and Tamburrini [24]).

Each active e neuron e_i fires on various elements of this subnet: (i) on the memory neuron (m neuron) m_i with which it is connected, whose role is that of storing the information, coming through e_i, that fact f_i has been detected; (ii) on the fact neuron (neuron) f_i; (iii) on itself, sending a self-inhibitory signal immediately after activation (e neurons do not retain memory of past events); and (iv) on an incoherence signaling neuron ic_{ij}. Its threshold is $1 - \varepsilon$, with $0 < \varepsilon < 1$.

There are n *memory neurons* (m neurons) (Fig. 19), which retain memory of the facts that have been detected and given as input to the system. The memory neuron m_i is connected to the entry neuron e_i. Once activated by e_i, m_i remains active by means of a self-sustaining mechanism that enables it to "remember" that fact f_i has been detected and to signal this information, when restarting the net, to the restart neuron r_i. m_i also fires on neurons in the explanation subnet. The m-neuron threshold is again $1 - \varepsilon$, with $0 < \varepsilon < 1$.

There are n *fact neurons* (f neurons) (Fig. 20), transmitting information to the abductive subnet about the facts given as input. An f neuron f_i can be activated by the e neuron e_i with which it is connected and can send out signals to a collection WM_{f_i} of working-memory neurons (wm neurons) of the abductive subnet. The coupling coefficients between f neurons and wm neurons are crucial variable parameters of the net, which must be determined by the expert, possibly in the manner described in Burattini and Tamburrini [24].

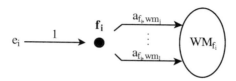

Figure 20 $f_i \equiv \{\{\langle e_i, 1 \rangle\}, \delta^0, 1 - \varepsilon\}$.

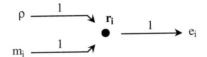

Figure 21 $r_i \equiv \{\{\langle \rho, 1 \rangle, \langle m_i, 1 \rangle\}, \delta^0, 2 - \varepsilon\}.$

There are *n restart neurons* (*r* neurons) (Fig. 21), already mentioned in connection with the entry and memory neurons, whose task is that of restarting the
network when new information is obtained via the predictive subnet. They receive
messages from the *m* neurons and from a special restart neuron ρ, while they send
out impulses to the entry neurons.

Finally, there are *k incoherence neurons* (*ic* neurons) (Fig. 22). For each pair
e_i, e_j, whose simultaneous presence is ruled out by the human experts, an *ic*
neuron is created which becomes active when both e_i and e_j are activated. This
neuron signals the presence of an incoherence to the user.

2. Abductive Subnet

The facts presented as input to the system may have different significance with
respect to the problem of eliciting a hypothesis. Therefore, a suitable weight must
be assigned by the expert, or by the procedure described in [26], to connections
between the appropriate pairs of neurons "representing" facts and hypotheses.
Given these weighed relations, the abductive subnet analyzes the information
flowing from the data given as input under the form of excitatory impulses, checking whether any hypothesis can be elicited as an explanation for some of the observed facts. If this is possible, the hypothesis subnet is activated; otherwise, the
control is passed over to the predictive subnet.

In the abductive subnet, there are *m working-memory neurons* (*wm* neurons)
(Fig. 23), where *m* is the number of possible hypotheses in the given diagnostic
domain. The task is that of evaluating the relevance of the observed facts for eliciting one of the declared hypotheses. Each *wm*-neuron wm_i collects information
from the elements of a set $F_{wm_i} = \{f_1, \ldots, f_k\}$ of f neurons of the evidence

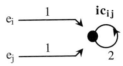

Figure 22 $ic_{ij} \equiv \{\{\langle e_i, 1 \rangle, \langle e_i, 1 \rangle, \langle ic_{ij}, 2 \rangle\}, \delta^0, 2 - \varepsilon\}.$

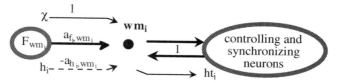

Figure 23 $wm_i \equiv \{\{\langle f_{m_j} a_{f_{jwm_i}} \rangle, \langle h_i, a_{h_k wm_i} \rangle, \langle \chi, 1 \rangle \langle \text{control}_{xi}, 1 \rangle\}, \delta^{wm_i}, \tau_{wm_i}\}.$

subnet, where $1 \leqslant k \leqslant n$. wm_i has a threshold set by the expert and corresponding to the minimum amount of information needed to elicit hypothesis h_i. Its decay law, δ_{wm_i}, differs from δ_0. In addition to messages from the f neurons of the evidence subnet, each wm neuron also receives inhibitory impulses from a special clearing neuron and from some control neurons. When a wm neuron receives impulses from the connected f neurons at time t, it starts firing at time $t + 1$ on neurons of the predictive and hypothesis subnets. It keeps on firing until, by its decay law, the excitation falls below the threshold value. The impulses sent at time t toward the hypothesis subnet reach at $t + 1$ the connected neurons in that subnet, whereas the impulses sent toward the predictive subnet are delayed by some "delay" neurons. This retardation mechanism has been introduced to enable the system to verify first whether the available information is sufficient to elicit explanatory hypotheses, so that the predictive subnet can be activated at a later time, only if this condition is not verified.

Finally, there is also a controlling subnet, whose role is that of terminating the process when no more hypotheses can be investigated and of activating the subnet outputting the list of hypotheses, ranked according to the excitation level reached by their "representing" neurons.

3. Hypothesis Subnet

If a set H of hypotheses has been elicited, two different situations may arise:

(i) The hypotheses in H explain all the facts given as input. H is declared to the user and stored for later use in the justification subnet. The hypothesis subnet inhibits activation of the predictive subnet and thus the hypothesis selection process terminates.

(ii) Some facts given as input are not explained by any hypothesis in H; H is stored, the hypothesis subnet does not inhibit the predictive process about facts unexplained by H. The predictive subnet is activated in order to raise questions about hypotheses not in H that may explain these facts.

In this subnet, there are m *hypothesis-triggering neurons* (*ht* neurons) (Fig. 24), one for each hypothesis declared by the expert. The ht neuron ht_i re-

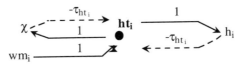

Figure 24 $ht_i \equiv \{\{\langle wm_i, 1\rangle, \langle \chi, -\tau_{ht_i}\rangle\}, \delta^{ht_i}, \tau_{ht_i}\}.$

ceives impulses from the *wm* neuron $\boldsymbol{wm_i}$ of the abductive subnet. It also receives from and sends out impulses to the clear neuron χ. An *ht* neuron has a suitably high threshold value representing the minimum support value needed to trigger the associated hypothesis. This threshold can be reached only if the associated *wm* neuron fires for several consecutive instants of time, and, in turn, the number of consecutive firings of $\boldsymbol{wm_i}$ is proportional to the amount of excitation received from the set F_{wm_i}. By this mechanism, $\boldsymbol{wm_i}$ expresses the amount of support given by the observed facts in F_{wm_i} to hypothesis h_i and $\boldsymbol{ht_i}$ "decides" whether this evidence is sufficient to elicit $\boldsymbol{h_i}$ (see Burattini *et al.* [26] on the minimum value of the support): if the sum of the impulses coming from $\boldsymbol{wm_i}$ reaches its threshold, the *ht* neuron $\boldsymbol{ht_i}$ sends impulses to the corresponding hypothesis neuron $\boldsymbol{h_i}$.

There are also *m hypothesis neurons* (*h* neurons) (Fig. 25), one for each hypothesis relative to the given domain. If neuron $\boldsymbol{h_i}$ is activated by neuron $\boldsymbol{ht_i}$, this indicates that hypothesis h_i is proposed by the system as an explanation for some observed facts. The *h* neurons store all explanatory hypotheses advanced by the system. This is achieved, once an *h* neuron is activated, by a self-excitatory mechanism which keeps its level of excitation above the threshold. An *h* neuron also sends excitatory impulses to neurons of the explanation subnet and inhibitory impulses to neurons of the predictive subnet. This subnet outputs a partially ordered list of selected hypotheses (if any). The intended meaning of this partial order is a nonexclusive preference order for the selected hypotheses. This order is obtained by comparing the excitation levels of the *ht* neurons representing such hypotheses.

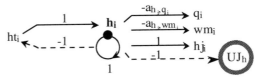

Figure 25 $h_i \equiv \{\{\langle h_i, 1\rangle, \langle ht_i, 1\rangle\}, \delta^0, 1 - \varepsilon\}.$

4. Predictive Subnet

From now on, we will omit most implementation details in describing the main types of neurons, concentrating on their role in the process of hypothesis selection. The predictive subnet is managed by the abductive and hypothesis subnets, and is activated when further investigation is needed to explain all observed facts. This subnet is, as it were, a mirror image of the abductive subnet: the links from evidence to hypotheses in the latter are inverted in the predictive subnet in order to infer which evidence might be present if a specific diagnostic hypothesis were correct.

Upon activation of this subnet, two preliminary tasks must be performed: (i) selecting a hypothesis (making sure that it is not in the set H, if any such set has been selected in the hypothesis subnet) to be probed first and (ii) asking a question that may contribute to test the selected hypothesis. Also at this stage, the heuristic strategies enabling the system to raise appropriate questions must be implemented by the expert, possibly with the help of a procedure described in [26].

Each question formulated in the predictive subnet is about the presence of a certain observable phenomenon relevant to test the hypothesis under examination (e.g., "Does patient x show symptom y?"). Such questions admit three types of answers: "yes," "no," or "I don't know." If the answer is "yes", the new information is added to the initial input and the system is restarted on the augmented set of data. If the answer is "no," the system has obtained negative information (e.g., "symptom y has not been observed") which is also added to the initial input if this negative information is explicitly represented as a fact in the system's knowledge base. The system is then restarted to evaluate the augmented input. In all other cases (the answer is "I don't know," or else is "no," but the corresponding negative information is not explicitly represented in the system's knowledge base), the system proceeds to ask a question about another fact relevant to assessing the hypothesis under examination.

When all questions relative to the selected hypothesis have been made, and some facts remain unexplained, the predictive subnet starts asking questions concerning the next lower ranked hypothesis. The predictive process terminates either when a set H of hypotheses explaining all observed facts is isolated or when there are no more questions to be made.

The predictive subnet is formed by three main layers (question neurons, fact-mirror neurons, and answer neurons), activated when one of the following situations occurs:

1. no explanatory hypothesis has been elicited;
2. the hypotheses elicited by the system do not explain all observed facts.

Case 1. In the query layer, there are m *question neurons* (q neurons). Their structure is analogous to that of the ht neurons (same threshold and excitation decay law) and their role is that of selecting a hypothesis to be tested by a question–

answer process. These neurons receive from the wm neurons the same impulses as the ht neurons, but suitably delayed. Because no ht neuron is activated by these impulses, a control neuron β sends uniform excitatory impulses to the q neurons, activating the q neuron $\boldsymbol{q}_i$ that first reaches its threshold. Intuitively, the activation of $\boldsymbol{q}_i$ signals that a question about the presence of a fact relevant to assess hypothesis h_i must be raised. However, once a q neuron $\boldsymbol{q}_i$ is activated by β, how is an appropriate question about the corresponding hypothesis h_i selected?

In the predictive subnet, there is a set of n *fact-mirror neurons* (fm neurons). These neurons can receive excitation from q neurons in the following way. Each q neuron $\boldsymbol{q}_i$ is connected to a subset FM_{q_i} of the set of fm neurons. The coupling coefficients between $\boldsymbol{q}_i$ and each element in FM_{q_i} must be determined by the human experts, possibly with the help of the algorithm described in [26]. When $\boldsymbol{q}_i$ fires, signaling that a question about hypothesis h_i must be raised, it sends excitatory impulses to the elements of FM_{q_i}. Only one of the fm neurons in FM_{q_i} fires in response to the impulses coming from $\boldsymbol{q}_i$ and triggers a question about h_i addressed to the user. This fm neuron, let us call it $\boldsymbol{fm}_s$, enables the system to raise the question: "Has fact f_s been detected?" The neuron $\boldsymbol{fm}_s$ is connected with three *answer neurons* (a neurons) $\boldsymbol{y}_{fm_s}$, $\boldsymbol{n}_{fm_s}$, and $\boldsymbol{d}_{fm_s}$, respectively representing the answers "yes," "no," and "I don't know." The user's answer will activate one of these a neurons.

Suppose first that the answer is "yes." The net is restarted by a restart neuron activated by $\boldsymbol{y}_{fm_s}$, augmenting the input of the previous run with the fact represented by the entry neuron $\boldsymbol{e}_s$.

Suppose now that the answer is "no." We have two subcases:

(a) If the negative information thus obtained is represented under the form of a fact in the system's knowledge base, then the behavior of the net is the same as in a "yes" answer situation.

(b) Otherwise, the process on $\boldsymbol{q}_i$—of asking another question relative to the hypothesis h_i—continues. If no more questions are available for that hypothesis, the question process will be applied to the next lower ranked hypothesis, say, h_j, via the q neuron $\boldsymbol{q}_j$.

Finally, let us suppose that the answer is "I don't know." The system will behave as in subcase (b).

Each fm neuron $\boldsymbol{fm}_j$ is inhibited by an m neuron (when the fact represented by $\boldsymbol{fm}_j$ has already been detected and therefore a question about its presence is not needed) or by a self-inhibitory impulse (when the user has already given one of the admissible answers to the question triggered by $\boldsymbol{fm}_j$). When $\boldsymbol{fm}_j$ is thus inhibited, it remains excluded from the query process, no matter how much excitation it receives from q neurons.

Case 2.   The hypotheses elicited by the system are stored by self-sustaining h neurons which send an inhibitory impulse to the corresponding q neurons. In this way, the query process can be activated only on hypotheses that the system has not elicited yet, via those q neurons that are not inhibited by h neurons. Indeed, the q neuron q_i receives inhibitory impulses from h_i, only if h_i has been previously activated, indicating that hypothesis h_i has already been elicited and additional investigation is not needed.

If no more questions to ask and hypotheses to probe are available, the system terminates the query process, declaring to the user the list of ranked hypotheses that it was able to elicit and the list of facts that are possibly left unexplained by these hypotheses.

5. Justification Subnet

This subnet is currently capable of answering only one type of question ("On the basis of which facts was a certain explanatory hypothesis advanced?"), in addition to providing the list of all unexplained facts, whenever this is the case. However, neurally implemented extensions of this module enabling the system to provide a more detailed justification of advanced hypotheses, by tracing the inferential steps leading up to the selection of those hypotheses, are clearly possible.

The user activates this subnet by asking which facts support a certain hypothesis. Such interaction may be thought of as a backward mechanism which, starting from the input hypothesis, looks for all observed facts inducing its elicitation. The answer is worked out by two layers of neurons interacting with the m neurons and the h neurons.

6. Control Neurons

Clearly, a network of neurons performing the abduction–prediction cycle for hypothesis selection in the way described here must be controlled and synchronized: this process must be performed in a certain number of *sequential* steps, in each of which the available information is elaborated in *parallel*. Also this controlling and synchronizing function is performed by various neural elements. The role of some controlling neurons was made explicit in the preceding description of the evidence, abductive, predictive and hypothesis subnets.

IX. CONCLUSIONS

We have shown how to simulate, by means of a localist approach to neural representation and processing, various symbolic, rule-based reasoning. Some of these technical tools have been applied to designing rule-based expert systems, such as the one presented in Section VIII, which exploit the massively parallel processing capacities of neural networks while retaining, in virtue of the localist approach, the full justification capacities of conventional symbolic systems. These localist networks, however, remain brittle much in the way of conventional symbolic systems. Moreover, some numerical computations involved in quantitative uncertain reasoning require very large networks. These limitations suggest the opportunity of extending the present approach toward unified neurosymbolic systems which combine localist networks with distributed representations, as well as toward strictly hybrid systems. Accordingly, one is led, solely on the basis of a balanced assessment of limitations and potentialities of the present approach, to distinguishing between three different strategies for neurosymbolic integration:

  (i) using specialized and structured localist networks for symbolic reasoning, both crisp and uncertain;

  (ii) combining localist networks for symbolic processing with distributed neural networks, the latter representing individual pieces of knowledge as distributed patterns across a large number of neural units;

  (iii) combining separate localist networks and conventional symbolic systems for symbolic processing.

Additional distinctions, partly overlapping with this classification, can be made in the setting of more general analyses of possible approaches to neurosymbolic integration (see, e.g., Sun and Bookman [27] and Hilario [28], and the references therein).

In this chapter, we have mostly worked within approach (i), by presenting specialized, localist networks for various types of crisp or uncertain symbolic reasoning. We have also addressed the problem of learning in a strictly nonconnectionist fashion, by means of a localist network with adjustable weights, which can perform an elementary form of chunking and add new rules to a preexisting production system.

Approach (iii) was implicitly considered in our treatment of quantitative uncertain reasoning. In this case, the parallel execution afforded by the localist neural implementations may not be beneficial in terms of computation time because of the heavy numerical computations that have to be performed by various subnets. On account of this fact, we suggested replacing these subnets by conventional symbolic processors computing the same function. The resulting system is strictly hybrid, but the parallel architecture originally introduced by means of the purely neural system is largely preserved in the structure of the new system.

Approach (ii) is well suited for domains in which a cooperation between neural learning and rule-based reasoning is needed to solve specific problems. For example, De Gregorio [29] analyzes an object classification problem from visual data which is difficult to solve by merely training a neural net, and finds an adequate solution by means of a hybrid system performing an abduction–prediction cycle (as described in Section VIII.A), with the following division of labor between the neural and the symbolic reasoning modules. The neural net, instead of attempting a direct classification of the viewed object, is trained to provide a classification of selected visual clues at particular locations of the image. Then the clues are symbolically coded and make possible the selective activation of production rules in the symbolic reasoning module. The action parts of these rules code possible classifications for the viewed objects. If the clues obtained so far are insufficient to arrive at a particular classification, the symbolic reasoning module can ask for new clues from the neural module in a hypothesis-driven query mode. This cycle (again an instance of the abduction–prediction process) is iterated until the system settles on a unique classification. The knowledge base of the symbolic reasoning module is a set of production rules which is equivalent to a set of rules of the form examined in Section III. Therefore, this module can be replaced by a localist network as described in Section IV. The new system combines a localist network for symbolic processing with a distributed neural network for classification of visual clues, and therefore falls squarely within approach (ii) to neurosymbolic integration. The massively parallel processing capacities of neural networks are exploited in the symbolic processing module, whereas both parallel processing and noise-tolerant learning and classification are put to work in the perceptual module. In addition to providing an interesting engineering solution, this system suggests the potential interest of approach (ii) for the computational modeling of high-level perception (such as, e.g., high-level vision; see Ullman [30] and Kosslyn and Koenig [31]), where bottom-up perceptual processing and top-down interpretative reasoning tightly interact.

ACKNOWLEDGMENTS

We are most grateful to Massimo De Gregorio and Andrea Pasconcino for many invaluable discussions which contributed to shaping various materials presented in this chapter. Moreover, Pasconcino developed the software for the chunking system and improved an earlier version of the forward chaining network. Aldo Filosa and Umberto Giani contributed in a crucial way to designing the knowledge base and the inferential strategies for the nonneural version of the diagnostic system in Section VIII. Parts of this chapter were adapted from Aiello *et al.* [32, 33], and Burattini and Tamburrini [24]. © 1992, 1995 Wiley, New York.

REFERENCES

[1] E. R. Caianiello. *J. Theoret. Biol.* 2:204–235, 1961.

[2] M. R. Genesereth and N. J. Nilsson. *Logical Foundations of Artificial Intelligence.* Morgan Kaufmann, Los Altos, CA, 1987.

[3] J. W. Grzymala-Busse. *Managing Uncertainty in Expert Systems.* Kluwer Academic, Dordrecht, 1991.

[4] E. Burattini, A. Pasconcino, and C. Tamburrini. *Mathware Soft Comput.* 2:85–116, 1995.

174

[5] G. G. Towell and J. W. Shavlik. *Artificial Intell.* 70:119–165, 1994.

[6] A. Newell. *Unified Theories of Cognition.* Harvard University Press, Cambridge,

[7] M. Tambe, A. Newell, and P. S. Rosenbloom. *Machine Learning* 5:299–348, 1990.

[8] D. E. Rumelhart and J. L. McClelland, Eds. *Parallel Distributed Processing: Explorations in Microstructure of Cognition* Vol. I. MIT Press, Cambridge, MA, 1986.

[9] T. Kohonen. *Associative Memory: A System-Theoretical Approach.* Springer-Verlag, Heidelberg, 1977.

[10] T. Kohonen. In *Proceedings of the Sixth International Conference on Pattern Recognition* (M. Lang, Ed.), pp. 114–125. IEEE Computer Society Press, Silver Spring, MD, 1982.

[11] A. Pasconcino. Sistemi Esperti e Reti Neuroniche: realizzazione di un sistema composito parallelo. Tesi di Laurea, Università degli Studi di Napoli "Federico II," 1994.

[12] R. Sun. *Integrating Rules and Connectionism for Robust Commonsense Reasoning.* Wiley, New York, 1994.

[13] M. L. Minsky. In *The Psychology of Computer Vision* (P. Winston, Ed.), pp. 34–57. McGraw-Hill, New York, 1974.

[14] L. G. Valiant. *Circuits of the Mind.* Oxford University Press, Oxford, 1994.

[15] E. Sandewall. *Machine Intelligence* (B. Meltzer and D. Michie, Eds.), Vol. 7, pp. 195–204. Wiley, New York, 1972.

[16] I. Kramosil. In *Proceedings IJCAI-75*, pp. 53–56, 1975.

[17] R. Reiter and G. Criscuolo. In *Proceedings IJCAI-81*, pp. 270–276, 1981.

[18] J. R. Josephson and S. G. Josephson. *Abductive Inference.* Cambridge University Press, Cambridge, 1996.

[19] Y. Peng and J. A. Reggia. *Abductive Inference Models for Diagnostic Problem-Solving.* Springer-Verlag, Heidelberg, 1990.

[20] J. A. Reggia, D. S. Nau, and P. Y. Wang. *Internat. J. Man Machine Studies* 19:437–460, 1983.

[21] S. B. Ahuja. Y. S. Woo, and A. Schwartz. *Internat. J. Intelligent Systems* 4:155–180, 1989.

[22] R. Reiter. *Artificial Intell.* 32:81–132, 1987.

[23] L. Console and P. Torasso. In *Proceedings of the Ninth ECAI*, Stockholm, pp. 160–166, 1990.

[24] E. Burattini and G. Tamburrini. *Internat. J. Intelligent Systems* 7:521–545, 1992.

[25] E. Burattini, G. Criscuolo, A. Filosa, U. Giani, and F. Mele. In *Proceedings of the Ninth European Meeting on Cybernetics and Systems Research*, Vienna, pp. 467–473, 1988.

[26] E. Burattini, M. De Glas, and M. De Gregorio. In *IEEE International Conference on Systems, Man, and Cybernetics*, Chicago, pp. 272–278, 1992.

[27] R. Sun and L. A. Bookman, Eds. *Computational Architectures Integrating Neural and Symbolic Processes.* Kluwer Academic, Dordrecht, 1995.

[28] M. Hilario. In *Connectionist–Symbolic Integration: From Unified to Hybrid Approaches, Working Notes* (R. Sun and F. Alexandre, Eds.), pp. 1–6, 1995.

[29] M. De Gregorio. *Mathware Soft Comput.* 3:271–279, 1996.

[30] S. Ullman. *High-Level Vision.* MIT Press, Cambridge, MA, 1996.

[31] S. M. Kosslyn and O. Koenig. *Wet Mind, The New Cognitive Neuroscience.* Free Press, New York, 1995.

[32] A. Aiello, E. Burattini, and G. Tamburrini. *Internat. J. Intelligent Systems* 10:735–749, 1995.

[33] A. Aiello, E. Burattini, and G. Tamburrini. *Internat. J. Intelligent Systems* 10:751–769, 1995.

[34] E. Burattini and M. De Gregorio. *Inform. Decision Technol.* 19:471–481, 1994.

Construction of Rule-Based Intelligent Systems

Graham P. Fletcher
Department of Computer Sciences
University of Glamorgan
Wales CF37 1DL, United Kingdom

Chris J. Hinde
Department of Computer Sciences
University of Glamorgan
Wales CF37 1DL, United Kingdom

I. INTRODUCTION

Engineers are finding new and different applications for neural networks every day. These new applications are exposing many limitations in our current techniques. Perhaps one of the more important of these limitations is the trust that can be placed on a neural network, or more correctly the hypothesis that it has constructed. With a localized paradigm it is sometimes possible to assign meanings to the neurons manually. However, as the networks grow in size beyond several different layers, this can become very difficult.

Hinton [1] expresses the following view:

> the problem is to devise effective ways of representing complex structures in connectionist networks without sacrificing the ability to learn the representations. My own view is that connectionists are still a very long way from solving this problem.

It is relatively straightforward to transform a propositional rule base into a neural network. However, the transformation in the other direction has proved a much harder problem to solve. This chapter explains techniques that allow neurons, and thus networks, to be expressed as a set of rules. These rules can then be used within a rule-based system, turning the neural network into an important tool in the construction of rule-based intelligent systems.

Fuzzy Logic and Expert Systems Applications

The rules that have been extracted, as well as forming a rule-based implementation of the network, have further important uses. They also represent information about the internal structures that build up the hypothesis, and, as such, can form the basis of a verification system. This chapter also considers how the rules can be used for this purpose.

II. REPRESENTATION OF A NEURON

Feedforward neural networks can be built to use many types of input parameters. Real-valued or continuous inputs are the most difficult to deal with. Consider the example in Fig. 3; there are only two input parameters, but the desired hypothesis requires eight decision planes. These decision planes are implemented in the first layer, known as the quantization layer, of any neural network that is built to implement this hypothesis (Fig. 1). The weights and biases of these nodes represent the equations of the decision planes and cannot be simplified. Some of the individual decision planes are shown in Fig. 2.

In the idealized neuron model, thresholding means that all transformations beyond the first are Boolean. The quantization layer has split the input space into

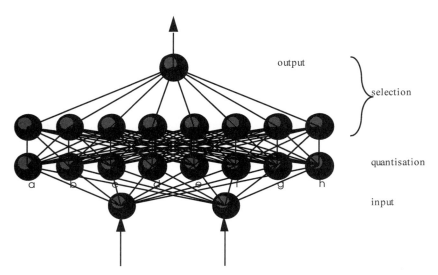

Figure 1 A two-input network with eight quantization neurons to change real-valued inputs to Booleans and a two-layer selection section terminating in a single output neuron.

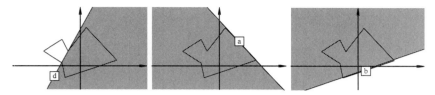

Figure 2 Showing the separation of the input plane into halves by individual neurons in the quantization layer. The halves are then combined to form the region in Fig. 3.

regions which are then combined logically by the subsequent layers. In the example, the shape is described by

$$(a \wedge b \wedge c \wedge d \wedge e) \vee (\overline{d} \wedge f \wedge g \wedge h).$$

As the quantization layer cannot be simplified, rule extraction has to be concerned with the second and subsequent layers of the network. The inequalities that represent the quantization layer can be stored and substituted into the rules that are extracted from the rest of the network. The resulting expression can be quite clear.

Boolean inputs to a network effective remove the quantization layer, simplifying the process by allowing Boolean rule analysis to begin at the first layer of

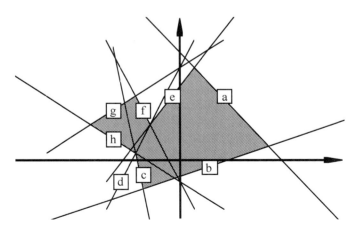

Figure 3 Lines representing the quantization neurons in Fig. 1 separating the plane into various half planes which are then selected using the subsequent selection layers resulting in the shaded region shown.

neurons. Therefore a network with Boolean inputs would give a rule of the form

$$\text{If } (a \wedge b) \vee (\overline{c \wedge a}),$$

then ...

compared to the type of rules produced from networks with real or continuous inputs

$$\text{If } [3.8a + 7b - 4 > 0] \wedge [3.1b - 2c + 9 > 0],$$

then ....

So far an assumption has been made that a neuron implements a crisp decision plane; in other words, it is a perceptron rather than the more common sigmoid-based neuron. The use of a sigmoid function changes the view of thresholding. However, with a fully trained network, the sigmoid function will be a very close approximation to a threshold and therefore there is no need to differentiate between the different types of neurons. There are two reasons for this close approximation:

- As the networks harden the weights grow in size which makes the range of inputs for which the output is not close to Boolean very small.
- If simulated annealing is used to speed up training, then as the temperature approaches zero the activation curve becomes very close to a threshold.

Although a threshold neuron can always be represented as a Boolean function, the form of this function cannot be restricted to a representation where each input variable is mentioned only once. The interest in such representations stems from scaleability; the search space for the correct Boolean function grows exponentially as the number of inputs rises. If a set of operators could be found that represents any neuron while mentioning each operand only once, then an effective algorithm could be produced for finding a concise representation. The problem is intractable [2] in that not only does the search space grow exponentially with the number of inputs the Boolean representation can also grow exponentially. In many cases, however, where there is a concise Boolean representation for the neuron this can be found efficiently. Where the Boolean representation of the neuron becomes large and unintelligible, it is clear that a Boolean representation is inappropriate and other representations must be sought [3].

Previous investigations into this problem used a piecemeal approach, splitting off the Boolean problem space into a class of operators which are referred to as O and A operators as they are generalizations of OR and AND.

The problem space covered by the operators was shown to be complete up to three dimensions by Mihalaros [4] and therefore to provide a means of representing all possible neurons with an input dimension of three or less. This work by Mihalaros presented no complete analysis of how to find the O operator that matched a general neuron, or even if all possible neurons in higher dimensions were covered. Perhaps more significantly and what encouraged further work is that the traditionally hard problem of n-input parity neural networks can be simply represented using O operators [5].

Although initially promising, Fletcher and Hinde [3] showed that O operators are not complete for four or more dimensions and, as such, cannot in general be used to analyze networks in higher dimensions. Where it is possible to use O operators, they provide a concise and clear representation and, as such, are ideal for large networks with a low connectivity. Unfortunately, as the number of dimensions increases, the proportion covered by O and A operators falls until they are virtually useless.

O and A operators are incomplete; they do not have the flexibility to represent all possible neurons. The aim of finding some form of representation that mentions each input once is, in the authors' opinion, basically flawed. A more general technique must be adopted. The best understood technique for representing logical functions is traditional Boolean logic. Although the representation for a complex neuron can become unacceptably long, Boolean logic can still provide a powerful tool. The aim of the next section is to produce a system for analyzing neurons in terms of a Boolean rule.

III. CONVERTING NEURAL NETWORKS TO BOOLEAN FUNCTIONS

A neural network performs a logical transformation of a set of inputs to a set of outputs. The range and domain will contain tuples that consist of real and/or Boolean values. The whole problem is to represent the transformation in some concise and meaningful way. So far in this chapter we have been looking at how this "network transfer function" could be represented. For the purposes of this section, real-valued inputs will be ignored. In effect, the nodes in the quantization layer of networks that use real inputs are treated as the inputs to a Boolean network; thus all networks are Boolean.

A correct Boolean representation of the network transfer function is logically very simple to calculate. A truth table could be calculated by applying every input pattern to the network. This truth table could then be turned into a Boolean function using an algorithmic implementation of Karnaugh maps [6]. Both stages of this method have a time complexity of $O(2^n)$. While logically very simple, these processes have prohibitively large time and space complexities. So it is necessary to break the problem into much smaller problems. The transfer function of the network is built up from the transfer functions of the individual neurons. If the function of the network is intractable to calculate directly, then it can be derived by finding the solutions for all of the neurons that make the network.

The possible functions that can be represented by a single neuron fall into distinct groups. All of the elements of a group are identical except for the negation of literals. Further, the groups will contain exactly one "natural" neuron and possibly some "real" ones, where we define natural and real to be:

- *Natural*: All the weights and the bias of the neuron are positive; that is, the transfer function is effectively

$$W_1 * \text{In}_1 + W_2 * \text{In}_2 + \cdots + W_n * \text{In}_n > B$$

and

$$W_1 \geq 0, \ W_2 \geq 0, \ldots, W_n \geq 0, \qquad B \geq 0.$$

- *Real*: At least one of the weights or the bias of the neuron is negative; that is, the transfer function is effectively

$$W_1 * \text{In}_1 + W_2 * \text{In}_2 + \cdots + W_n * \text{In}_n > B$$

and

$$W_1 < 0 \text{ or } W_2 < 0 \text{ or} \ldots \text{ or } W_n < 0 \text{ or } B < 0.$$

All the following representations are neurons in the same group. The first is the natural neuron; subsequent ones represent real neurons from the same group.

$(I_1 \wedge I_3) \vee (I_1 \wedge I_4) \vee (I_2 \wedge I_3 \wedge I_4)$	Natural,
$(I_1 \wedge \neg I_3) \vee (I_1 \wedge I_4) \vee (I_2 \wedge \neg I_3 \wedge I_4)$	Real,
$(I_1 \wedge I_3) \vee (I_1 \wedge I_4) \vee (\neg I_2 \wedge I_3 \wedge I_4)$	Real,
$(I_1 \wedge I_3) \vee (I_1 \wedge \neg I_4) \vee (I_2 \wedge I_3 \wedge \neg I_4)$	Real.

A. BOOLEAN REPRESENTATION
OF A NATURAL NEURON

Natural neurons can be turned into Boolean functions with relatively little effort. As none of the inputs is negatively weighted, all we need to find are the minimum sets of inputs required to overcome the bias. This problem is similar to (but much simpler than) the classic knapsack [2, 7]. Instead of trying to find a set of inputs that exactly fit the bias, the required answer is all the sets of inputs that just exceed the bias. For example, the neuron in Fig. 4 has four inputs with weights of 9, 4, 6, and 7, and a bias of 14. Input I_1 & Input I_4 just exceeds the bias as $9 + 7 > 14$, but Input I_1, Input I_4 & Input I_2 is too large as removing I_2 would still leave the total above 14. (Input I_1 & Input I_4) is therefore part of the answer and (Input I_1 & Input I_4 & Input I_2) is not.

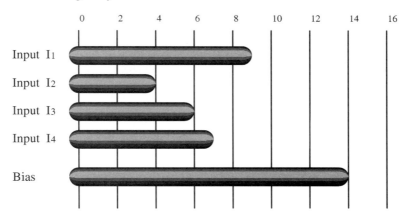

Figure 4 Representing a natural neuron as a knapsack problem where the input "parcels" are fitted into the bias "package."

The complete set of solutions or infimums for the problem shown in Fig. 4 is $(I_1 \& I_3)$, $(I_1 \& I_4)$, and $(I_2, I_3 \& I_4)$. This represents the complete Boolean function of the neuron. In the conventional Boolean format it would be written as

$$(I_1 \wedge I_3)/(I_1 \wedge I_4)/(I_2 \wedge I_3 \wedge I_4).$$

An algorithm to produce one part of the result is given as a PROLOG program shown in Fig. 5. By backtracking the procedure, it is possible to derive each part in turn until all the infimums have been produced.

The algorithm so far described will return a solution that is logically correct but larger than necessary. As well as the infimums that make up part of the answer, the algorithm may also return sets of inputs more specific than infemums. For example, if $(I_1 \wedge I_3)$ is an infimum, then it is also possible for the algorithm to

```
knapsack(_Unused_inputs,Bias,Answer,Answer):-
    Bias =< 0.
knapsack(Unused_inputs,Bias,SubAnswer,Answer):-
    Bias > 0,
    append(_,[(Input_Name,Input_Weight)|Inputs], Unused_inputs),
    New_Bias is Bias - Input_Weight,
    knapsack(Inputs,New_Bias,[Input_Name|SubAnswer],Answer).
```

Figure 5 PROLOG representation of the "knapsack" algorithm used to extract Boolean rules from a set of weights associated with a neuron.

return $(I_1 \wedge I_3 \wedge I_4)$. While logically this makes no difference, we require a concise and meaningful answer. This added clutter is therefore undesirable. The removal of these extra terms can be achieved by sorting the weights into decreasing size. A further advantage of this is that sorting is a good heuristic for reducing average processing time to find the solution. Let the input to the algorithm be weights $W_1, W_2, W_3, \ldots, W_n$ and the bias value B.

The mistakes that cause the answer to be nonminimal occur when there is a term that could be removed and still overcome the bias, that is,

$$W_1 + W_2 + W_3 + \cdots + W_m > B, \tag{1}$$

$$W_2 + W_3 + \cdots + W_m > B. \tag{2}$$

Because the algorithm stops as soon as the bias is achieved,

$$W_1 + W_2 + W_3 + \cdots + W_m - W_m < B. \tag{3}$$

If the weights were in sorted order, then

$$W_1 > W_m. \tag{4}$$

Substituting (4) into (3) gives

$$\begin{aligned} W_1 + W_2 + W_3 + \cdots + W_m - W_1 < B, \\ W_2 + W_3 + \cdots + W_m < B, \end{aligned} \tag{5}$$

which contradicts (2). Therefore if the weights are sorted the two requirements for added complication in the answer cannot occur.

B. BOOLEAN REPRESENTATION OF A REAL NEURON

The transfer function of a real neuron is much harder to calculate directly than that of a natural neuron. Earlier in the chapter we stated that "All of the elements of a group are identical except for the negation of literals." Therefore finding the representation for the natural neuron in the same group as the real neuron provides a solution that can be modified to represent the original problem by negating some of the inputs. This means that the target of matching a real neuron to a Boolean representation has been reduced to two separate problems:

- How to find the Boolean representation of a natural neuron. This problem has already been covered.
- How to convert the solution for a natural neuron so that it matches one of the real neurons in the same group.

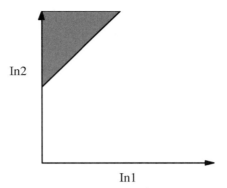

Figure 6 Two-dimensional neuron with one negative and one positive weighted input, represented as a hyperplane.

The shaded region in Fig. 6 is represented by the equation $In_2 - In_1 \geq 0.5$. To convert this representation of a real neuron into a natural form, it is necessary to remove the negative weight without affecting its Boolean characteristics. The negative weight associated with $Input_1$ can be removed by moving the origin of the axis to position $(1, 0)$ and then rotating them. The equation for the graph using the new axes (Fig. 7) is $In_2 + In_3 \geq 1.5$, where $In_3 = \neg In_1$. If a Boolean operator, F, is found for this new neuron, then all instances of In_3 in F could be replaced by $\neg In_1$ to give a correct solution for the original problem.

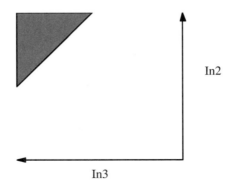

Figure 7 Transformed hyperplane with all positive weights.

C. Examples of Boolean Function Derivation

A neural network to calculate the function "three-input parity" is shown in Fig. 8. The Boolean function for each neuron is calculated separately

$$O = H, \qquad\qquad F = \neg(\neg D \vee \neg E),$$
$$H = G \vee F, \qquad\qquad D = \neg(A \wedge B) \vee \neg(A \wedge C) \vee \neg(B \wedge C),$$
$$G = \neg(\neg A \vee \neg B \vee \neg C), \qquad E = A \vee B \vee C.$$

Combining these gives

$$O = \neg(\neg A \vee \neg B \vee \neg C) \vee \neg(\neg(\neg(A \wedge B) \vee \neg(A \wedge C) \vee \neg(B \wedge C)) \vee \neg(A \vee B \vee C)).$$

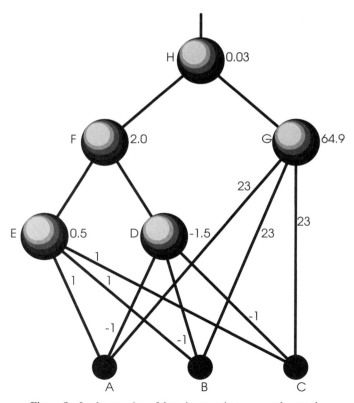

Figure 8 Implementation of three-input parity as a neural network.

IV. EXAMPLE APPLICATION OF BOOLEAN RULE EXTRACTION

This application is the dispensing of adhesive in the manufacture of "mixed technology" printed circuit boards (PCBs) in which through hole and surface mount components are present on the same board. The surface mount components are secured to the board, prior to a wave soldering operation, by a small (0.0002 to 0.005 c^3 depending on the component) amount of adhesive. The amount of adhesive dispensed is critically dependent on several process environment variables (e.g., temperature, humidity, erratic thixotropic behavior of the adhesive, air bubbles in the flow, and variations in the PCB substrate).

The dispensing unit consists of a syringe of adhesive coupled to a pressure control unit. The unit is made up of a solenoid valve, pressure regulator, temperature sensor, and a pressure transducer to monitor the variation of pressure within the syringe. The dispensing unit is fixed to a SEIKO RT3000 robot which moves the syringe to locations of the PCB where the adhesive has to be dispensed. Feedback data collection is carried out by an image processing system (Imaging Technology ITI151) coupled to a Pulnix TM-460 CCD camera incorporating a magnifying optical system.

The original software was developed using the MUSE real-time artificial intelligence (AI) toolkit. MUSE is a hybrid modular system supporting a range of knowledge representation paradigms: PopTalk, a procedural language with object-oriented programming extensions, a forward chaining rule language, a backward chaining language, data-directed programming through the use of demons, and flexible relation supporting general relations between objects.

Particular support for real-time operation includes agenda-based priority scheduling, interrupt handling, and fast data capture.

Messom *et al.* [8] produced a neural network system for controlling the adhesive dispensing machine (Fig. 9). This is the sort of problem that has typically been tackled by the use of a rule induction package. The trained neural network should therefore be equivalent to a set of rules that could have been learned by such a package.

The first layer of neurons receives real-valued inputs, but produce outputs that are very nearly bipolar. These neurons are effectively quantifying the input region, and are therefore called quantization nodes. The actions of the quantization nodes are described as a set of inequalities. This step does not simplify the information but does display it in a manner that is much more natural to read than a set of weights on a diagram. Examples of the rules resulting from these inequalities are shown in Fig. 10.

The remainder of the network receives inputs that are close to bipolar and delivers bipolar outputs, and, as such, they can be said to be implementing Boolean transfer functions. The results from expressing these transfer functions as rules

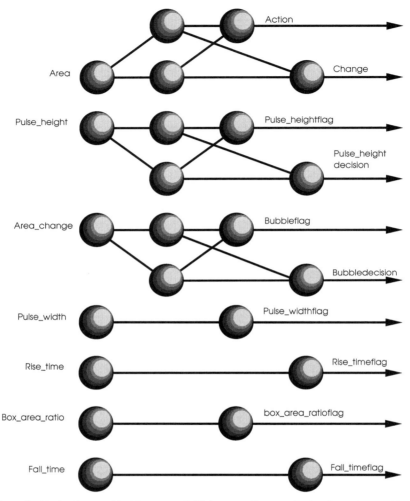

Figure 9   Net implemented by Messom *et al.* [8] for controlling an adhesive dispensing machine.

are very similar in format to the first-layer inequalities. Examples of the derived rules for the subsequent layers are shown in Fig. 11.

It is now possible to directly implement this set of rules in a rule base without altering the action of the system. For clarity, some general simplification of the Boolean rules is required. Furthermore, substituting the inequalities produces much more natural-looking results as shown in Fig. 12.

IF rise_time < 1.23
THEN rise_timeflag
ELSE ¬rise_timeflag

IF fall_time > 1.1
THEN fall_timeflag
ELSE ¬fall_timeflag

IF pulse_width < 1.0429
THEN pulse_widthflag
ELSE ¬pulse_widthflag

IF boxarearatio > 0.8520
THEN boxarearatioflag
ELSE ¬boxarearatioflag

IF area > 1.0258
THEN midnode1
ELSE ¬midnode1

IF area > 0.9722
THEN midnode2
ELSE ¬midnode2

IF pulse_height < 1.025
THEN midnode3
ELSE ¬midnode3

IF pulse_height > 0.975
THEN midnode4
ELSE ¬midnode4

IF area_change < 0.5500
THEN midnode5
ELSE ¬midnode5

IF area_change > 0.4499
THEN midnode6
ELSE ¬midnode6

Figure 10 Examples of the rules resulting from the inequalities arising from the first layer of the net implemented by Messom *et al.* [8] for controlling an adhesive dispensing machine.

The relative ease with which the transformation, from control network to rule set, can be made illustrates the usefulness of the interpretation system for medium-sized control networks. This type of result is most useful to check that the hypothesis is reasonable. It also allows the network to explain its actions; something that they cannot classically do.

V. NETWORK DESIGN, PRUNING, AND WEIGHT DECAY

As the number of inputs to a neuron grows so does the length of the Boolean description. In the worst case the number of conjunctions in the disjunctive normal form grows at a rate of 2^{n-1} where there are n inputs to the neuron. The time taken to derive a rule is proportional to the number of conjunctions. The rules become intractable to compute for neurons with more that 40 inputs, and meaningless for a human far earlier. The Boolean rule depicted in Fig. 13 has several thousand conjunctions, only the first few are shown.

IF (midnode1 ∨ ¬(midnode2))
THEN action
ELSE ¬action

IF ¬((midnode1 ∨ midnode2))
THEN change
ELSE ¬action

IF ¬((¬(midnode4) ∨
 ¬(midnode3)))
THEN pulse_heightflag
ELSE pulse_heightflag

IF ¬((midnode3 ∨
 ¬(midnode4)))
THEN pulse_heightdecision
ELSE ¬pulse_heightdecision

IF ¬((¬(midnode6) ∨
 ¬(midnode5)))
THEN bubbleflag
ELSE ¬bubbleflag

IF ¬((¬(midnode6)∨
 midnode5))
THEN bubbledecision
ELSE ¬bubbledecision

IF ¬((¬(midnode6) ∨
 ¬(midnode5)))
THEN bubbleflag
ELSE ¬bubbleflag

IF ¬((¬(midnode6) ∨
 midnode5))
THEN bubbledecision
ELSE ¬bubbledecision

Figure 11 Examples of the rules resulting from the conversion of the second layer of the net to Boolean expressions.

As the representation grows so quickly, we cannot allow the use of complex networks. Therefore we must develop some method of producing a network that correctly represents our problem, but simple enough that the analysis produces meaningful rules like the ones for the adhesive dispensing machine discussed earlier and unlike the one shown in Fig. 13. The appropriateness of a Boolean rule representation for a neural network reflects the appropriateness of a Boolean rule representation in the problem domain.

A. NETWORK DESIGN

The first and most obvious way to achieve a network of the correct complexity is to design its structure by hand. Neural networks are ideal for use in process control applications. By designing the network by hand, we are allowing the controller to be informed of certain of the process characteristics and to learn the rest. Rule extraction can be used to convert the neural controller back to a more traditional technology, but after it has learned the balance of the control characteristics.

IF rise_time < 1.25	IF fall_time < 1.1
THEN rise_timeflag	THEN fall_timeflag
ELSE ¬rise_timeflag	ELSE ¬fall_timeflag

IF rise_time < 1.25	IF fall_time < 1.1
THEN rise_timeflag	THEN fall_timeflag
ELSE ¬rise_timeflag	ELSE ¬fall_timeflag

IF pulse_width < 1.0429	IF boxarearatio > 0.8520
THEN pulse_widthflag	THEN boxarearatioflag
ELSE ¬pulse_widthflag	ELSE ¬boxarearatioflag

IF rise_time < 1.25
THEN rise_timeflag
ELSE ¬rise_timeflag

IF rise_time < 1.25
THEN rise_timeflag
ELSE ¬rise_timeflag

IF pulse_width < 1.0429
THEN pulse_widthflag
ELSE ¬pulse_widthflag

IF area_change > 0.5500
THEN bubbledecision
ELSE ¬bubbledecision

IF (area_change > 0.4499 ∧
   area_change < 0.5500)
THEN bubbleflag
ELSE ¬bubbleflag

IF (pulse_height > 0.975 ∧
   pulse_height < 1.025)
THEN pulse_heightflag
ELSE ¬pulse_heightflag

IF fall_time < 1.1
THEN fall_timeflag
ELSE ¬fall_timeflag

IF fall_time < 1.1
THEN fall_timeflag
ELSE ¬fall_timeflag

IF boxarearatio > 0.8520
THEN boxarearatioflag
ELSE ¬boxarearatioflag

IF (area > 1.02581 ∨ area <
   0.9722)
THEN action
ELSE ¬action

IF pulse_height > 1.025
THEN pulse_heightdecision
ELSE ¬ pulse_heightdecision

IF area < 0.9722
THEN change
ELSE ¬change

Figure 12 Overall rules resulting from amalgamating the inequalities shown in Fig. 10 into the Boolean expressions shown in Fig. 11.

(ip20 ∧ ip19 ∧ ¬ip18 ∧ ¬ip17 ∧ ¬ip12 ∧ ¬ip11 ∧ ip10 ∧ ip9 ∧ ip4 ∧ ¬ip3 ∧ ¬ip2 ∧ ¬ip1)

or

(¬ip26 ∧ ip25 ∧ ip19 ∧ ¬ip18 ∧ ¬ip17 ∧ ¬ip12 ∧ ¬ip11 ∧ ip10 ∧ ip9 ∧ ip4 ∧ ¬ip2 ∧ ¬ip1)

or

(¬ip26 ∧ ip25 ∧ ip7 ∧ ip10 ∧ ip5 ∧ ip4 ∧ ¬ip3 ∧ ¬ip2 ∧ ¬ip22)

or

(¬ip26 ∧ ip21 ∧ ip19 ∧ ¬ip18 ∧ ¬ip17 ∧ ¬ip12 ∧ ¬ip11 ∧ ip10 ∧ ip9)

or

(¬ip21 ∧ ip13 ∧ ip12 ∧ ip11 ∧ ¬ip10 ∧ ip9 ∧ ip4 ∧ ¬ip3 ∧ ip2 ∧ ip1)

⋮   ⋮   ⋮   ⋮

Figure 13 Showing the unintelligibility of a Boolean rule derived from a digitized set of images fed to a neural network in a pattern recognition application.

B. System Investigation

All complex processes will be influenced by many system variables. The first problem in any design is to identify these influential system variables from the mass of data available. This is achieved by an investigation of the relevance of the algebraic combinations of process variables. This is derived from a combination of polynomial and adaptive linear neurons, which are known as polynomial adaptive linear neurons or PADALINEs.

The basic idea is that new inputs can be formed from functions of the basic input nodes. If the function being modeled is the sum of several elementary functions, then the PADALINE can discover the correct input parameters. For a full description of PADALINEs and their usage, refer to Hinde [9].

C. Segmentation of System Variables

Before the relevant system variables can be combined logically, they must, in effect, be digitized. This is achieved by a quantization layer of neurons. Quantization takes several forms: thresholding (a) or regions (b), and it may also be one or several variables (c). Figure 14 shows the various strategies that may be employed to transform continuous real-valued inputs into Boolean attributes.

D. Boolean Structure

The final part of the network is the Boolean transformation that is applied to the selected region. All control problems require specific action to be taken when the process system is in a specific state. The controller has to provide suitable outputs when the inputs are within a specific region. The Loughborough control architecture developed by Messom [5] solves these problems. For more information on the design of this type of network, refer to Messom *et al.* [8].

Network design works well for systems where the underlying structure is understood. In these situations a designer can build the known portion of the problem structure into the control network. The learning by the network and subsequent analysis is refining or honing the designer's prior knowledge into an intelligent rule-based system.

E. Pruning and Weight Decay

There are many domains where the underlying structure of the problem is not well understood. In these situations it is not possible to adequately design the network topology. Our goal is still to produce a minimal network that correctly

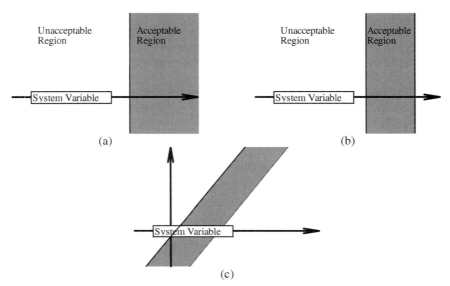

Figure 14 Quantization takes several forms: thresholding (a) or regions (b), and it may also be one or several variables (c). A parallel pair of opposite facing neurons is capable of performing the tasks shown in (b) and (c). (a) represents a condition that a variable must be greater than a certain value.

models the problem, but we must develop a new methodology. All systems can be modeled using large generic networks. For example, any system built up solely of n Boolean inputs and one output can be always be modeled using n fully connected internal nodes in a single layer. Many of today's standard textbooks give capacity results for standard classes of problems. So building a network that is capable of representing our required hypothesis is not a problem. However, many problems do not require the full capacity of the standard network topologies and we wanted a network of minimal size. The solution is to start with a standard network and to remove neurons and links between neurons dynamically during learning, thus delivering a result markedly smaller than the standard solution.

For the purposes of conversion to rule-based systems, the number of neurons in a network does not pose a major problem. The complexity of interconnection between the neurons is what dictates the size of the resultant rules. Therefore we require a technique to remove as many interconnections as possible.

When the weight assigned to an interconnection is very small, it may be removed from the network with negligible effect. We therefore want to encourage as many of the connections as possible to tend toward zero weight. Standard back propagation minimizes the error function

$$e = \sum_{All\_Examples} [Example - Actual]^2.$$

By expanding this error function so that a penalty function is incurred for the size of the weights in the network, back propagation will attempt to eliminate as many weights as possible:

$$e = \alpha \cdot \sum_{All\_Examples} [Example - Actual]^2 + (1 - \alpha) \cdot \sum_{All\_weights} f(weight).$$

The value of the constant α reflects the relative importance of accuracy versus compactness. An α near 1.0 will produce a network that is as correct as possible, at the expense of compactness. An α near 0.0 will produce a very small network, but one that is less accurate.

A suitable function for the penalty function was proposed by Setiono and Liu [10]

$$f(w) = \frac{w^2}{1 + 10w^2} + \frac{w^2}{1 * 10^5}.$$

We now have a method of training a network so that the solution has many connections with small weights, but which ones should be deleted?

Let the usefulness of a connection be defined as the largest value it passes forward over all of the examples. If the usefulness of a connection is below a predefined threshold, then it can be deleted. Similarly, the usefulness of a neuron can be defined as the largest value it has in response to any example. If the usefulness of a neuron falls below the same threshold, then it too can be deleted.

Networks are constructed by starting with a standard network, training until all of the classifications made by the system are correct and then deleting all connections and neurons whose usefulness is too low. These deletions will lower the accuracy of the network, so retraining is required. The loop is repeated until no more links or neurons can be removed. For more information on this technique, refer to Setiono and Liu [10].

VI. SIMPLIFYING THE DERIVED RULE BASE

So far in this chapter we have been considering how to produce a rule-based system that exactly mirrors the action of the network in every detail. The "knapsack/Boolean function" extraction process produces rules that have exactly the same functions as the neurons. For example, the neuron depicted in Fig. 15 would produce $(A \wedge B) \vee (A \wedge C) \vee (C \wedge B)$. If the neuron had been trained from the

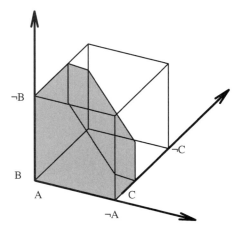

Figure 15 Showing the neural representation of the function $(A \wedge B) \vee (A \wedge C) \vee (C \wedge B)$.

incomplete example shown in Fig. 16, then there are two further, equally correct hypotheses that could have been learned instead (see Fig. 21).

The choice between the three correct hypotheses is completely arbitrary. As all are consistent with the data and are therefore arguably correct, we could substitute the rules derived from any of these with the rules from the original neuron. If we

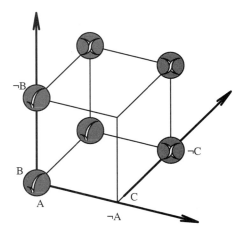

Figure 16 Showing the training set which may be used to represent the function $(A \wedge B) \vee (A \wedge C) \vee (C \wedge B)$. Notice that there is no data point at $(\neg A \wedge B \wedge C)$.

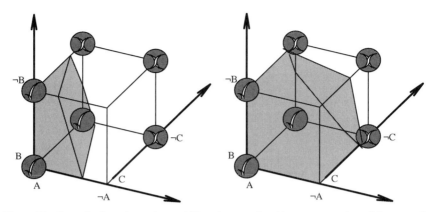

Figure 17   Two other hypotheses that could have been produced in response to the training examples shown in Fig. 16.

are using the derived rules to produce a rule-based intelligent system, then any of these neurons could be used instead of the one learned. The three derived rules are

$$(A \wedge B) \vee (A \wedge C) \vee (C \wedge B),$$
$$(A \wedge B) \vee (C \wedge B),$$
$$(A \wedge B) \vee C.$$

Consider Occam's razor

> The most likely hypothesis is the simplest one that is consistent with all observations.

Then the shortest and simplest of the three possible solutions should be used. This has two advantages. First, it is more likely to be correct and, as importantly, the resulting rule base will be much easier to read. Setiono and Liu [10] give one method for achieving this. In their system the complete set of input–output training pairs is calculated for each neuron. In their words they then

> Find the minimum number of attributes in the attribute list that uniquely differentiate the items.

This can be achieved using any standard symbolic rule extraction method. In essence, the neural training and weight decay is used to break down the data into smaller interconnected data sets. Each of these is then analyzed independently. The smaller rule sets could then be recombined to produce a complete representation. In the following example, a network has been trained to recognize three input exclusive or using the training data shown in Fig. 18.

IN1	IN2	IN3	OUT
0	0	0	0
0	0	1	1
0	1	0	1
0	1	1	0
1	0	0	1
1	0	1	0
1	1	0	0
1	1	1	1

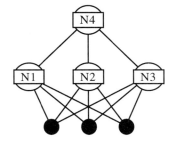

Figure 18   Training set and associated neural network for a three-input exclusive or, or even parity.

The network learning algorithm produces the minimum size network required to model this function. Four sets of training data associated with the four neurons in the network can be extracted from the network and they are summarized in Fig. 19. The first six relate the original input data to the outputs of the three nodes in the hidden layer, the seventh relates the inputs and output of the final node.

Rule extraction from these training sets gives the rules shown in Fig. 20. The three inputs are used along with the outputs of nodes 1–3 to calculate their rules. The outputs of the three nodes in the second layer and the output to node 4 are used to calculate a rule for node 4. Neural training has given the structure of the rule base, that is, the number of rules and how they interact. Traditional symbolic rule extraction produces the bodies of the rules.

This method of splitting the extraction of rules from data between the neural network and the symbolic rule induction works well. The neural training produces subhypotheses that are represented by the nodes in the hidden layers and the symbolic rule extraction produces good concise rules. However, the rule extraction can be a very complex process, and by starting with just the input–output

IN1	IN2	IN3	N1	N2	N3	N4
0	0	0	0	0	0	0
0	0	1	1	0	0	1
0	1	0	1	0	0	1
0	1	1	1	1	0	0
1	0	0	1	0	0	1
1	0	1	1	1	0	0
1	1	0	1	1	0	0
1	1	1	1	1	1	1

Figure 19   Four sets of training data associated with the four neurons in the network shown in Fig. 18 can be extracted from the network.

If '*More than none input fire*'
then
    Node1
else
        Node1

If '*More than one input fire*'
then
    Node2
else
        ¬Node2

If '*More than two input fire*'
then
    Node3
else
        ¬Node3

If *(Node1 ∧ ¬Node2)∨(Node4)*
then
    Output
else
        ¬Output

Figure 20   Results of symbolic rule extraction applied to the data generated from the network shown in Fig. 18 and tabulated in Fig. 19.

data pairs for the neuron we have discarded information. Symbolic rule extraction assigns a weight to each of the variables. In the example shown in Fig. 19, the weight attached to input 1 when building the rule for node 2 is

$$W_a = \left\{ \begin{array}{l} (Number\_of\_examples\_where[In1 = N1]) - \\ (Number\_of\_examples\_where[In1 \neq N1]) \end{array} \right\}$$
$$= 6 - 2$$
$$= 4.$$

The weight refers to the ability of the input variable to predict the answer. As I_1 is the best predictor, the examples should be split firstly on I_1. Symbolic rule extraction is based on these weights calculated from the examples. These weights already exist within the neural system as the connection weights. Return to the knapsack algorithm given earlier in Fig. 5 and reproduced here in Fig. 21 for clarity. Previously, we were searching for all sets of inputs whose weights just exceed the bias. However, we can extend the algorithm to take account of the

```
knapsack(_Unused_inputs,Bias,Answer,Answer):-
   Bias =<0.
knapsack(Unused_inputs,Bias,SubAnswer,Answer):-
   Bias > 0,
   append(_,[(Input_Name,Input_Weight)|Inputs], Unused_inputs),
   New_Bias is Bias - Input_Weight,
   knapsack(Inputs,New_Bias,[Input_Name|SubAnswer],Answer).
```

Figure 21   Original knapsack algorithm.

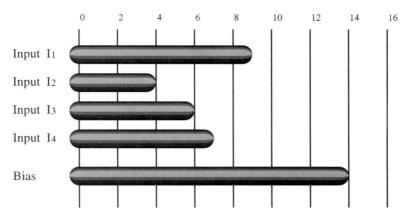

Figure 22   Weights assigned to inputs I_1, I_2, I_3, and I_4 together with the bias shown as a bar graph.

training examples. For example, given the weights in Fig. 22, I_1 would be the first input considered. Normally, the weight for I_1 would be removed from the bias giving a new bias of 5. The search would then continue for inputs to go with I_1 that can overcome this new bias. Given the training examples with I_1 true, we have three cases

1. All the examples are positive. We need search no further; as far as the examples are concerned, I_1 is sufficient to indicate a positive response even though the bias value has not been overcome.
2. All the examples are negative. There is no point searching further as the examples tell us that the current set of variables should never give a positive response. Therefore the current variables should be changed.
3. The examples contain both positive and negative examples. The algorithm needs to continue by adding further variables until either the bias is overcome or one of the first two cases applies.

The algorithm incorporating these observations is given in PROLOG in Fig. 23.

VII. EXAMPLE OF THE CONSTRUCTION OF A RULE-BASED INTELLIGENT SYSTEM

In this section a problem is introduced and a rule base derived using the neural techniques illustrated in the preceding sections.

A number of published studies have evaluated the application of artificial neural networks in the area of medical prediction. This example is based on the identification of renal transplant recipients who risk developing cytomegalovirus

```
{The Bias has been overcome}
knapsack(_Unused_inputs,Bias,_Examples,Answer,Answer):-
   Bias =<0.

{There are no false examples}
knapsack(_Unused_inputs,_Bias,Examples,Answer,Answer):-
   \+(append(_,[(Inputs,Outputs)|_],Examples),
      Outputs = -1).

{There are both positive and negative examples}
knapsack([First_Input|Other_Inputs],Bias,Examples,SubAnswer,
                                              Answer):-
   Bias > 0,
   findall((Name,1),append(_,[(Name,1)|_],Examples),Pos_examples),
   findall((Name,-1),append(_,[(Name,-1)|_],Examples),Neg_examples),
   Pos_examples =\= [],
   Neg_examples =\= [],

   {First input is set to true}
   (Input_Name,Input_Weight) = First_Input
   New_Bias is Bias - Input_Weight,
   knapsack(Other_Inputs,New_Bias,Pos_examples,
                                  [Input_Name|SubAnswer],Answer).

knapsack([First_Input|Other_Inputs],Bias,Examples,SubAnswer,
                                              Answer):-
   Bias > 0,
   findall((Name,1),append(_,[(Name,1)|_],Examples),Pos_examples),
   findall((Name,-1),append(_,[(Name,-1)|_],Examples),Neg_examples),
   Pos_examples =\= [],
   Neg_examples =\= [],

   {first input is set to false}
   (Input_Name,Input_Weight) = First_Input
   knapsack(Other_Inputs, Bias,Neg_examples,SubAnswer,Answer).
```

Figure 23   New "knapsack" algorithm in PROLOG incorporating the changes outlined.

(CMV) disease. CMV infection may be present in a patient prior to transplantation or it may be introduced into a patient through a CMV-infected donor organ. The infection can lead to the development of CMV disease and this is a significant cause of morbidity and mortality among immunocompromised renal recipients. The data set represents 548 renal transplants that took place at the Cardiff Royal Infirmary, Wales, between 1986 and 1994.

Definitions

Pretransplant Cytomegalovirus Infection

CMV(+) is recorded if IgG antibodies are present in a pretransplant blood sample, and is indicative of the presence of, or previous exposure to, CMV infection.

Posttransplant Cytomegalovirus Infection

For a CMV(−) recipient, posttransplant CMV(+) indicates the appearance of IgG antibodies in the blood. For all recipients, the appearance of IgM antibodies, the detection of CMV antigen, or the presence of CMV culture from blood or urine is recorded as posttransplant CMV(+).

Posttransplant Cytomegalovirus Disease

The patient develops fever, pneumonia, gastrointestinal diarrhea, renal insufficiency, raised alanine amino transferase.

Cytomegalovirus Prophylaxis

Donor is CMV(+) and recipient is pretransplant CMV(−). This currently forms the basis for treating with Sandoglobulin as part of the posttransplant therapy.

Human Leukocyte Antigen Mismatch Grade

Scale for measuring the match between tissue types. 0—excellent match; 6—complete mismatch.

Panel Reactive Antibodies

The percentage of a random panel of cells to which the recipient has antibodies and which cause a positive reaction. This percentage gives an indication of the number of possible donors to which the recipient would be sensitized.

The parameters recorded for each patient are shown in Fig. 24.

The initial network was constructed with 5 real-valued inputs and 11 Boolean inputs. There was a single hidden layer of 8 neurons feeding through to a single output. The network is illustrated in Fig. 25. Using the network training algorithms previously discussed, the network shown in Fig. 26 was constructed. This network classified the appearance of CMV disease with an accuracy of 85%.

Extracting the rules from this network gives the following contraindicators to CMV disease. Any two of these contraindicators, except for the pair 1 and 4, suggest that the patient is in a low-risk group.

1. Pretransplant CMV status of donor	B
2. Pretransplant CMV status of recipient	B
3. Transplant type (kidney or kidney and pancreas)	B
4. Donor age	R
5. Recipient age	R
6. Donor sex	B
7. Recipient sex	B
8. Recipient diabetes	B
9. CMV prophylaxis	B
10. Had previous transplant	B
11. Donor source	B
12. HLA mismatch	R
13. PRA latest	R
14. PRA highest	R
15. Number of rejections	B
16. Posttransplant CMV infection	B

Figure 24   Definitions of the 16 inputs to the network to classify whether CMV disease would appear in the patient. The R or B denotes whether the input is a real continuous-valued input or a Boolean input.

Contraindicator 1

No previous CMV in Donor∧
(No previous Tx ∨ No CMV Postinfection ∨ Recent Diabetes)

Contraindicator 2

Recip Female ∧ (Previous CMV in Donor ∨ K&P Tx ∨ PRA Highest > 31.7)

or

Recip Male ∧ Previous CMV in Donor ∧ (K&S Tx ∨ PRA Highest > 31.7)

Contraindicator 3

No CMV Postinfection ∧ No previous Tx ∧ No CMV Prophylaxis∧
(No Previous CMV in Donor ∨ PRA latest < 14.01)

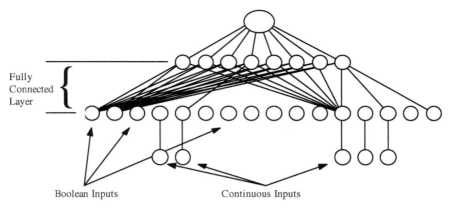

Figure 25   Initial network used to classify the appearance of CMV disease. The network is fully connected but not all the connecting arcs are shown.

or

        Recept Female ∧ K Tx ∧

                (No previous Tx ∨ No CMV Postinfection ∨ No CMV Prophylaxis)

or

                Recept Female ∧ No previous CMV in Donor

or

        Recept Male ∧ No previous CMV in Donor ∧ PRA latest < 14.01.

Contraindicator 4

            No previous CMV in Donor ∧ No CMV Postinfection

Contraindicator 5

              Previous CMV in Donor ∧ PRA latest > 14.01

The rules represent knowledge about the domain, and could now be used to implement a rule-based system to identify potential problem patients. The conversion

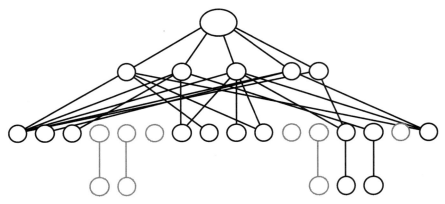

Figure 26 Final network used to classify the appearance of CMV disease. The network is now not fully connected and the arcs remaining are shown. Note that some of the input nodes are now not connected as they have no influence on the output.

of the neural solution into the rule base has many potential profits. Of these, perhaps the most important is verification. The knowledge has been converted into a form that could now be taken back to the hospital.

VIII. USING RULE EXTRACTION TO VERIFY THE NETWORK

So far this chapter has argued for, and demonstrated how, to extract Boolean rules from neural networks. One of the major uses envisioned is the verification of the original neural network. The pole-balancing problem is used to demonstrate the usefulness of the network-derived rules in verifying the action of networks.

The pole-balancing problem is an example of applied adaptive control and has become a standard tutorial problem. The control system must balance a pole on a motorized cart by moving the cart forward and back in a confined space. The implementation of most interest here is the neural network [11] shown in Fig. 27, which was successful in balancing the pole under a variety of circumstances. The inputs to this system are Boolean, so there is no quantization layer and no need for inequalities in the rule set. One of the objectives of the research was to study the usefulness of qualitative inputs to a neural network.

The analysis of the pole-balancing net resulted in the following rule. Careful analysis of the rule will make it possible to find, and therefore correct, the errors in the network.

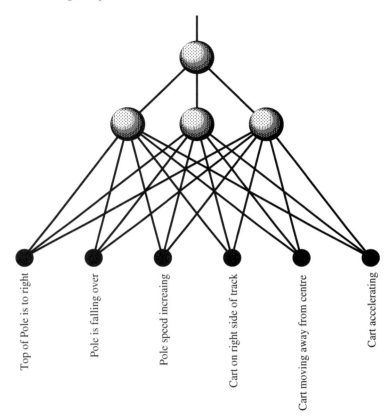

Figure 27 Neural network developed by Zhang and Grant [11] to solve the pole-balancing problem.

IF   (Top of pole is to right ∧ ¬Pole is falling over ∧
¬Pole speed increasing ∧ Cart accelerating)

∨

(Top of pole is to right ∧ ¬Pole is falling over ∧
¬Pole speed increasing ∧ Cart on right side of track ∧
Cart moving away from centre)

∨

(Top of pole is to right ∧ Pole is falling over ∧
Pole speed increasing)

THEN Apply right force

ELSE Apply left force

Simple image enhancement techniques can be used to simplify the rules, resulting in the underlying functions of the network and a list of exceptions. The underlying function can be verified and the exceptions indicate possible problems with the network hypothesis.

A. APPLYING SIMPLE IMAGE ENHANCEMENT TECHNIQUES TO RULES

The rule represented as a Karnaugh map in Fig. 28 has a fairly complex Boolean expression:

$$(I_1 \wedge I_3 \wedge I_4) \vee (\neg I_2 \wedge \neg I_3 \wedge I_4) \vee (I_2 \wedge \neg I_1 \wedge \neg I_3) \vee (\neg I_1 \wedge \neg I_2 \wedge I_3 \wedge I_4).$$

The map gives a better understanding of the simple basic concept. The aim of using image enhancement techniques is to allow high-dimensional problems to be represented in their original form but with much of the complexity removed. Enhancement will alter the rules, maintaining the main underlying objectives while removing superfluous small cases. This makes the analysis of the network hypotheses tractable for much larger networks. For example, the Boolean expression represented in Fig. 28 could be simplified to become the one represented in Fig. 29.

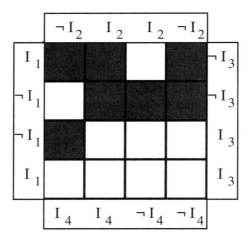

Figure 28 Function with a very simple basic concept but a relatively complex Boolean representation.

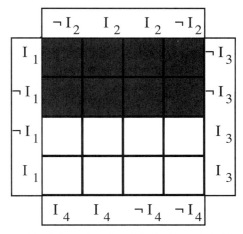

Figure 29 Enhanced version of the Boolean rule in Fig. 28.

There are two separate processes required to enhance a Boolean rule. The first reduces it by removing burrs. A burr is an area on the Karnaugh map that is separate from or smaller than the main body of the function. The second process is the reverse, removing burrs from the inverse function fills any holes on the rules. The removal of burrs from a Boolean rule can be achieved easily if the Boolean expression is in a minimal disjunction of conjunctions. The example in Fig. 28 was represented as

$$(I_1 \wedge I_3 \wedge I_4) \vee (\neg I_2 \wedge \neg I_3 \wedge I_4) \vee (I_2 \wedge \neg I_1 \wedge \neg I_3) \vee (\neg I_1 \wedge \neg I_2 \wedge I_3 \wedge I_4).$$

The Boolean representation for the version after the burrs have been removed is

$$(I_1 \wedge I_3 \wedge I_4) \vee (\neg I_2 \wedge \neg I_3 \wedge I_4) \vee (I_2 \wedge \neg I_1 \wedge \neg I_3).$$

Each of the conjunctions in the Boolean equation represents an area on the Karnaugh map. The size of this area is inversely proportional to the length of the conjunction. By removing the longer conjunctions from the Boolean rule, it is possible to remove the smaller areas on the map; thus removing the burrs from the rule. The burr is removed from the Boolean by removing the longest of the conjunctions.

Hole filling is an equally simple task. In effect, hole filling is burr removal of the inverse function, which is exactly how it is implemented.

I1 - Top of pole to right   I4 - Cart on right side of track
I2 - Pole is falling over   I5 - Cart moving away from center
I3 - Pole speed increasing   I6 - Cart accelerating

IF $(I1 \wedge \neg I2 \wedge \neg I3)$ IF $(M1 \vee M2)$
THEN   THEN
            M2 = True   Apply right force to cart
ELSE   ELSE
                    M2 = False   Apply left force to cart
IF $(I1 \wedge \neg I2 \wedge I3) \vee (I1 \wedge \neg I2 \wedge I4) \vee (I1 \wedge I3 \wedge I4) \vee$
            $(I1 \wedge I3 \wedge I5) \vee (I1 \wedge \neg I2 \wedge I6) \vee (I1 \wedge I3 \wedge I6)$
THEN
            M1 = True
ELSE
            M1 = False

Figure 30   Raw rules extracted from the network trained to balance a pole on a cart.

B. USING ENHANCEMENT TO EXPLAIN THE ACTION OF THE POLE-BALANCING NETWORK

By using the simple image enhancement techniques described previously on the rules derived for the pole-balancing network, it is possible to reveal information about the internal hypothesis. The rules for the network in their raw form are shown in Fig. 30. Repeatedly enhancing these rules gives the three versions shown in Figs. 31–33. Each is enhanced one step further than the preceding copy. Substituting the simplest rules for M1 & M2 shown in Fig. 33 into the top-level rule gives the rule shown in Fig. 34.

The basic main rule (Fig. 34) is saying follow the top of the pole. This is clearly the correct basic rule. At the higher level of complexity shown in Fig. 32, we can check that the shape of the more important input variables is correct using

IF $(I1 \wedge \neg I2 \wedge I3) \vee (I1 \wedge \neg I2 \wedge I4) \vee$ IF $(I1 \wedge \neg I2 \wedge I3)$
            $(I1 \wedge \neg I2 \wedge I6) \vee (I1 \wedge I3)$ THEN
THEN   M2 = True
            M1 = True   ELSE
ELSE   M2 = False
            M1 = False

Figure 31   Rule shown in Fig. 30 enhanced one stage.

IF   (I1 ∧ ¬I2) ∨ (I1 ∧ I3)   IF   (I1 ∧ ¬I2 ∧ I3)
THEN   THEN
        M1 = True   M2 = True
ELSE   ELSE
        M1 = False   M2 = False

Figure 32   Rule shown in Fig. 31 enhanced one further stage.

IF   I1   IF   True
THEN   THEN
        M1 = True   M2 = True
ELSE   ELSE
        M1 = False   M2 = False

Figure 33   Rule shown in Fig. 32 enhanced yet again.

IF Top of the pole is to right
THEN
            Apply right force
ELSE
            Apply left force

Figure 34   Rule shown in Fig. 35 with the basic input values and output names substituted to aid readability and understanding.

IF   (Top of pole is to right ∧ Pole speed increasing) ∨
        (Top of pole is to right ∧ Pole is falling over)
THEN
            Apply right force
ELSE
            Apply left force

Figure 35   Rule shown in Fig. 32 with the basic input values and output names substituted to aid readability and understanding.

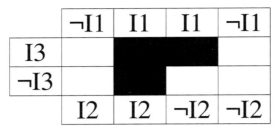

Figure 36   Map of the rule after some simplification. This shows an asymmetrical response.

a Karnaugh map. The rule for applying force is shown in Fig. 35. The map for this rule is shown in Fig. 36. The map shows that the rule need not be symmetric. This means that the network will respond differently to the same situation on different sides. Although the network balanced the pole under test conditions, it cannot be exactly correct. The reasons for moving the cart to the right should mirror the reasons for moving the cart to the left; so one of the rules must be wrong.

The part of the rule missed by the network corresponds to accelerating right when the top is to the left so as to slow down the movement of the pole if it is about to overshoot the center line. It is not surprising this has been missed as it is not a common case. However, the training could be modified to force this example to occur.

The benefits of being able to extract the underlying hypothesis of a neural network are the ability to understand the hypothesis of the network either to gain new insights into the behavior of the mechanism being modeled or to check that the hypothesis embodied in the network makes sense.

IX. CONCLUSIONS

This chapter has shown how trained neural networks can be transformed into understandable Boolean rule-based systems. It has also made observations about when this might be appropriate and when it may be unwise. Transforming neural networks into rule-based systems is an effective way of inducing rules from examples, although it is clearly not the only way. Transforming neural networks can expose deficiencies in the network which further training can rectify, thus taking neural networks away from and art form and brining it into line with more conventional software validation methods.

REFERENCES

[1] G. E. Hinton. Preface. *Artificial Intell.* 46:1–5, 1990.

[2] M. R. Garey and D. S. Johnson. *Computers and Intractability*. Freeman, New York, 1979.

[3] G. P. Fletcher and C. J. Hinde. Using neural networks as a tool for constructing rule based systems. *Knowledge-Based Systems* 8:183–189, 1995.

[4] M. Mihalaros. Studying the interpretation of feedforward neural networks using logical functions. M.S. Thesis, Loughborough University, Uunited Kingdom, 1992.

[5] C. H. Messom. Engineering reliable neural network systems. Ph.D. Thesis, Loughborough University, United Kingdom, 1992.

[6] R. D. Dowsing, V. J. Rayward-Smith, and C. D. Walter. *A First Course in Formal Logic and Its Application in Computer Science*. Blackwell Scientific, Oxford, 1986.

[7] R. M. Karp. Reducibility among comnbinatorial problems. In *Complexity of computer computations* (R. E. Miller and J. W. Thatcher, Eds.), pp. 85–103. Plenum, New York, 1972.

[8] C. H. Messom, C. J. Hinde, A. A. West, and D. J. Williams. Designing neural networks for manufacturing process control systems. In *Proceedings of the International Symposium on Intelligent Control. IEEE*, New York, 1992.

[9] C. J. Hinde. Heuristic techniques applied to an industrial situation. Ph.D. Thesis Brunel University, 1974.

[10] R. Setiono and H. Liu. Symbolic representation of neural networks. *Computer* pp. 71–78, 1996.

[11] B. Zhang and E. Grant. A neural net approach to autonomous machine learning of pole balancing. In *Proceedings ELUL'89* p. 123. IEEE, New York, 1989.

Expert Systems in Soft Computing Paradigm

Sankar K. Pal
Machine Intelligence Unit
Indian Statistical Institute
Calcutta 700 035, India

Sushmita Mitra
Machine Intelligence Unit
Indian Statistical Institute
Calcutta 700 035, India

I. INTRODUCTION

There has recently been a spurt of activity to integrate different computing paradigms, such as fuzzy set theory, neural networks, genetic algorithms, and rough set theory, under the heading *soft computing* [1–3], for generating more efficient hybrid systems. The purpose of soft computing is to provide flexible information processing capability for handling real-life ambiguous situations by exploiting the tolerance for imprecision, uncertainty, approximate reasoning, and partial truth to achieve tractability, robustness, and low cost [4]. The guiding principle is to devise methods of computation which lead to an acceptable solution at low cost by seeking an approximate solution to an imprecisely/precisely formulated problem.

One such integration that has been made by several researchers during the last five to seven years is *neuro-fuzzy computing* [5, 6], where the merits of fuzzy set theory [7, 8] and artificial neural networks (ANNs) [9–12] are fused to improve the performance in decision-making systems. The integration promises to provide both generic (parallelism, fault tolerance, adaptivity, and uncertainty management) and application-specific advantages to handle real-life problems. In many cases these models perform better than either a neural network or a fuzzy system considered individually. Neuro-fuzzy hybridization is performed broadly in two ways: a neural network equipped with the capability of handling fuzzy information (termed fuzzy-neural network), and a fuzzy system augmented by

Fuzzy Logic and Expert Systems Applications

211

neural networks to enhance some of its characteristics like flexibility, speed, and adaptibility (termed neural-fuzzy system).

Other hybridizations include the *genetic-neural* [13, 14], *fuzzy-genetic* [15], *neuro-fuzzy-genetic* [16], *rough-fuzzy* [17], and *rough-neuro-fuzzy* [18] approaches, where the characteristics of genetic algorithms (GAs) [19, 20] and rough sets [21, 22] are being exploited. Such applications are relatively new as compared to the neuro-fuzzy approaches. The primary role of GAs here is to provide techniques for efficient searching and optimization, whereas that of rough sets is the management of uncertainty and knowledge extraction. These hybrid paradigms are suitable for solving complex real-world problems for which only one tool may not be adequate. In other words, during hybridization the individual tools act synergetically (not competitively) to increase the application domain of each other when used in a soft computing paradigm.

In this chapter we discuss the issue of designing expert systems in a soft computing environment. As the knowledge base of an expert system is a repository of human knowledge and because some of these may be imprecise in nature, this may often result in a collection of rules and facts which, for the most part, are neither totally certain nor totally consistent. The expert system is also likely to be required to infer from premises that are imprecise, incomplete, or not totally reliable. The uncertainty of information in the knowledge base of the question-answering system thus induces some uncertainty in the validity of its conclusions [23]. Hence a basic problem in the design of expert systems is the analysis of the transmitted uncertainty from the premises to the conclusion and the association of a certainty factor [24]. Fuzzy expert systems [24, 25], incorporating the concept of fuzzy sets at various stages, help to a reasonable extent in the management of uncertainty in such situations.

Artificial neural networks (ANNs) are also used in designing expert systems. Such models are called connectionist expert systems [26], and they use the set of connection weights of a *trained* neural net for encoding the knowledge base for the problem under consideration. The use of ANNs helps in (a) incorporating parallelism and (b) tackling optimization problems in the knowledge base space. These models are usually suitable in data-rich environments and seem to be capable of overcoming the problem of the *knowledge acquisition bottleneck* of traditional expert systems. They help in minimizing human interaction and the associated inherent bias during the phase of knowledge base formation (which is time consuming in the case of traditional models), and they also reduce the possibility of generating contradictory rules. Powerful learning techniques exist for generating connectionist architectures from training samples. This enables us to automate the construction of knowledge bases for *classification-type* expert systems. When the connection weights of a trained fuzzy neural net are used as the knowledge base, we call the model a neuro-fuzzy expert system. This enables one to accommodate the merits of neuro-fuzzy computing in expert system design.

Generally, ANNs consider a fixed topology of neurons connected by links in a predefined manner. These connection weights are usually initialized by small random values. Knowledge-based networks [27, 28] constitute a special class of ANNs that consider crude domain knowledge to generate the initial network architecture which is later refined in the presence of training data. This process helps in reducing the searching space and time while the network traces the optimal solution. Node growing and link pruning are also performed to generate the optimal network architecture. A knowledge-based network can be used for designing a knowledge-based connectionist expert system. Rough sets, known to be effective in knowledge reduction, have very recently been used in extracting domain knowledge for encoding knowledge-based networks [18]. GAs, being efficient search techniques, have been utilized for optimizing the network parameters [14]. Both these tools hold promise in generating efficient knowledge-based connectionist expert systems in the framework of rough-neuro-genetic or rough-neuro-fuzzy-genetic computing.

The block diagrams of the basic modules of an expert system, fuzzy expert system, fuzzy neural net, connectionist expert system, neuro-fuzzy expert system, and knowledge-based connectionist expert system are shown in Fig. 1. As stated previously, a fuzzy neural net constitutes the knowledge base of a neuro-fuzzy expert system. (Note that this excludes other possible integrations, such as bringing the concept of ANN into the framework of a fuzzy expert system.) Whereas the rules are collected by knowledge engineers for designing the knowledge base of a traditional expert system (or fuzzy expert system), the connectionist models use the trained link weights of the neural net/fuzzy neural net to automatically generate the rules, either for later use in a traditional version or for providing justification in the case of an inferred decision. This automates and also speeds up the knowledge acquisition process. The use of fuzzy neural nets helps in the handling of uncertainty at various levels (e.g., input, output, learning, and neuronal) and generates fuzzy rules capable of more realistically representing real-life situations. The knowledge-based connectionist expert systems, on the other hand, initially encode crude domain knowledge among the connection weights of the neural net, thereby speeding up the training phase and generating better performance. Refined rules are later extracted from the less redundant trained network.

Section II is devoted to the general problems of expert system design, and the relevance of fuzzy sets, connectionist models, and neuro-fuzzy computing along this line. The utility of knowledge-based networks and the feasibility of using other soft computing tools, for example, GAs and rough sets in this context, are also described. A survey on connectionist expert systems (without fuzzy) is included in Section III, for the convenience of the reader. In Section IV we provide a review on existing models of neuro-fuzzy expert systems, keeping in mind the rich literature currently available in this field. A comparative study is provided in tabular form. A brief discussion on other hybrid models is provided in Section V.

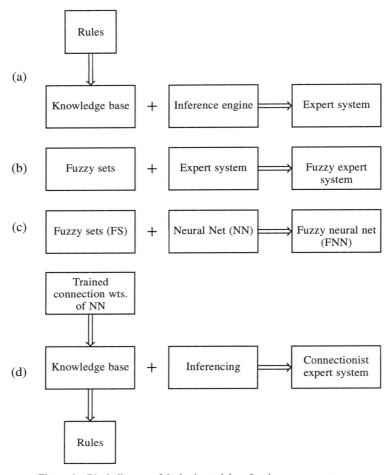

Figure 1   Block diagram of the basic modules of various expert systems.

II. EXPERT SYSTEMS: SOME PROBLEMS AND RELEVANCE OF SOFT COMPUTING

The major components of an expert system [29] are the *knowledge base, inference engine*, and *user interface*. The knowledge base contains the expert-level information necessary to solve problems in a specific domain. This information is generally represented in the form of a set of rules, although frames [30], semantic nets [31], and belief networks [32] are also in vogue. We shall consider

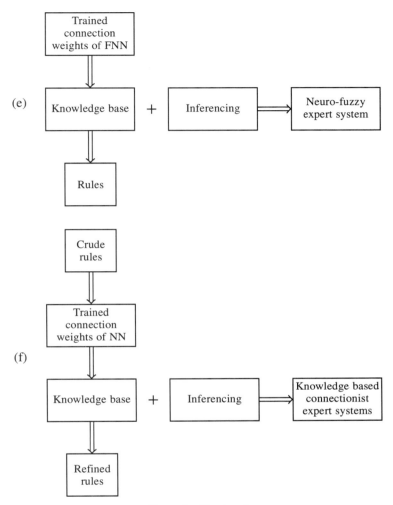

Figure 1 *(Continued)*

rule-based systems in this discussion. Knowledge bases, being domain specific, are nontransferable. The inference engine interacts both with the knowledge base and a *working memory* (that records *facts* about the current problem and is updated with the availability of new information). Pattern matching occurs between the *rules* in the knowledge base and the *facts* in the working memory to select the relevant rules applicable. Note that when no matching occurs, no rule is selected, whereas when multiple rules apply, *conflict resolution* strategies are used to select the *most specific* one. The same inference engine can be used with different knowledge bases.

We provide here, first of all, a mathematical formulation of expert systems. This is followed by a discussion on fuzzy logic and its role in the management of uncertainties, the relevance of connectionist models, the need for neuro-fuzzy computing, and the utility of knowledge-based networks.

Let us consider finding a decision consisting of a sequence of hypotheses optimizing some criteria in an environment characterized by available information. Let $\mathbf{D}$ be a candidate decision consisting of n decision elements d_i, where each decision element d_i belongs to a finite, discrete set D^i:

$$\mathbf{D} = (d_1, d_2, \ldots, d_n), \qquad d_i \in D^i.$$

As a link between the decision and the available information, a number N of measurements (or observations) are available

$$m_i: D \to \mathcal{M}: D \to m_i(D), \qquad i = 1, \ldots, N,$$

where $\mathcal{M}$ is the measurement space. Heuristic functions are used for rating the different candidate decisions according to these measurements. These ratings describe how well (or how likely) a decision (and its associated measurement) fits in with the environment

$$h_i: \mathcal{M} \to \mathcal{R}: m_i \to h_i(m_i), \qquad i = 1, \ldots, N, \tag{1}$$

where $\mathcal{R}$ is the space of the possible rating values (mostly a subset of the real numbers). Each heuristic can be considered as a piece of knowledge, usually coming from an expert, and is used for partially assessing the quality of the decision. Heuristics are combined to form a global rating r, which is a measure of the quality of the decision

$$r = O\big[h_1\big(m_1(\mathbf{D})\big), h_2\big(m_2(\mathbf{D})\big), \ldots, h_N\big(m_N(\mathbf{D})\big)\big], \tag{2}$$

where O is the combination operator across all heuristics.

A. ROLE OF FUZZY LOGIC

Fuzzy logic is based on the theory of fuzzy sets and, unlike classical logic, it aims at modeling the imprecise (or inexact) modes of reasoning and thought processes (with linguistic variables) that play an essential role in the remarkable human ability to make rational decisions in an environment of uncertainty and imprecision. This ability depends, in turn, on our ability to infer an approximate answer to a question based on a store of knowledge that is inexact, incomplete, or not totally reliable. In fuzzy logic everything, including truth, is a matter of degree

[24]. Zadeh has developed a theory of approximate reasoning based on fuzzy set theory. By approximate reasoning we refer to a type of reasoning that is neither very exact nor very inexact. This theory aims at modeling the human reasoning and thinking process with linguistic variables [33] in order to handle both soft and hard data, as well as various types of uncertainties. Many aspects of the underlying concept have been incorporated in designing decision-making systems [34]. Because fuzzy sets are a generalization of classical set theory, the embedding of conventional models into a larger setting endows fuzzy models with greater flexibility to capture various aspects of incompleteness or imperfection (i.e., deficiencies) in whatever information and data are available about a real process. Let us now explain the role of fuzzy logic in the management of uncertainty in expert systems.

The knowledge base of an expert system contains *human* knowledge, most of which is imprecise and qualitative. To describe situations where the boundary between competing hypotheses is vaguely defined, human experts use terms such as *very likely, likely, more or less likely, low, medium, high*, etc. Encoding this sort of expertise by probabilities results in the loss of information about this vagueness or imprecision. Using linguistic variables for such terms enables a knowledge engineer to capture the essence of the experts' experience and judgment without attempting to overquantify intuition. Moreover, facts about the world are rarely known with certainty. Conventional rule-based systems, with two-valued logic, usually evade this issue of partial matching.

In conventional statistical designs, the input patterns are quantitatively exact to within the resolution of the sensors used to collect them. However, real processes also may possess imprecise or incomplete input features. In such cases it may become convenient to use linguistic variables and hedges [35] like *low, medium, high, very, more or less*, etc. to augment or even replace numerical input feature information. Any input feature value can be described in terms of some combination of membership values in the linguistic property sets *low, medium*, and *high*.

The importance of fuzzy logic to the management of uncertainty in expert systems mainly lies in its ability to deal with fuzzy quantifiers and modifiers. Fuzzy logical systems allow a proposition or conclusion to range over fuzzy subsets (like *very true, more or less true, likely true*, etc.) of truth-value sets characterized by their possibility distributions. Fuzzy modifiers like *not, very, more or less, extremely, slightly, much, a little*, etc. can also be represented. A fuzzy *certainty factor* is associated with the conclusion to analyze the transmission and cumulation of uncertainty from the premises to the conclusion. Deduction of conclusions from observations and rules in the knowledge base is made using either *truth value restriction* or *compositional rule of inference*. Hence, partial match can occur between the antecedent of a rule and a fact supplied by the user.

In short, fuzzy logic or reasoning [24] provides a natural conceptual framework for knowledge representation and inferencing from knowledge bases that are imprecise, incomplete, or not totally reliable. The advantage of using fuzzy reasoning is that it can yield an approximate answer even when probabilistic theories are not applicable, as the latter often require idealized assumptions such as the independence of evidence and the mutual exclusiveness and exhaustiveness of hypotheses.

The range of the space of the measurement values in Eq. (1) can now be divided into a number of classes, each characterized by a membership function and a linguistic variable describing how well it fits the hypothesis that the candidate is the solution to the problem. Mathematically,

$$h_i: \mathcal{M} \to [0,1]^K: m \to h_i^K(m) = \left(\mu_i^1(m), \mu_i^2(m), \ldots, \mu_i^K(m)\right), \qquad (3)$$

where K indicates the number of classes. Each linguistic term is a fuzzy set which designates a category partially qualifying a candidate solution in the sense of the considered heuristic (e.g., very likely, likely, not unlikely, etc.). The set of heuristics forms a knowledge base of fuzzy rules whose antecedents are related to the measurements or observations and whose consequent part determines the fuzzy (partial) quality of the decision.

Fuzzy rule-based systems can be incorporated in fuzzy expert systems. Such a system can be expressed by a set of fuzzy inference rules. In each rule, there is a premise and a consequence. The premise is described by a fuzzy proposition and the consequence can be a fuzzy conclusion. A typical fuzzy inference rule for an N-input K-output system can be expressed as

$$\text{If } x_1 \text{ is } A_{i1}, x_2 \text{ is } A_{i2}, \ldots, x_N \text{ is } A_{iN},$$
$$\text{then } y_1 \text{ is } B_{i1}, y_2 \text{ is } B_{i2}, \ldots, y_M \text{ is } B_{iK},$$

where $\mathbf{X} = \{x_j, j = 1, 2, \ldots, N\} \in \mathcal{R}^N$ are the inputs to the fuzzy system, $\mathbf{Y} = \{y_j, j = 1, 2, \ldots, K\} \in \mathcal{R}^K$ are the outputs, and $A_{ij}, \ j = 1, 2, \ldots, N$, and $B_{ij}, \ j = 1, 2, \ldots, K$, are fuzzy subsets, such that $B_{ij} = A_{i1} \odot A_{i2} \odot \cdots \odot A_{iN}$ and $\odot$ is a fuzzy compositional operator. Thus a fuzzy rule-based system implements a mapping $\mathcal{R}^N \to \mathcal{R}^K$. Fuzzy inference methods are algorithms that deduce results from the inference rules and the presented inputs. Note that the consequent part of the rules can also be represented by scalars or membership values $\mu_i^j(m), \ j = 1, 2, \ldots, K$, where m refers to the measurement of the input variable x. These rules constitute the knowledge base of the fuzzy expert system.

The various approaches in fuzzy inferencing for expert systems include the approximate analogical reasoning based on similarity measures by Turksen and Zhong [36], the problem reduction method of Ishizuka *et al.* [37], modeling of physicians' decision processes by Esogbue and Elder [38], and inferencing in the framework of *inflammatory protein variations* by Sanchez and Bartolin [39] (using weighting). Wang and Mendel [40] developed a slightly different method for creating a fuzzy rule base made up of a combination of rules generated from numerical examples and linguistic rules supplied by human experts. The input and output domain spaces are divided into a number of linguistic subspaces. Human intervention is sought to assign degrees to the rules, and conflicts are resolved by selecting those rules yielding the maximum of a computed measure corresponding to each linguistic subspace. For other details on fuzzy expert systems, one may refer to the standard literature [23, 25].

B. USE OF CONNECTIONIST MODELS

The various uncertainty management schemes of traditional expert systems share some common problems. For example, a willing human expert able to accurately quantify expertise is needed. The transfer of the knowledge takes place gradually through many interviews between the expert and the system, and is therefore very time consuming. Usually humans are prone to be easily biased and thus the quality of knowledge extracted from the experts depends greatly on the methods used for assessment. Moreover, large knowledge bases need to be searched quickly and it is also very important to check that this knowledge base remains consistent as more information is accumulated. It would therefore be welcome if knowledge assessment could be automated by freeing it from human intervention, thereby avoiding human bias and subjectivity.

It is worth mentioning that the most difficult, time-consuming, and expensive task in building an expert system is constructing and debugging its knowledge base. In practice, the knowledge base construction can be said to be the *only* real task in building an expert system considering the proliferating presence of *expert shells*. Several approaches have been explored for easing this knowledge acquisition bottleneck.

Connectionist expert systems [26] offer an alternative approach both to the knowledge base construction and to the inferencing phase, providing interaction with the user accompanied by justification(s) of the conclusion(s) reached. Rules are not required to be supplied by humans. Instead, the connection weights of a trained neural network encode among themselves, in a distributed fashion, the information conveyed by the input–output combinations of the training set. The problems faced by traditional expert systems regarding the difficulties in normalizing across different experts' scales, conversion from human expressions to numerical terms, bias of the expert(s), generation of contradictory rules by the experts, etc., may be overcome here. The use of the learning technique of neural networks enables the model to extract the information inherent in the data (which is not utilized in traditional models) and allows dynamical adjustments to changes in the environment. It also enables one to handle a complicated environment for which either no mathematical model exists or, even if it exists, is so strongly nonlinear that a design method does not exist. Besides, the various characteristics of neural nets, namely, generalization, tolerance to noise, graceful degradation at the border of the domain of expertise, ability to discover new relations between variables, etc., are built in and hence can be exploited by the connectionist expert systems. A detailed review is provided in Section III.

Let us now provide a mathematical formulation of a layered neural network that can be used for constructing a connectionist expert system. A neuron can be depicted as an information-processing element which receives an n-dimensional input vector

$$\mathbf{X}(t) = \left[x_1(t), x_2(t), \ldots, x_n(t)\right] \in \mathcal{R}^n \tag{4}$$

and yields a scalar neural output $y(t) \in \mathcal{R}^1$ at instant t (which can correspond to a pattern presentation in one epoch). The input vector, $\mathbf{X}(t) \in \mathcal{R}^n$, represents the signals being transmitted from the n-neighboring neurons (including the self-feedback signal) and/or the outputs (measurements) from the sensory neurons. Mathematically, the information-processing ability of a neuron can be represented as a nonlinear mapping operation

$$\mathbf{X}(t) \in \mathcal{R}^n \rightarrow y(t) \in \mathcal{R}^1. \tag{5}$$

A confluence operation $\otimes$ essentially provides a measure of similarity between the neural input vector $\mathbf{X}(t)$ (new information) and the synaptic weight vector $\mathbf{W}(t)$ (accumulated knowledge base). Generally summation and product operations are used in this stage. A nonlinear activation function then performs a nonlinear mapping on the similarity measure through a nonlinear activation function $\psi[\cdot]$. Hence

$$y(t) = \psi\left[\mathbf{W}(t) \otimes \mathbf{X}(t)\right]. \tag{6}$$

A neural network can be viewed as a collection of such neurons connected to each other according to a specific topology. It therefore performs a mapping from the n-dimensional input space (input layer) to a K-dimensional output space (output layer) such that

$$\mathbf{X}(t) \in \mathcal{R}^n \rightarrow \mathbf{Y}(t) \in \mathcal{R}^K, \tag{7}$$

where K refers to the number of output classes in case of a classifier. The supervised learning uses a collection of N input–output training pairs $\{(\mathbf{X}(t), \mathbf{D}(t)), \ t = 1, \ldots, N\}$, where $\mathbf{X}(t) \in \mathcal{R}^n$ and $\mathbf{D}(t) \in \mathcal{R}^K$ are the input pattern and desired output, respectively. The objective is to optimize a cost function

$$E_N = \sum_{t=1}^{N} E_t = \sum_{t=1}^{N} d\big(\mathbf{Y}(t), \mathbf{D}(t)\big), \tag{8}$$

where $d(\cdot)$ is a distance in $\mathcal{R}^K$ and $\mathbf{Y}(t)$ is the computed output given by Eq. (7). A common choice, which simplifies the mathematical analysis, is that of considering the distance induced by an L_p norm $(1 \leq p \leq \infty)$. The error-based gradient-descent learning algorithm for weight updating is represented as

$$\mathbf{W}(t + 1) = \mathbf{W}(t) + \eta \triangle \mathbf{W}(t) \tag{9}$$

for the N_w connection weights of the neural net.

Connectionist expert systems use the connection weights $\mathbf{W}$ of the trained neural network [Eq. (9)] to form the knowledge base. The magnitudes of these connection weights are used to generate rules to justify any decision. The maximum weighted paths from the output layer to the input layer are used in the process [26, 41]. Note that in traditional expert systems the knowledge base is formulated in terms of rules by interaction with the experts. On the other hand, here the rules may be automatically extracted from the trained connection weights that form the knowledge base. This procedure will be discussed in more detail in Sections III and IV.B.

C. NEED FOR INTEGRATING FUZZY LOGIC AND NEURAL NETWORKS

Both neural networks and fuzzy systems are trainable dynamic systems that estimate input–output functions. They estimate a function without any mathematical model and *learn from experience* with sample data. A fuzzy system adaptively infers and modifies its fuzzy associations from representative numerical samples. Neural networks, on the other hand, can *blindly* generate and refine fuzzy rules from training data [42]. Fuzzy systems and neural networks also differ in how they estimate sampled functions, the kind of samples used, and how they represent and store these samples. Fuzzy systems estimate functions with fuzzy set samples (A_i, B_i), whereas neural systems use numerical point samples (x_i, y_i), where both kinds of samples reside in the input–output product space $X \times Y$. Hence the input–output mapping corresponds to $f: X \to Y$ in both cases.

Fuzzy theory is considered to be advantageous in the logical field, and in handling higher-order processing easily. The higher flexibility is a characteristic feature of neural nets produced by learning, and hence this suits data-driven processing better [43].

For the last few years, researchers all over the world [5, 6, 44–46] have been trying to combine the merits of fuzzy and neural approaches under the heading *neuro-fuzzy computing* for building more intelligent decision-making systems. This enables one to incorporate the generic advantages of artificial neural networks like massive parallelism, robustness, and learning in data-rich environments into the expert system model. The modeling of imprecise and qualitative knowledge as well as the transmission of uncertainty are possible through the use of fuzzy logic. Besides this generic advantage, the neuro-fuzzy approach provides some application-specific merits in the following way. For example, in the case of classification-type connectionist expert systems, one is typically interested in exploiting the capability of neural nets in generating the required (linearly non-separable) decision regions. The uncertainties involved in the input description and output decision are also taken care of by the concept of fuzzy sets. It is observed that in certain cases a neuro-fuzzy model performs better than either a neural network or a fuzzy system considered individually [47, 48].

Keeping in mind Eqs. (4)–(9) defining a neural net, let us now provide a mathematical formulation of a layered fuzzy neural net that can be used for designing a neuro-fuzzy expert system. A fuzzy neural network can incorporate fuzziness at the input–output level, in the connection weights, in the confluence operation, or in the activation function. Let the fuzzy input and output vectors be represented as $\widehat{\mathbf{X}}$ and $\widehat{\mathbf{Y}}$, respectively, where these correspond to fuzzy numbers or intervals or the augmented space consisting of linguistic terms. Similarly, the connection weight vector may be represented as $\widehat{\mathbf{W}}$. Arithmetic operations like fuzzy addition and fuzzy multiplication can be used in the new confluence operation $\widehat{\otimes}$. The nonlinear activation function $\widehat{\psi}$ can incorporate fuzzy logic operations like *and, or,* and *not*. Hence the resultant mapping from the $\widehat{n}$-dimensional input space to the $\widehat{K}$-dimensional output space becomes

$$\widehat{\mathbf{X}}(t) \in \mathcal{R}^{\widehat{n}} \to \widehat{\mathbf{Y}}(t) \in \mathcal{R}^{\widehat{K}}, \tag{10}$$

where a single fuzzy neuron implements the nonlinear operation

$$\hat{y}(t) = \widehat{\psi}\big[\widehat{\mathbf{W}}(t)\widehat{\otimes}\widehat{\mathbf{X}}(t)\big]. \tag{11}$$

The training data $\{(\widehat{\mathbf{X}}(t), \widehat{\mathbf{D}}(t)), \widehat{\mathbf{X}}(t) \in \mathcal{R}^{\widehat{n}},\ \widehat{\mathbf{D}}(t) \in \mathcal{R}^{\widehat{K}},\ t = 1, \ldots, N\}$ is used to optimize the cost function

$$\widehat{E}_N = \sum_{t=1}^{N} \hat{d}\big(\widehat{\mathbf{Y}}(t), \widehat{\mathbf{D}}(t)\big), \tag{12}$$

where $\hat{d}(\cdot)$ is a distance in $\mathcal{R}^{\widehat{K}}$. The learning algorithm now becomes

$$\widehat{\mathbf{W}}(t+1) = \widehat{\mathbf{W}}(t) + \eta \triangle \widehat{\mathbf{W}}(t) \tag{13}$$

for the $\widehat{N}_w$ connection weights of the fuzzy neural net.

Neuro-fuzzy expert systems use the connection weights $\widehat{W}$ of the fuzzy neural net [Eq. (13)] to form the corresponding knowledge base. The connection weights encode the knowledge base of the problem during training by using the training set $\{\widehat{\mathbf{X}}(t), \widehat{\mathbf{D}}(t)|t = 1, \ldots, N\}$, where the implemented mapping is $\mathcal{R}^{\widehat{n}} \to \mathcal{R}^{\widehat{K}}$. Note that the antecedent $\widehat{\mathbf{X}}(t)$ and the consequent $\widehat{\mathbf{D}}(t)$ may involve linguistic terms, or fuzzy intervals/numbers, or fuzzy membership values in [0, 1]. Fuzzy rules may be extracted using the connection weights of the network by backtracking along the maximum weighted paths [41].

D. UTILITY OF KNOWLEDGE-BASED NETWORKS

Recently, there have been some attempts to improve the performance of expert systems by using knowledge-based networks (KBNs) which use the domain knowledge to determine the initial structure of the network. This process helps in reducing the searching space and time while the network traces the optimal solution. Such a model has the capability of outperforming a standard multilayer perceptron (MLP) as well as other related algorithms including symbolic and numerical ones [27, 28]. However, in the absence of knowledge, one has to resort to a purely data-driven mode of learning as in simple connectionist expert models. When the initial knowledge fails to explain many instances, additional hidden units and connections need to be added (often empirically). The initial encoded knowledge may be refined with experience by performing learning in the data environment. The resulting networks generally involve less redundancy in their topology.

Let us provide here a mathematical formulation in line with the modeling in Eqs. (4)–(9). The knowledge-based nets implement a mapping

$$\mathbf{X}'(t) \in \mathcal{R}^{n'} \rightarrow \mathbf{Y}(t) \in \mathcal{R}^{K} \tag{14}$$

from the n'-dimensional input space to the K-dimensional output space, where $n' \leq n$ and

$$y(t) = \psi\left[\mathbf{W}'(t) \otimes \mathbf{X}'(t)\right]. \tag{15}$$

The training data $\{(\mathbf{X}'(t), \mathbf{D}(t)), \ \mathbf{X}'(t) \in \mathcal{R}^{n'}, \ \mathbf{D}(t) \in \mathcal{R}^{K}, t = 1, \ldots, N\}$ is used to optimize the cost function E_N of Eq. (8). The learning algorithm becomes

$$\mathbf{W}'(t+1) = \mathbf{W}'(t) + \eta \triangle \mathbf{W}'(t) \tag{16}$$

for the N'_w connection weights such that $N'_w \leq N_w$ of Eq. (9).

1. Incorporating Fuzziness

Some attempts on using fuzzy sets for the design of knowledge-based systems have also been recently reported. Analogous to the idea of Eqs. (10)–(13), the mapping from the $\hat{n}'$-dimensional input space to the $\widehat{K}$-dimensional output space can be represented here as

$$\widehat{\mathbf{X}}'(t) \in \mathcal{R}^{\hat{n}'} \rightarrow \widehat{\mathbf{Y}}(t) \in \mathcal{R}^{\widehat{K}}, \tag{17}$$

where $\hat{n}' \leq \hat{n}$ and

$$\hat{y}(t) = \widehat{\psi}\left[\widehat{\mathbf{W}}'(t)\widehat{\otimes}\widehat{\mathbf{X}}'(t)\right]. \tag{18}$$

The training data $\{(\widehat{\mathbf{X}}'(t), \widehat{\mathbf{D}}(t)), \ \widehat{\mathbf{X}}'(t) \in \mathcal{R}^{\hat{n}'}, \ \widehat{\mathbf{D}}(t) \in \mathcal{R}^{\widehat{K}}, \ t = 1, \ldots, N\}$ is used to optimize the cost function $\widehat{E}_N$ of Eq. (12). The learning algorithm becomes

$$\widehat{\mathbf{W}}'(t+1) = \widehat{\mathbf{W}}'(t) + \eta \triangle \widehat{\mathbf{W}}'(t) \tag{19}$$

for the $\widehat{N}'_w$ connection weights such that $\widehat{N}'_w \leq \widehat{N}_w$ of Eq. (13).

2. Using Rough Sets and Genetic Algorithms

One of the major problems in connectionist/neuro-fuzzy expert system design is the choice of the optimal network structure. This has an important bearing on any performance evaluation. Moreover, the models are generally very much data dependent and the appropriate network size also depends on the available training data. Various methodologies developed for selecting the optimal network structure include growing and pruning of nodes/links, employing genetic search, and embedding initial knowledge in the network topology. The last approach has been investigated to some extent in knowledge-based networks. The soft computing tools, used effectively in this connection, are rough sets [21, 22] and genetic algorithms [19, 20].

The theory of rough sets [21] has recently emerged as another major mathematical approach for managing uncertainty that arises from inexact, noisy, or incomplete information. It has been investigated in the context of expert systems, decision support systems, machine learning, inductive learning, and various other areas of application. It is found to be particularly effective in the area of knowledge reduction. The focus of rough set theory is on the ambiguity caused by limited discernibility of objects in the domain of discourse. The intention is to approximate a *rough* (imprecise) concept in the domain of discourse by a pair of *exact* concepts, called the lower and upper approximations. These exact concepts are determined by an *indiscernibility* relation on the domain, which, in turn, may be induced by a given set of *attributes* ascribed to the objects of the domain. These approximations are used to define the notions of *discernibility matrices*, *discernibility functions* [49], *reducts*, and *dependency factors* [21], all of which play a fundamental role in the reduction of knowledge.

Genetic algorithms (GAs) [19, 20] are randomized search and optimization techniques guided by the principles of evolution and natural genetics. They are efficient, adaptive, and robust search processes, producing near optimal solutions and have a large amount of implicit parallelism. The algorithm starts the search from an initial population of chromosomes, encoded as bit strings, and applies several genetic operators like selection, crossover, and mutation (over a sequence of generations) to finally arrive at a globally optimal solution based on a fitness function. Unlike conventional search techniques, GAs work simultaneously on multiple points in the search space. Owing to their stochastic character, they have a very low chance of getting stuck at local minima. The criterion of "survival of the fittest" provides evolutionary pressure for populations to grow with increasingly fit individuals.

Before we describe various neuro-fuzzy and other hybrid expert systems in Sections IV and V, let us provide a brief survey on existing connectionist (non-fuzzy) expert systems, including those using knowledge-based networks, for the convenience of the reader. Note that all the hybrid models to be described here have their origin in connectionist expert systems.

III. CONNECTIONIST EXPERT SYSTEMS: A REVIEW

Here we consider a few of the existing layered connectionist expert systems modeled by Gallant [26], Saito and Nakano [50], Lacher *et al.* [51], and Poli *et al.* [52]. The inputs and outputs consist of *crisp* variables in all cases. Generally the symptoms are represented by the input nodes, whereas the diseases and possible treatments correspond to the intermediate and/or output nodes [26, 50]. The linear discriminant network of [26] (dealing with *sacrophagal* problems) is generated from the dependency information regarding the variables, which is provided by the expert in the form of an adjacency matrix. This is then trained by the simple *pocket algorithm*. The absence of hidden nodes and nonlinearity limit the utility of the system in modeling complex decision surfaces. The multilayer network in [50] is designed for detecting *headache*. A patient responds to a questionnaire regarding his or her perceived symptoms and these constitute the input to the network.

Lacher *et al.* [51] have designed event-driven, acyclic networks of neural objects called expert networks. The network is built under the *commercial shell M.1*. There are regular nodes and operation nodes (for conjunction and negation). Input weights are hard wired, whereas the output weights of a node are adaptive. Antecedents of a disjunction in a rule are simplified to generate a set of individual rules before formulating the initial network architecture. The backpropagation algorithm is modified to work in the event-driven environment, where both forward and backward signals propagate in *data flow* fashion. The form of the rules (coarse knowledge) is tuned with the associated certainty factors (fine knowledge) and the resultant network trained for better performance.

A novel approach to designing a *modular* connectionist expert system, called *Hypernet*, has been reported by Poli *et al.* [52]. The feedforward network consists of a reference-generating module, a drug compatibility module, and a therapy-selecting module in order to simulate the physician's reasoning as closely as possible. The user-friendly system provides a graphics interface for easy handling as well as verification of decisions. The model is implemented for diagnosing and treating hypertension. The performance is good owing to the embedded modularity of the network.

Rule generation is also possible for the models in [26, 50]. In [50] the doctor is supplied with information regarding possible diagnoses based on output node values. Relation factors, estimating the strength of the relationship between symptom(s) and disease(s), are extracted from the network and used to help doctors. Rules are generated from the changes in levels of input and output units; the connection weights are not involved in the process. These rules are then used to allow the patient to confirm the symptoms initially provided by him or her to the system, in order to eliminate noise from the answers. The model in [26] incorporates inferencing/forward chaining, confidence estimation, backward chaining, and explanation of conclusions by *if–then* rules. To generate a rule, the attributes with greater inference strength (magnitude of connection weights) are selected and a conjunction of the more significant premises is formed to justify the output concept. Here, the user can also be queried to supplement incomplete input information.

Ishikawa [53] demonstrates the extraction of rules from a network trained by *structural learning with forgetting* with mushroom data. The nonredundant network architecture, so generated, is examined to detect the regularities in the training data. Omlin and Lee Giles [54] use trained discrete-time recurrent neural networks to correctly classify strings of a regular language. Rules defining the learned grammar can be extracted from networks in the form of deterministic finite-state automata (DFAs) by applying clustering algorithms in the output space of recurrent state neurons. A heuristic is used to choose among the consistent DFAs the model which best approximates the learned regular grammar.

An MLP-based model for the identification of electroencephalogram (EEG) power spectra of rats in depression has recently been reported by Mitra *et al.* [55]. The input consists of frequency, represented both as individual values and as nonoverlapping bands, normalized in the range [0, 1]. The output refers to the control and depressed states. It has been observed that the role of exercise reverses the effect of stress. Rules have also been generated in terms of the linguistic labels *small* and *large* corresponding to the relative values of the features. Note that this is slightly different from the *crisp* rules, indicating the presence or absence of certain features (symptoms) as in [26, 50].

The knowledge-based models discussed here [27, 28, 56] involve *crisp* inputs and outputs. The initial domain knowledge, in the form of rules, is mapped into the multilayer feedforward network topology using binary link weights to maintain the semantics. Yin and Liang [56] have employed a *gradually-augmented-node* learning algorithm to incrementally build a dynamic knowledge base capable of both acquiring new knowledge and relearning existing information. The rules are explicitly represented among the *condition nodes, rule nodes*, and *action nodes* and the algorithm gradually builds the multilayer feedforward network. This connectionist incremental expert model is used as an *animal identification system* whose network structure is changed dynamically according to the new environment or through human intervention. In Fu's model [27] hidden units and additional connections are introduced appropriately when the network performance stagnates during training using backpropagation. Weight decay, pruning of weights, and clustering of hidden units are incorporated to improve the generalization of the network.

Towell and Shavlik [28] have designed a hybrid learning system for problems from molecular biology. Disjunctive rules are rewritten as multiple conjunctive rules while building the network structure. Nodes and links are incorporated, upon instructions from the user, to augment the knowledge-based module. Expansion of the network guided by both the domain theory and training data has been reported by Opitz and Shavlik [57]. Dynamic addition of hidden nodes is made by heuristically searching through the space of possible network topologies, in a manner analogous to the adding of rules and conjuncts to the symbolic rule base.

A way of using the knowledge of the trained neural model to extract the revised rules for the problem domain is described in [27, 58]. Meaningful rules can be extracted from the knowledge-based network in refined form by employing clustering, averaging, elimination, optimization, and simplification [58]. The algorithm considers groups of links as equivalence classes, thereby generating a bound on the number of rules rather than establishing a ceiling on the number of antecedents. Note that this approach differs from that in [50], where a breadth-first search is employed to exhaustively find those input settings that cause the weighted sum to exceed the bias at a node.

IV. NEURO-FUZZY EXPERT SYSTEMS

This section provides a review on neuro-fuzzy models for inferencing and rule generation, with the objective of generating expert systems. A comparative analysis of the basic features of these models with those of the traditional and connectionist (nonfuzzy) versions is provided in Table I.

A. WAYS OF INTEGRATION

The state of the art for the various techniques of combining neural networks and fuzzy sets involves synthesis at various levels. We categorize the different fusion methodologies, made so far, as follows [59].

1. *Incorporating fuzziness into the neural network framework.* This involves fuzzifying the input data, assigning fuzzy labels to the training samples,

Table I

Comparative Study of Various Expert Systems

	Expert system	Connectionist expert system	Neuro-fuzzy expert system	Knowledge-based connectionist/neuro-fuzzy expert system
Knowledge base	Knowledge acquisition and representation in the form of rules, frames, semantic nets, or belief networks	Connection weights of trained neural net that were initialized with small random values	Connection weights of trained fuzzy neural net that were initialized with small random values	Connection weights of trained nonfuzzy/fuzzy neural net that were initialized with crude domain knowledge in rule form with binary link weights [27, 28, 56, 80–83], *a priori* class information and distribution of pattern points [85]
Knowledge refinement	Addition of new knowledge (say, as new rules)	Empirical addition of hidden nodes/links	Empirical addition of hidden nodes/links	Network optimization using growing and pruning of nodes/links, based on training data and additional knowledge [27, 28, 56, 57, 85]
Inferencing	Matching facts with the existing knowledge base	Presentation of crisp input, forward pass, and generation of crisp output	Presentation of fuzzy input, forward pass, and generation of fuzzy output	Presentation of input, forward pass, and generation of output
Rule generation	—	Crisp rules obtained during backward pass using changes in levels of input and output units [50], magnitude of connection weights [26, 55]	Fuzzy rules obtained during backward pass using node activations and link weights [41, 70–75]	Rules obtained during backward pass [27, 58, 81]; negative rules also possible [85]

possibly fuzzifying the learning procedure, and obtaining neural network outputs in terms of fuzzy sets [60, 61].

2. *Designing neural networks guided by fuzzy logic formalism.* Neural networks are designed to implement fuzzy logic and fuzzy decision making, and to realize membership functions representing fuzzy sets [62, 63].

3. *Changing the basic characteristics of the neurons.* Neurons are designed to perform various operations used in fuzzy set theory (like fuzzy union, intersection, aggregation represented by *and, or,* and hybrid operators) instead of the standard multiplication and addition operations [64, 65].
4. *Making the individual neurons fuzzy.* The input and output of the neurons are fuzzy sets and the activity of the networks involving the fuzzy neurons is also a fuzzy process [66].
5. *Using measures of fuzziness as the error or instability of a network.* The fuzziness/uncertainty measures of a fuzzy set are used to model the error or instability or energy function of the neural-network-based system [67].

As the existing neuro-fuzzy expert systems fall under categories 1 and 3 only, we shall not be concerned with the remaining groups (dealing mainly with classification or control problems) in this discussion.

B. VARIOUS METHODOLOGIES

Neuro-fuzzy expert systems use the connection weights of trained fuzzy neural nets for encoding the knowledge base, thereby enabling one to incorporate the advantages of fuzzy set theory into the connectionist expert system model. Besides the generic advantages of neural networks and fuzzy systems, like parallelism, robustness, adaptivity, and handling of uncertainty, one can incorporate their application-specific merits in this paradigm. For example, the capability of neural nets in generating linearly nonseparable decision regions can be exploited. Moreover, the modeling of uncertainty in the input description and output decision can be tackled by the concept of fuzzy sets. As an illustration of the characteristics of neuro-fuzzy expert systems, the models by Hayashi [68], Hudson *et al.* [69], Sanchez [61], Mitra and Pal [41], and Romaniuk and Hall [70] are described here. Note that while the last model falls under category 3 of the fusion methodologies, the remaining models pertain to category 1.

Yoshida *et al.* [71] have *defuzzified* real-life fuzzy data, using the *level set* representation, to produce the *crisp* inputs $\{+1, -1, 0\}$ required by the *distributed single-layer perceptron-based* model trained with the *pocket algorithm* for diagnosing *hepatobiliary disorders*. All contradictory training data are excluded, as these cannot be tackled by the model. In Hayashi's extension [68], the input layer consists of both fuzzy and crisp cell groups, whereas the output is modeled only by fuzzy cell groups. The crisp cell groups are represented by m cells taking on two values in $\{(+1, +1, \ldots, +1), (-1, -1, \ldots, -1)\}$. Fuzzy cell groups, on the other hand, use binary m-dimensional vectors, each taking on values in $\{+1, -1\}$. Linguistic relative importance terms like *very important* and *moderately important* are allowed in each proposition; linguistic truth values like *completely true, true, possibly true, unknown, possibly false, false,* and *completely false* are also assigned by the domain experts depending on the output values. Multiple correct pattern classes, using different linguistic truth values, are possible.

Hudson *et al.* [69] use input nodes that simply represent the data values for signs, symptoms, and test results (may be continuous or discrete). The interactive nodes account for the interactions which may occur between these parameters. A feedforward neural network model is used for detecting *carcinoma of the lung*. Information is extracted directly from the accumulated data and then combined with a rule-based expert system incorporating approximate reasoning techniques. The learning method is an adaptation of the *potential function* approach to pattern recognition and is used to determine the weighting factors as well as the relative strengths of rules for two-class problems.

Sanchez [61] has associated two types of connection weights, namely, primary linguistic weights and secondary numerical weights, to generate the knowledge base for a *biomedical application* (*inflammatory protein variations*) using a feed-forward network. Triangular membership functions like *negative large, negative medium, negative small, approximately zero, positive small, positive medium,* and *positive large*; or, *decreased, normal,* and *increased* account for the linguistic weights, whereas the quantitative weights lie in the range [0, 1]. The linguistic weights are tuned according to the information provided from the input–output examples, whereas the numeric weights and the network topology are determined by solving *fuzzy relation equations*.

A *cell recruitment* learning algorithm, capable of forgetting previously learned facts by learning new information, has been employed by Romaniuk and Hall [70] to build a fuzzy connectionist expert system for determining the *creditworthiness of credit applicants*. The network consists of *positive* and *negative collector cells* along with *unknown* and *intermediate* cells and can handle *fuzzy* or *uncertain* data. Fuzzy functions like *maximum, minimum,* and *negation* are applied at the neuronal levels depending on the corresponding bias values. This incremental learning algorithm can be used either in conjunction with an existing knowledge base or alone.

Extraction of fuzzy *if–then* production rules is possible in [70–72], using a top-down traversal involving analysis of the node activations, their bias, and the associated link weights. Rhee and Krishnapuram [72] have reported a method for rule generation from minimal approximate fuzzy aggregation networks. They estimate the linguistic labels and the corresponding triangular membership functions for the input features from the training data. Hybrid operators with compensatory behavior, whose parameters can be learned during gradient descent to estimate the type of aggregation, are employed at the neuronal level. Pruning of redundant features and/or hidden nodes helps in generating appropriate rules in terms of *and–or* operators that are represented by these hybrid functions.

Mitra and Pal [41] have reported the use of a fuzzy MLP for classification and rule generation. The input is represented in terms of π-functions corresponding to the linguistic properties *low*, *medium*, and *high*. Handling of inputs in numeric, linguistic, and set forms is possible. The output is in terms of fuzzy class membership values and enables efficient handling of overlapping pattern classes. The antecedent parts of rules are generated by backtracking along the maximum-weighted connection paths of the trained network. The consequent part is determined from a certainty measure which expresses the confidence (belief) of an output decision. The node excitations corresponding to a test pattern determines the appropriate *if–then* parts of a rule generated to justify an inferred decision. Note that this investigation provides a basic module for designing a classification-type connectionist expert system. The rules thus obtained can also constitute the knowledge base of a traditional expert system in the same application domain. Here (unlike the other models) both the antecedent and the consequent parts of these rules are provided in linguistic (or natural) form. Linguistic hedges/modifiers like *very, more, or less* and *not* can be represented as antecedent clauses.

Consider the simple three-layered network given in Fig. 2 demonstrating a simple rule generation instance regarding class 1 [41]. A sample set of connection weights w_{ji}^h, input activation y_i^0, and the corresponding linguistic labels are depicted in the figure. The solid and dotted–dashed paths (that have been selected) terminate at input neurons i_s and i_n, respectively. The dashed lines indicate the

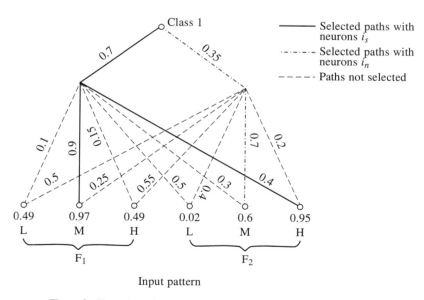

Figure 2   Example to demonstrate rule generation scheme by backtracking.

paths not selected, using the w_{ji}^h and y_i^h values during backtracking. We select only those maximum weighted paths from the output to the input layer, such that all neurons lying along them have $y_i^h > 0.5$. Let the certainty measure for the output neuron under consideration be 0.7. This corresponds to the label *likely* in the consequent part. Then the rule generated by the model in this case to justify its conclusion regarding class 1 would be

If F_1 is *very medium AND* F_2 is *high*, then *likely* class 1.

We generate clauses for an *if–then* rule until the *net path weights* wet_{i0} satisfy the relation

$$\sum_{i_s} wet_{i_s^0} > 2 \sum_{i_n} wet_{i_n^0}.$$

Here the *net path weights* are found to be 2.7 ($= 1.6 + 1.1$) and 1.05 for the i_s and i_n neurons, respectively, such that $2.7 > 2 * 1.05$. The modifier *very* (corresponding to F_1) is obtained by selecting the one having the *minimum* distance from the input vector. Similarly, in the case of F_2, modifiers are required using this minimum distance criterion.

The user can be queried in case of unknown or unavailable input features. Handling of missing or incomplete inputs is also possible. Applications have been made for vowel recognition and detection of *Kala-azar* (a tropical disease). This has been extended in [73] to design a neuro-fuzzy expert system for diagnosing *hepatobiliary disorders*. Here the linguistic labels at the input can be automatically tuned from the training data.

Another interesting application has also been reported [74] using the unsupervised, self-organizing Kohonen net. This approach is completely different from the fuzzy Kohonen net, in unsupervised mode, as reported in [62, 63]. The network has been modified to incorporate linguistic π-functions and contextual class information at the input, thereby enabling it to function under partial supervision. Unlike the other methods (involving layered feedforward nets under full supervision), this fuzzy version of the Kohonen net has been effectively used for classification, querying, and rule generation. Note that the three models [41, 73, 74] fall under category 1 of the fusion methodology.

A fourth model, using logical *and–or* functions (in terms of *product-probabilistic sum* and *max-min*) at the neuronal level, has been reported [75]. This is grouped under category 3. It has been observed that more meaningful rules (in terms of *and–or* clauses) can be generated here in case of simpler problems, although the classification performance is better in case of the more generalized sigmoidal function of [41, 73].

It is worth mentioning that all these models incorporate overlapping linguistic labels, represented by π-functions, at the input. This is different from the approach of Keller *et al.* [76] where trapezoidal possibility distributions, sampled at discrete points, are used to represent fuzzy linguistic terms and modifiers. The concept of class membership helps the models to tackle overlapping and fuzzy pattern classes. This approach is an extension of the work of Keller and Hunt [60] for multiclass problems using multilayer networks. Another approach for fuzzification at input and output has been reported by Ishibuchi *et al.* [77] using interval vectors. Although this is different, it will not be elaborated here as it does not cover the domain of connectionist expert system design or rule generation. Moreover, the conventional triangular membership functions used in control problems are also slightly different from the π-functions. It is to be noted that the triangular functions can be used in place of the more general continuous π-functions if desired.

C. USING FUZZY KNOWLEDGE-BASED NETWORKS

A brief survey of this field is provided here based on the studies of Masuoka *et al.* [78], Kasabov [79], Kosko [80], Machado and Rocha [81], Pedrycz and Rocha [82], and Hirota and Pedrycz [83]. The first three approaches fall under category 1 of the fusion methodologies, whereas the rest can be grouped under category 3.

Knowledge extracted from experts in the form of membership functions and fuzzy rules (in *and–or* form) is used to build and preweight the neural net structure which is then tuned using training data. The model by Masuoka *et al.* [78] consists of the input variable membership net, the rule net, and the output variable net. Kasabov [79] uses three neural subnets, namely, production memory, working memory, and variable binding space to encode the production rules, which can later be updated. A fuzzy signed digraph with feedback, termed the fuzzy cognitive map, has been used by Kosko [80] to represent knowledge. An additive combination of augmented connection matrices is employed to include the views of a number of experts for generating the knowledge network.

Machado and Rocha [81] have used a connectionist knowledge base involving fuzzy numbers at the input layer, fuzzy *and* at the hidden layers, and fuzzy *or* at the output layer. The hidden layers chunk input evidence into clusters of information for representing regular patterns of the environment. The output layer computes the degree of possibility of each hypothesis. The initial network architecture is generated using *knowledge graphs* elicited from experts by the application of the knowledge acquisition technique of [84]. The experts express their knowledge about each hypothesis of the problem domain by selecting an appropriate set of evidence and building an acyclic weighted *and–or* graph to describe how these must be combined to support decision making.

Pedrycz and Rocha [82] have used basic aggregation neurons (*and/or*) and referential processing units (matching, dominance, and inclusion neurons) to design knowledge-based networks. The inhibitory and excitatory characteristics are captured by embodying direct and complemented input signals and fully supervised learning is employed. Another related approach by Hirota and Pedrycz [83] has incorporated the use of fuzzy clustering for developing the geometric constructs leading to the design of knowledge-based networks.

Most of these models are mainly concerned with the encoding of initial knowledge by a fuzzy neural network followed by refinement during training. Extraction of fuzzy rules in this framework has been attempted in [78, 79, 81]. Inference, inquiry, and explanation are possible during consultation in [81]. Mitra *et al.* [85] have recently designed a knowledge-based neuro-fuzzy system for classification and rule generation. This approach falls under category 1 of the fusion methodologies. Here crude initial domain knowledge is encoded among the connection weights using the *a priori* class information (and their complements) and the distribution of pattern points in the feature space. An accurate estimation of the links connecting the output, and hidden layers (in terms of the preceding layer link weights and node activations) is provided. The input, output, and learning scheme are similar to that in [41]. Node growing and link pruning are incorporated to generate the optimal network architecture. Inferencing, querying, and rule generation are demonstrated (as in [41]) for recognizing vowels and diagnosing *hepatobiliary disorders*. Negative rules, indicative of cases where a pattern does not belong to a class, can also be generated. This is specially suitable in the ambiguous cases where positive rules (dealing with the belongingness of a pattern to a particular class) cannot be obtained. The performance of the knowledge-based net is seen to be superior as compared to the models incorporating no initial knowledge.

V. OTHER HYBRID MODELS

The relevance of rough sets and genetic algorithms to the design of expert systems has been described in Section II.D.2. As mentioned before, the literature on various approaches along this line is scarce as compared to neuro-fuzzy expert systems. However, we provide here some of the attempts recently reported in this area. A few methods related to expert system design are also described.

A. ROUGH SETS

Many have looked into the implementation of decision rules extracted from operation data using rough-set formalism, especially in problems of machine learning from examples and control theory [22]. In the context of neural networks, an attempt at such an implementation has been made by Yasdi [86]. The intention was to use rough sets as a tool for structuring the neural networks. The methodology consisted of generating rules from training examples by rough-set learning, and mapping the dependency factors of the rules into a single layer of connection weights of a four-layered neural network. The input and output layers involved fixed binary weights. *Max, min,* and *or* operators were applied at the hidden nodes. Application of rough sets in neurocomputing has also been made in [87]. However, in this method, rough sets were used for knowledge discovery at the level of data acquisition (i.e., in preprocessing of the feature vectors) and not for structuring the network.

Banerjee *et al.* [18] have proposed an integration of rough sets and fuzzy neural networks for designing a knowledge-based system. Rough-set-theoretic techniques are utilized for extracting crude domain knowledge that is encoded among the connection weights. Methods are derived to model (i) convex decision regions with single-object representatives and (ii) arbitrary decision regions with multiple-object representatives. A three-layered (fully adaptive) fuzzy MLP is considered. The feature space gives the condition attributes and the output classes the decision attributes, resulting in a decision table. This table, however, may be transformed, keeping the complexity of the network to be constructed in mind. Rules are then generated from the (transformed) table by computing relative reducts. The dependency factors of these rules are encoded as the initial connection weights of the fuzzy MLP, propagating their effect in a top-down manner in proportion to the fan-in at any particular neuron. The network is next trained to refine its weight values. The effectiveness of the model is demonstrated on both real-life and artificial data. The knowledge encoding procedure, unlike most other methods [27, 28], involves a nonbinary weighting mechanism based on a detailed and systematic estimation of the available domain information. It may be noted that the optimal number of hidden nodes is automatically determined from the syntax of the generated rules.

Figure 3 illustrates an example demonstrating the knowledge encoding procedure [18] for class c_2 using a three-layered network. Let us consider the reduct set $B = (L_1 \wedge M_1 \wedge M_3)$. Then the discernibility functions $f_D^{x_i}$ (in conjunctive normal form) for the six classes $i = 1, \ldots, 6$, obtained from the discernibility matrix, are

$$f_D^{x_1} = L_1 \wedge (M_1 \vee M_3), \qquad f_D^{x_2} = L_1 \wedge (M_1 \vee M_3), \qquad f_D^{x_3} = M_1 \wedge M_3,$$
$$f_D^{x_4} = L_1 \wedge M_1 \wedge M_3, \qquad f_D^{x_5} = M_1 \wedge M_3, \qquad f_D^{x_6} = L_1 \wedge M_1 \wedge M_3.$$

The dependency factors df_i for the resulting rules $r_i, i = 1, \ldots, 6$, are 2/3, 2/3, 1, 1, 1, 1. These factors are encoded as the initial connection weights of the fuzzy MLP. Consider rule r_2, namely, $L_1 \wedge (M_1 \vee M_3) \to c_2$, with dependency factor $df_2 = 2/3$. Here we require two hidden nodes corresponding to class c_2 to model the operator $\wedge$. The two links from the output node representing class c_2 to these two hidden nodes are assigned weights of $df_2/2$ to keep the weights

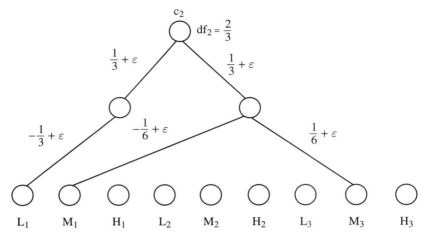

Figure 3 Example to demonstrate the initial weight encoding scheme using rough-set-theoretic techniques.

equally distributed. The signs of the weights are set to positive (negative) according to the values 1 (0) of the corresponding entries in the attribute value table. The attributes M_1 and M_3, connected by the operator $\vee$, are combined at one hidden node with link weights of $-df_2/4$, $df_2/4$, respectively, whereas the link weight for attribute L_1 is clamped to $-df_2/2$ (because there is no further bifurcation). All other connection weights are assigned very small random weights ϵ, lying in the range $[-0.005, +0.005]$. The resultant network is finally refined during training using a set of labeled samples.

B. GENETIC ALGORITHMS

Genetic algorithms have found various applications in fields like pattern recognition, image processing, and neural networks [88]. They have been used in determining the optimal set of connection weights [14] as well as the optimal topology of a layered neural network [89, 90]. These hold significance for designing connectionist expert systems. Pal and Bhandari [14] incorporated a new concept of nonlinear selection for creating mating pools and a weighted error as a fitness function. A fixed-topology MLP was used to determine the optimal solution for selecting a decision boundary for the pattern recognition problem. Maniezzo [89] used variable-length chromosomes, incorporating the concept of presence and absence bits, for encoding various topologies of an MLP. The concept of a GA simplex was also introduced. In another investigation, Pal and Bhandari [16] have demonstrated a way of integrating fuzzy sets, ANNs, and GAs for automatic selection of cloning templates when a *cellular neural network* is used in extracting object regions from noisy images. Fuzzy geometrical properties of image were used as the basis of the fitness function.

Opitz and Shavlik [90] have used the domain theory of Towell and Shavlik [28] (with random perturbation) to create an initial population of knowledge-based nets. Crossover and mutation operators are specifically designed to function on these networks. The algorithm uses these genetic operators to search the topology space in order to find networks, which are then trained using backpropagation.

VI. CONCLUSIONS

The problem of designing an expert system in the light of soft computing has been addressed. The relevance, characteristics, and merits of integrating different soft computing tools such as fuzzy sets, artificial neural networks, genetic algorithms, and rough sets in various forms have been described, with greater emphasis on neuro-fuzzy computing. Neuro-fuzzy models have been found to incorporate both the generic and the application-specific merits of neural networks as well as fuzzy systems. This has resulted in the generation of more intelligent decision-making systems. We have also included a brief survey on connectionist expert systems (without incorporating fuzzy sets) for the convenience of the reader. The use of knowledge-based networks has been discussed as one of the latest entrants in this field. A comparative study of the various methodologies has been provided in tabular form. Recent attempts at using rough sets for knowledge encoding and genetic algorithms for finding optimal net parameters have also been mentioned.

REFERENCES

[1] *Proceedings of the Third Workshop on Rough Sets and Soft Computing* (San Jose), 1994.
[2] *Proceedings of the Fourth International Conference on Soft Computing* (Iizuka, Japan), 1996.
[3] S. K. Pal and N. R. Pal. Soft computing: goals, tools and feasibility. *J. Inst. Electron. Telecomm. Engineers* 42:195–204, 1996.
[4] L. A. Zadeh. Fuzzy logic, neural networks, and soft computing. *Comm. ACM* 37:77–84, 1994.
[5] J. C. Bezdek and S. K. Pal, Eds. *Fuzzy Models for Pattern Recognition: Methods that Search for Structures in Data.* IEEE Press, New York, 1992.
[6] *Proceedings of the IEEE International Conference on Fuzzy Systems*, 1996.
[7] L. A. Zadeh. Fuzzy sets. *Inform. Control* 8:338–353, 1965.
[8] G. J. Klir and B. Yuan. *Fuzzy Sets and Fuzzy Logic: Theory and Applications.* Prentice–Hall, Englewood Cliffs, NJ, 1995.

[9] D. E. Rumelhart and J. L. McClelland, Eds. *Parallel Distributed Processing: Explorations in the Microstructures of Cognition*, Vol. 1. MIT Press, Cambridge, MA, 1986.

[10] R. P. Lippmann. An introduction to computing with neural nets. *IEEE Acoustics Speech Signal Process. Mag.* 4:4–22, 1987.

[11] T. Kohonen. *Self-Organization and Associative Memory*. Springer-Verlag, Berlin, 1989.

[12] J. Hertz, A. Krogh, and R. G. Palmer. *Introduction to the Theory of Neural Computation*. Addison–Wesley, Reading, MA, 1994.

[13] H. Muhlenbein. Limitations of multi-layer perceptron networks—step towards genetic neural networks. *Parallel Comput.* 14:249–260, 1990.

[14] S. K. Pal and D. Bhandari. Selection of optimum set of weights in a layered network using genetic algorithms. *Inform. Sci.* 80:213–234, 1994.

[15] A. Homaifar and E. McCormick. Simultaneous design of membership functions and rule sets for fuzzy controllers using genetic algorithms. *IEEE Trans. Fuzzy Systems* 3:129–139, 1995.

[16] S. K. Pal and D. Bhandari. Genetic algorithms with fuzzy fitness function for object extraction using cellular neural networks. *Fuzzy Sets Systems* 65:129–139, 1994.

[17] M. Banerjee and S. K. Pal. Roughness of a fuzzy set. *Inform. Sci.* 93:235–246, 1996.

[18] M. Banerjee, S. Mitra, and S. K. Pal. Rough fuzzy MLP: knowledge encoding and classification. *IEEE Trans. Neural Networks*, to appear.

[19] D. E. Goldberg. *Genetic Algorithms in Search, Optimization and Machine Learning*. Addison–Wesley, Reading, MA, 1989.

[20] Z. Michalewicz. *Genetic Algorithms + Data Structures = Evolutionary Programs*. Springer-Verlag, Berlin, 1994.

[21] Z. Pawlak. *Rough Sets, Theoretical Aspects of Reasoning about Data*. Kluwer Academic, Dordrecht, 1991.

[22] R. Slowiński, Ed. *Intelligent Decision Support, Handbook of Applications and Advances of the Rough Sets Theory*. Kluwer Academic, Dordrecht, 1992.

[23] H.-J. Zimmermann. *Fuzzy Sets, Decision Making and Expert Systems*. Kluwer Academic, Boston, 1987.

[24] L. A. Zadeh. The role of fuzzy logic in the management of uncertainty in expert systems. *Fuzzy Sets Systems* 11:199–227, 1983.

[25] A. Kandel, Ed. *Fuzzy Expert Systems*. CRC Press, Boca Raton, 1991.

[26] S. I. Gallant. Connectionist expert systems. *Comm. ACM* 31:152–169, 1988.

[27] L. M. Fu. Knowledge-based connectionism for revising domain theories. *IEEE Trans. Systems Man Cybernet.* 23:173–182, 1993.

[28] G. G. Towell and J. W. Shavlik. Knowledge-based artificial neural networks. *Artificial Intell.* 70:119–165, 1994.

[29] F. Hayes-Roth, D. A. Waterman, and D. B. Lenat. *Building Expert Systems*. Addison–Wesley, London, 1983.

[30] M. Minsky. A framework for representing knowledge. In *The Psychology of Computer Vision* (P. Winston, Ed.). McGraw–Hill, New York, 1975.

[31] R. Quillian. Semantic memory. In *Semantic Information Processing* (M. Minsky, Ed.). MIT Press, Cambridge, MA, 1968.

[32] J. Pearl. Distributed revision of composite beliefs. *Artificial Intell.* 33:173–215, 1987.

[33] L. A. Zadeh. The concept of a linguistic variable and its application to approximate reasoning, 1, 2, and 3. *Inform. Sci.* 8, 8, 9:199–249, 301–357, 43–80, 1975.

[34] M. M. Gupta, A. Kandel, W. Bandler, and J. B. Kiszka, Eds. *Approximate Reasoning in Expert Systems*. North-Holland, Amsterdam, 1985.

[35] S. K. Pal and D. Dutta Majumder. *Fuzzy Mathematical Approach to Pattern Recognition*. Wiley (Halsted Press), New York, 1986.

[36] I. B. Turksen and Z. Zhong. An approximate analogical reasoning schema based on similarity measures and interval-valued fuzzy sets. *Fuzzy Sets Systems* 34:323–346, 1990.

[37] M. Ishizuka, K. S. Fu, and J. T. P. Yao. Inference procedures under uncertainty for the problem-reduction method. *Inform. Sci.* 28:179–206, 1982.

[38] A. O. Esogbue and R. C. Elder. Fuzzy sets and the modelling of physician decision processes, I: The initial interview – information gathering session. *Fuzzy Sets Systems* 2:279–291, 1979.

[39] E. Sanchez and R. Bartolin. Fuzzy inference and medical diagnosis, a case study. *Biomedical Fuzzy Systems Bull.* 1:4–21, 1990.

[40] L. X. Wang and J. M. Mendel. Generating fuzzy rules by learning from examples. *IEEE Trans. Systems Man Cybernet.* 22:1414–1427, 1992.

[41] S. Mitra and S. K. Pal. Fuzzy multi-layer perceptron, inferencing and rule generation. *IEEE Trans. Neural Networks* 6:51–63, 1995.

[42] B. Kosko. *Neural Networks and Fuzzy Systems*. Prentice–Hall, Englewood Cliffs, NJ, 1991.

[43] H. Takagi. Fusion technology of fuzzy theory and neural network—survey and future directions. In *Proceedings of the 1990 International Conference on Fuzzy Logic and Neural Networks* (Iizuka, Japan) pp. 13–26, 1990.

[44] Y. H. Pao. *Adaptive Pattern Recognition and Neural Networks*. Addison–Wesley, Reading, MA, 1989.

[45] M. M. Gupta and D. H. Rao. On the principles of fuzzy neural networks. *Fuzzy Sets Systems* 3:1–18, 1994.

[46] J. J. Buckley and Y. Hayashi. Fuzzy neural networks: a survey. *Fuzzy Sets Systems* 3:1–13, 1994.

[47] S. K. Pal and S. Mitra. Multi-layer perceptron, fuzzy sets and classification. *IEEE Trans. Neural Networks* 3:683–697, 1992.

[48] S. K. Pal and D. P. Mandal. Linguistic recognition system based on approximate reasoning. *Inform. Sci.* 61:135–161, 1992.

[49] A. Skowron and C. Rauszer. The discernibility matrices and functions in information systems. In *Intelligent Decision Support, Handbook of Applications and Advances of the Rough Sets Theory* (R. Slowiński, Ed.), pp. 331–362. Kluwer Academic, Dordrecht, 1992.

[50] K. Saito and R. Nakano. Medical diagnostic expert system based on PDP model. In *Proceedings of the IEEE International Conference on Neural Networks* (San Diego), pp. I.255–I.262, 1988.

[51] R. C. Lacher, S. I. Hruska, and D. C. Kuncicky. Back-propagation learning in expert networks. *IEEE Trans. Neural Networks* 3:62–72, 1992.

[52] R. Poli, S. Cagnoni, R. Livi, G. Coppini, and G. Valli. A neural network expert system for diagnosing and treating hypertension. *IEEE Computer* 64–71, 1991.

[53] M. Ishikawa. Structural learning with forgetting. *Neural Networks* 9:509–521, 1996.

[54] C. W. Omlin and C. Lee Giles. Extraction of rules from discrete-time recurrent neural networks. *Neural Networks* 9:41–52, 1996.

[55] S. Mitra, S. N. Sarbadhikari, and S. K. Pal. An MLP-based model for identifying qEEG in depression. *Internat. J. Biomedical Comput.* 43:179–187, 1996.

[56] H. F. Yin and P. Liang. A connectionist incremental expert system combining production systems and associative memory. *Internat. J. Pattern Recognition Artificial Intell.* 5:523–544, 1991.

[57] D. W. Opitz and J. W. Shavlik. Heuristically expanding knowledge-based neural networks. In *Proceedings of the 13th International Joint Conference on Artificial Intelligence* (Chambery, France), pp. 1360–1365, 1993.

[58] G. G. Towell and J. W. Shavlik. Extracting refined rules from knowledge-based neural networks. *Machine Learning* 13:71–101, 1993.

[59] S. K. Pal and A. Ghosh. Neuro-fuzzy image processing: relevance and feasibility. In *Neural and Fuzzy Systems: The Emerging Science of Intelligence and Computing* (S. Mitra, W. Kraske, and M. M. Gupta, Eds.). SPIE Press, New York, 1993.

[60] J. K. Keller and D. J. Hunt. Incorporating fuzzy membership functions into the perceptron algorithm. *IEEE Trans. Pattern Anal. Machine Intell.* 7:693–699, 1985.
[61] E. Sanchez. Fuzzy connectionist expert systems. In *Proceedings of the 1990 International Conference on Fuzzy Logic and Neural Networks* (Iizuka, Japan), pp. 31–35, 1990.
[62] T. L. Huntsberger and P. Ajjimarangsee. Parallel self-organizing feature maps for unsupervised pattern recognition. *Internat. J. General Systems* 16:357–372, 1990
[63] J. C. Bezdek, E. C. Tsao, and N. R. Pal. Fuzzy Kohonen clustering networks. In *Proceedings of the First IEEE International Conference on Fuzzy Systems* (San Diego), pp. 1035–1043, 1992.
[64] J. M. Keller, R. Krishnapuram, and F. C.-H. Rhee. *Evidence aggregation networks for fuzzy logic inference. IEEE Trans. Neural Networks* 3:761–769, 1992.
[65] W. Pedrycz. Fuzzy neural networks with reference neurons as pattern classifiers. *IEEE Trans. Neural Networks* 3:770–775, 1992.
[66] S. C. Lee and E. T. Lee. Fuzzy neural networks. *Math. Biosci.* 23:151–177, 1975.
[67] A. Ghosh, N. R. Pal, and S. K. Pal. Self-organization for object extraction using multilayer neural network and fuzziness measures. *IEEE Trans. Fuzzy Systems* 1:54–68, 1993.
[68] Y. Hayashi. Neural expert system using fuzzy teaching input and its application to medical diagnosis. *Inform. Sci. Appl.* 1:47–58, 1994.
[69] D. L. Hudson, M. E. Cohen, and M. F. Anderson. Use of neural network techniques in a medical expert system. *Internat. J. Intell. Systems* 6:213–223, 1991.
[70] S. G. Romaniuk and L. O. Hall. Decision making on creditworthiness, using a fuzzy connectionist model. *Fuzzy Sets Systems* 48:15–22, 1992.
[71] K. Yoshida, Y. Hayashi, A. Imura, and N. Shimada. Fuzzy neural expert system for diagnosing hepatobiliary disorders. In *Proceedings of the 1990 International Conference on Fuzzy Logic and Neural Networks* (Iizuka, Japan), pp. 539–543, 1990.
[72] F. C. H. Rhee and R. Krishnapuram. Fuzzy rule generation methods for high-level computer vision. *Fuzzy Sets Systems* 60:245–258, 1993.
[73] S. Mitra. Fuzzy MLP based expert system for medical diagnosis. *Fuzzy Sets Systems* 65:285–296, 1994.
[74] S. Mitra and S. K. Pal. Fuzzy self organization, inferencing and rule generation. *IEEE Trans. Systems Man Cybernet.* 26:608–620, 1996.
[75] S. Mitra and S. K. Pal. Logical operation based fuzzy MLP for classification and rule generation. *Neural Networks* 7:353–373, 1994.
[76] J. M. Keller, R. R. Yager, and H. Tahani. Neural network implementation of fuzzy logic. *Fuzzy Sets Systems* 45:1–12, 1992.
[77] H. Ishibuchi, R. Fujioka, and H. Tanaka. Neural networks that learn from fuzzy if–then rules. *IEEE Trans. Fuzzy Systems* 1:85–97, 1993.
[78] R. Masuoka, N. Watanabe, A. Kawamura, Y. Owada, and K. Asakawa. Neuro-fuzzy system—fuzzy inference using a structured neural network. In *Proceedings of the 1990 International Conference on Fuzzy Logic and Neural Networks* (Iizuka, Japan), pp. 173–177, 1990.
[79] N. K. Kasabov. Adaptable neuro production systems. *Neurocomputing* 13:95–117, 1996.
[80] B. Kosko. Hidden patterns in combined and adaptive knowledge networks. *Internat. J. Approx. Reasoning* 2:377–393, 1988.
[81] R. J. Machado and A. F. Rocha. A hybrid architecture for connectionist expert systems. In *Intelligent Hybrid Systems* (A. Kandel and G. Langholz, Eds.). CRC Press, Boca Raton, 1992.
[82] W. Pedrycz and A. F. Rocha. Fuzzy-set based models of neurons and knowledge-based networks. *IEEE Trans. Fuzzy Systems* 1:254–266, 1993.
[83] K. Hirota and W. Pedrycz. Knowledge-based networks in classification problems. *Fuzzy Sets Systems* 59:271–279, 1993.
[84] B. F. Leao and A. F. Rocha. Proposed methodology for knowledge acquisition: a study on congenital heart disease diagnosis. *Methods Inform. Medicine* 29:30–40, 1990.

[85] S. Mitra, R. K. De, and S. K. Pal. Knowledge-based fuzzy MLP for classification and rule generation. *IEEE Trans. Neural Networks*, to appear.

[86] R. Yasdi. Combining rough sets learning and neural learning method to deal with uncertain and imprecise information. *Neurocomputing* 7:61–84, 1995.

[87] A. Czyzewski and A. Kaczmarek. Speech recognition systems based on rough sets and neural networks. In *Proceedings of the Third Workshop on Rough Sets and Soft Computing* (San Jose), pp. 97–100, 1994.

[88] S. K. Pal and P. P. Wang, Eds. *Genetic Algorithms for Pattern Recognition*. CRC Press, Boca Raton, 1996.

[89] V. Maniezzo. Genetic evolution of the topology and weight distribution of neural networks. *IEEE Trans. Neural Networks* 5:39–53, 1994.

[90] D. W. Opitz and J. W. Shavlik. Using genetic search to refine knowledge-based neural networks. In *Machine Learning: Proceedings of the 11th International Conference* (San Francisco), 1994.

Mean-Value-Based Functional Reasoning Techniques in the Development of Fuzzy Neural Network Control Systems

Keigo Watanabe

Faculty of Science and Engineering
Department of Mechanical Engineering
Saga University
1-Honjo-machi, Saga 840, Japan

Spyros G. Tzafestas

Department of Electrical and
Computer Engineering
Intelligent Robotics and
Automation Laboratory
National Technical University of Athens
Zografou, Athens 157 73, Greece

I. INTRODUCTION

Functional reasoning [1, 2] and simplified reasoning [3, 4], which are special cases of the so-called min-max-centroidal method [5, 6], have been proposed as fuzzy reasoning methods for treating fuzzy control and fuzzy modeling problems. These methods have the advantage that the fuzzy operation is simplified, because instead of using a membership function the conclusion part can be composed of a function of input data or be simply a constant value. However, it is not readily known to control engineers how to rationally design some parameters in the conclusion part using available control theories. One can refer either to a basic method [3, 4] in which a constant parameter is determined as a value on the support set when the membership function in the conclusion is assumed to be a singleton or to a method [7] in which a constant parameter is determined by averaging the mean

Fuzzy Logic and Expert Systems Applications

values for the membership functions in the antecedent with respect to the number of input data and which is finely adjusted for every control rule. However, it is unclear whether these methods are truly valid or not. As a more systematic design approach to the conventional fuzzy controller, a fuzzy model-based approach [8–11] and a variable structure system (VSS) or sliding mode approach [12–14] have been proposed.

As an alternative approach to the design of fuzzy controllers, neuro-fuzzy controllers (NFCs) or fuzzy neural network controllers (FNNCs) [15–24] are intensively studied, in which a fuzzy reasoning method such as those discussed previously is realized within a multilayered hierarchical neural network and the parameters that are represented by connection weights or involved in unit functions can be learned by using the actual data. The number of learning parameters or trials and errors for learning fuzzy control can be effectively reduced, if we can rationally design some parameters in the conclusion in advance.

In this chapter, we start by reviewing the conventional functional reasoning [1, 2] and simplified reasoning [3, 4] methods. Then, as a new reasoning method, we further introduce a mean-value-based functional reasoning [25, 26], in which the conclusion part consists of the mean values of the membership functions that are assigned to each input data. It is shown that if the conclusion is regarded as a VSS controller [27], then any parameter in the conclusion of the conventional functional reasoning can be rationally designed as the parameters for constructing r stable switching planes (or lines), while those of the mean-value-based functional reasoning can be designed as the parameters for constructing only one stable switching plane (or line). Here, r denotes the number of control rules. Next we describe a fuzzy Gaussian neural network (FGNN) [28–30] by applying the preceding fuzzy reasoning methods. It is then clarified that by using the mean-value-based functional reasoning the FGNN allows the number of learning parameters in the conclusion to be reduced drastically, compared with those of the conventional functional reasoning and simplified reasoning.

The chapter is organized as follows. The conventional functional reasoning, simplified reasoning, and mean-value-based functional reasoning methods are reviewed in Section II. In Section III, a design method for the conclusion of the preceding fuzzy reasoning schemes, based on VSS control theory, is described. In Section IV, three FGNNs are constructed using the conventional functional reasoning, simplified reasoning, and mean-value-based functional reasoning methods, and compared with each other, especially with regard to the number of learning parameters to be learned in the conclusion. In Section V, the effectiveness of the mean-value-based functional reasoning method is illustrated by designing and simulating a nonlearning fuzzy controller for controlling the attitude of a satellite. Finally, in Section VI, a fuzzy neural network controller based on mean-value-based functional reasoning is applied to the tracking control problem for a mobile robot with two independent driving wheels.

II. FUZZY REASONING SCHEMES

In this section, we review the conventional, the simplified, and the mean-value-based functional reasoning schemes.

A. INPUT-DATA-BASED FUNCTIONAL REASONING

The conventional or input-data-based functional reasoning [1, 2] method (see Fig. 1) is also called Sugeno's fuzzy reasoning method [3]. For n input variables $(x_1, \ldots, x_n)$ and p output variables $(u_1, \ldots, u_p)$ in the consequent part, the ith control rule R_i is described by

$$R_i: \text{If } x_1 = A_{i1} \text{ and} \cdots \text{and } x_n = A_{in},$$
$$\text{then } u_1 = f_{i1}(x_1, \ldots, x_n) \text{ and } \cdots \text{ and } u_p = f_{ip}(x_1, \ldots, x_n), \quad (1)$$

where A_{ij} denotes the fuzzy set in the antecedent associated with the jth input variable in the ith control rule, and $f_{ij}(x_1, \ldots, x_n)$ is the function associated with the jth variable in the conclusion of the ith control rule. Applying n confidences $\mu_{A_{i1}}(x_1), \ldots, \mu_{A_{in}}(x_n)$, the confidence in the antecedent h_i is, by definition, given by

$$h_i = \mu_{A_{i1}}(x_1) \cdot \mu_{A_{i2}}(x_2) \cdot \cdots \cdot \mu_{A_{in}}(x_n), \quad (2)$$

where "$\cdot$" denotes the algebraic product operation. Then the jth output consequent can be calculated as the following weighted mean of $f_{ij}(\cdot)$ with respect to the weight h_i:

$$u_j^* = \frac{\sum_{i=1}^r h_i f_{ij}(x_1, \ldots, x_n)}{\sum_{i=1}^r h_i}, \qquad j = 1, \ldots, p, \quad (3)$$

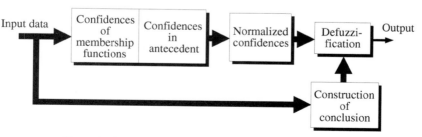

Figure 1   Concept of input-data-based functional fuzzy reasoning.

where r denotes the total number of control rules; if the number of membership functions (i.e., the number of labels) in the antecedent is ℓ, then, in general, $r = \ell^n$.

Note here that, in the single consequent case, the conclusion function is usually represented by a *linear* function:

$$u_1 = a_{0i} + a_{1i}x_1 + a_{2i}x_2 + \cdots + a_{ni}x_n. \tag{4}$$

As a special case of this representation, one can use the actual deviations of the membership functions in the antecedent as [23]:

$$u_1 = a_{0i} + a_{1i}(x_1 - c_{1i}) + \cdots + a_{ni}(x_n - c_{ni}), \tag{5}$$

where c_{ji} denotes the center value (e.g., the mean value of a Gaussian-like membership function) associated with the jth membership function in the antecedent of the ith control rule.

B. SIMPLIFIED REASONING

A further special case of input-data-based functional reasoning, called simplified reasoning [3, 4] (see Fig. 2), is based on the formula:

$$u_1 = a_{0i}. \tag{6}$$

The design parameters a_{ji} in the conclusion, as well as the scalers for the input data, significantly govern the performance of the unlearning fuzzy controller. An effective method for designing this parameter is not known at present. There exists an elementary method [7] in which the parameters are determined by averaging the mean values c_{ji} of the membership functions in the antecedent, with respect to the number of input data, and they are finely adjusted for every control rule by means of trial and error. In what follows, a more general design method for

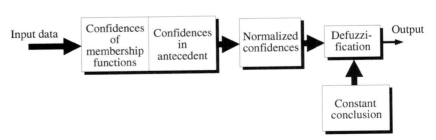

Figure 2   Concept of simplified fuzzy reasoning.

the parameter in the conclusion will be presented, which includes this elementary method as a special case.

C. MEAN-VALUE-BASED FUNCTIONAL REASONING

The confidence $\mu_{A_{ij}}(x_j)$ of the membership function indicates the degree of matching between the actual input data x_j and the hypothetical data distribution (membership function) on the support set $[-L, L]$ allocated by the control designer. It is then observed that the conclusion of the input-data-based functional reasoning given by (4) or of the simplified reasoning given by (6) does not include the allocation information for the membership function in the antecedent. For this allocation information, a mean-value-based functional reasoning has been proposed [25, 26], in which the mean values c_{ji} of each membership function in the antecedent are used in the conclusion as shown in Fig. 3. In this reasoning, the conclusion function in (1) is replaced by

$$u_1 = f_{i1}(c_{1i}, \ldots, c_{ni}) \text{ and } \cdots \text{ and } u_p = f_{ip}(c_{1i}, \ldots, c_{ni}) \tag{7}$$

and the output consequent is determined by

$$u_j^* = \frac{\sum_{i=1}^r h_i f_{ij}(c_{1i}, \ldots, c_{ni})}{\sum_{i=1}^r h_i}, \qquad j = 1, \ldots, p. \tag{8}$$

For the two-input data case (x_1, x_2), the output u_1 in a linear function can be represented by

$$u_1 = a_{0i} + a_{1i}c_{1i} + a_{2i}c_{2i} \tag{9}$$

or by the simpler form

$$u_1 = a_0 + a_1 c_{1i} + a_2 c_{2i} \tag{10}$$

considering that the mean-values c_{ji} are already dependent on the control rule.

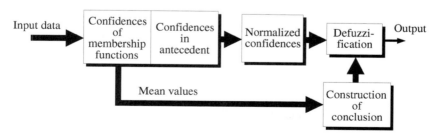

Figure 3   Concept of mean-value-based functional fuzzy reasoning.

III. DESIGN OF THE CONCLUSION PART IN FUNCTIONAL REASONING

In this section, a design technique for the conclusion of the three fuzzy reasoning methods stated previously through VSS control theory is described.

A. INPUT-DATA-BASED FUNCTIONAL REASONING

For a linear function of input-data-based functional reasoning with $a_{0i} \equiv 0$, the following hyperplane is defined:

$$\sigma_i \stackrel{\triangle}{=} S_i \mathbf{x}, \qquad i = 1, \ldots, r, \tag{11}$$

where $\mathbf{x}^T \stackrel{\triangle}{=} [x_1, x_2, \ldots, x_n]$ is regarded as a state vector and S_i denotes a $1 \times n$ design row vector that depends on the index i of the control rules. Now, consider the following VSS control law [27]:

$$u_1 = k_i \, \mathrm{sgn}(\sigma_i), \qquad k_i > 0. \tag{12}$$

Then taking a switching gain as $k_i = |\sigma_i|$ gives

$$u_1 = \sigma_i \tag{13}$$

because $|\sigma_i| \, \mathrm{sgn}(\sigma_i) = \sigma_i$. Thus, if the input data in the antecedent are regarded as state vectors, the design problem of the conclusion for input-data-based functional reasoning can be reduced to that of r stable switching planes (or lines) in (11).

B. MEAN-VALUE-BASED FUNCTIONAL REASONING

For a linear function of mean-value-based functional reasoning with $a_0 \equiv 0$ in the form of (10), we define the following switching plane:

$$\sigma_i \stackrel{\triangle}{=} S \mathbf{c}_i, \qquad i = 1, \ldots, r, \tag{14}$$

where $\mathbf{c}_i^T \stackrel{\triangle}{=} [c_{1i}, c_{2i}, \ldots, c_{ni}]$ is regarded as a state vector and S denotes a $1 \times n$ design row vector which is independent of the index i of the control rules. If we consider the same control law as used previously, then we have the same result as in (13). Note, however, that the design vector S and the mean-value vector $\mathbf{c}_i$ are constant, and hence $\lim_{\sigma_i \to 0} \sigma_i \dot{\sigma}_i \equiv 0$, because $\dot{\sigma}_i \equiv 0$. This implies that (12) and (14) do not satisfy the condition $\lim_{\sigma_i \to 0} \sigma_i \dot{\sigma}_i < 0$, which is required for the existence of a sliding mode at the neighborhood of the switching plane $\sigma_i = 0$. Therefore, Eqs. (12) and (14) do not generate an actual sliding mode as in a usual sliding mode control system, and they can be only formally regarded as VSS controllers.

Thus, it is seen that the design problem for the conclusion of mean-value-based functional reasoning can be reduced to that of only one stable switching plane (or line) in (14), if the mean value c_i on the membership functions in the antecedent is regarded as a state vector. The conclusion part of mean-value-based functional reasoning is completely constant, whereas that of input-data-based functional reasoning is time-varying and all r switching planes (or lines) in (11) must be designed to be stable; this yields an unrealistic design procedure if the number of control rules is very large. Furthermore, it is easy to see that the conclusion part of the simplified reasoning can also be directly replaced by the σ_i value which is computed from (14) off-line.

Because mean-value-based functional reasoning assumes that the mean values of the membership functions in the antecedent are utilized in the conclusion, the form of the membership function must be of an isosceles triangle, Gaussian-type, or an isosceles trapezoid, etc., which possess the information of the mean value. Therefore, in such a functional reasoning, the knowledge of an expert cannot necessarily be reflected in the determination of the form of the membership function in the antecedent, whereas, in traditional functional or simplified reasoning, a membership function of any form can be selected to reflect the knowledge of an expert in the form determination of the membership function in the antecedent.

IV. FUZZY GAUSSIAN NEURAL NETWORKS

In this section, three fuzzy Gaussian neural networks (FGNNs) are constructed using the reasonings stated previously, and compared with each other, especially in the number of learning parameters to be learned in the conclusion.

A. CONSTRUCTION

Figures 4–6 illustrate three FGNNs based on input-data-based fuzzy reasoning, simplified fuzzy reasoning, and mean-value-based fuzzy reasoning, respectively. Here, it is assumed that there are two inputs (x_1, x_2), a single output (u_1^*), and three labels for a Gaussian membership function in the antecedent part. Then the number of identifiable control rules is $r = 3^2$.

The variable within the curly brackets denotes a signal passing through the neural network, the circle symbol is the unit, w_{cj}^i is the connection weight that represents the center value for the jth Gaussian membership function of the ith input data, and the connection weight w_{dj}^i denotes the reciprocal value of the deviation from the center w_{cj}^i to which the jth Gaussian function of ith input data

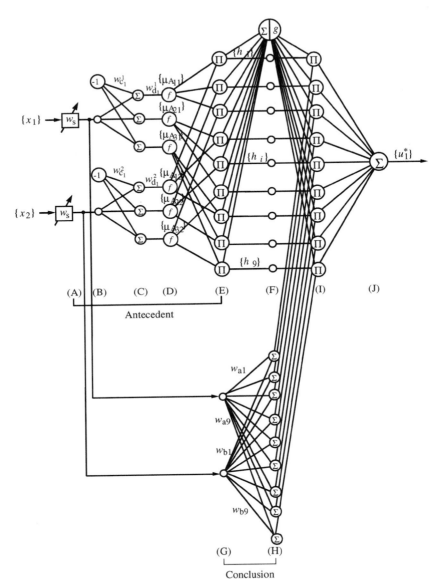

Figure 4 Fuzzy Gaussian neural network using input-data-based functional reasoning. Reprinted from K. Watanabe *et al.*, Fuzzy-neural network controllers using mean-value-based functional reasoning, *Neurocomputing* 9:39–61, 1995, with kind permission of Elsevier Science-NL, Sara Burgerhartstraat 25, 1055 KV Amsterdam, The Netherlands.

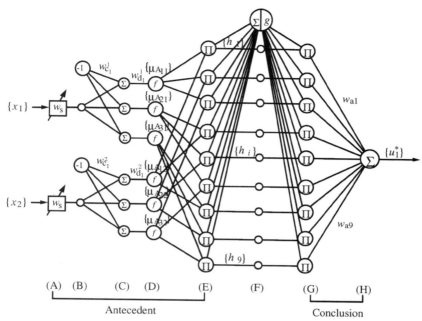

Figure 5 Fuzzy Gaussian neural network using simplified reasoning. Reprinted from K. Watanabe *et al.*, Fuzzy-neural network controllers using mean-value-based functional reasoning, *Neurocomputing* 9:39–61, 1995, with kind permission of Elsevier Science-NL, Sara Burgerhartstraat 25, 1055 KV Amsterdam, The Netherlands.

on the standardized support set has value 0.5. In addition, the unit with the symbol -1 generates the output of -1; the unit with the symbol $\sum$ outputs the summation of the inputs. Similarly, the unit with the symbol $\prod$ outputs the product of the inputs. The input–output relation at the unit with the symbol f is defined by the following Gaussian function:

$$f(x) = \exp\left(\ln(0.5) \cdot x^2\right) \qquad (15)$$

as a unit function. Furthermore, the unit with no symbols simply distributes the input to the output.

Layers $A–E$ in Figs. 4–6 correspond to the antecedent part of the fuzzy control rule, and layers G and H correspond to the conclusion part. The inputs x_i applied to layer A are scaled by using an adaptive input scaling technique [31]. At layer C, the connection weight $-w_{cj}^i$, which is a bias, is added to the scaled input, and it is multiplied by w_{dj}^i, which is an input to the Gaussian function at layer D. At layer E in all figures, we obtain the confidences h_i in the antecedent part for every

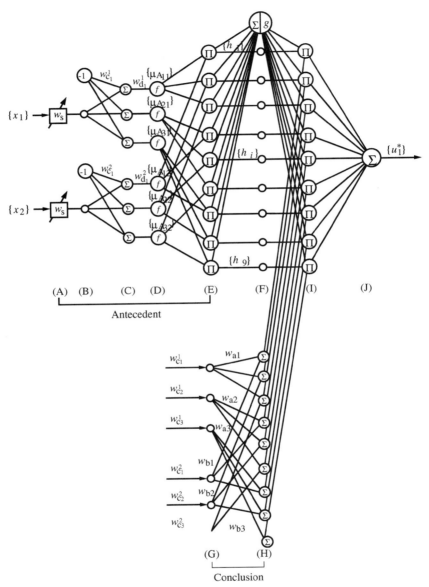

Figure 6 Fuzzy Gaussian neural network using mean-value-based functional reasoning. Reprinted from K. Watanabe *et al.*, Fuzzy-neural network controllers using mean-value-based functional reasoning, *Neurocomputing* 9:39–61, 1995, with kind permission of Elsevier Science-NL, Sara Burgerhartstraat 25, 1055 KV Amsterdam, The Netherlands.

control rule. At the first unit of layer F, a summation of inputs and the inverse calculation are performed. That is, the unit that has symbols $\sum$ and g generates the output through the following function:

$$g(x) = \frac{1}{x}, \qquad (16)$$

with a linear summed input.

At layers G and H in Fig. 5, we directly obtain the consequent as the weighted mean of w_{ai} with respect to the weight h_i, where w_{ai} denotes a constant parameter a_{0i} in the conclusion. As seen from Figs. 4 and 6, for input-data-based or mean-value-based functional reasoning, the conclusion part is calculated at these layers (G and H), and we finally have the similar consequent at layers I and J.

Note here that Horikawa *et al.* [15] have already developed some FNNs similar to those shown in Figs. 4 and 5 by using a sigmoidal function with range $[0, 1]$. However, to construct a pseudo-trapezoidal membership function in their approach, two sigmoidal functions with ranges $[0, 1]$ and $[-1, 0]$ must be superimposed. Therefore, their FNNs require additional intermediate layers to generate the membership function in the antecedent, because a pseudo-trapezoidal membership function is constructed by summing two sigmoidal functions with different signs. This also causes the number of units at the corresponding intermediate layer to grow as the number of fuzzy labels becomes larger. On the contrary, our approach does not suffer from this problem and also gives a reduced number of learning parameters in the conclusion, as will be discussed in the next subsection.

B. NUMBER OF LEARNING PARAMETERS

All of the preceding FGNNs have the same number of parameters to be learned in the antecedent. However, they have different numbers of parameters to be learned in the conclusion. By introducing two kinds of parameters α and β, Eq. (11) can be rewritten as

$$\sigma_i = \frac{1}{\beta_i} [\alpha_i \quad 1] \begin{bmatrix} x_1 \\ x_2 \end{bmatrix}, \qquad i = 1, \ldots, r, \qquad (17)$$

where α_i is the slope of a switching line and β_i is usually an averaging constant with respect to the number of inputs. Following this construction, the input-data-based functional reasoning approach must learn 49 w_{ai} and 49 w_{bi} parameters, if seven labels for each input are used. Here, $w_{ai} = \alpha_i / \beta_i$ and $w_{bi} = 1 / \beta_i$. Furthermore, if the parameter β_i is regarded as the averaging parameter with respect to the number of inputs, then it is required to learn 49 parameters w_{ai}, because we can fix the parameter β_i as 2, that is, $w_{bi} = 0.5$, $i = 1, \ldots, 49$. At this stage, both the input-data-based functional reasoning and the simplified reasoning methods have the same number of learning parameters to be learned in the conclusion. Note also that, to obtain all of the different learned parameters w_{ai} for $i = 1, \ldots, 49$, one must set all different initial parameters for w_{ai}, $i = 1, \ldots, 49$, because all units at layer H have the same "delta" quantities in the back-propagation algorithm.

Similarly, by introducing two kinds of parameters α and β in the mean-value-based functional reasoning, Eq. (14) can be rewritten as

$$\sigma_i = \frac{1}{\beta}[\alpha \quad 1]\begin{bmatrix} w_{cj}^1 \\ w_{ck}^2 \end{bmatrix}, \qquad j, k = 1, \ldots, 7. \tag{18}$$

Note here that both parameters α and β are independent of the index i of the control rules. Therefore, to learn their parameters, it is sufficient to learn the minimum number of w_{aj} and w_{bk} by using w_{cj}^1, $j = 1, \ldots, 7$, w_{ck}^2, $k = 1, \ldots, 7$, and δ_H. Here, $w_{aj} = \alpha/\beta$, $w_{bk} = 1/\beta$, and δ_H denotes the "delta" quantity for any unit in layer H. As shown in Fig. 7, the connection weights w_{aj} and w_{bj} between two jth units with respect to w_{cj}^1 and w_{cj}^2 at layer G and the lth unit at layer H, where $l = (j-1)*7+j$, $j = 1, \ldots, 7$, can be concretely learned. Note that the remaining unlearned connection weights, which have the same w_{cj}^1 and w_{cj}^2 as the learned connection weights, should be simply replaced by the learned ones: w_{aj} and w_{bj}.

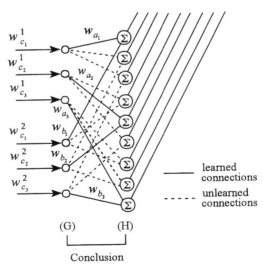

learned connections

----- unlearned connections

(G)   (H)

Conclusion

Figure 7   Learning method of parameters in the conclusion for mean-value-based functional reasoning. Reprinted from K. Watanabe *et al.*, Fuzzy-neural network controllers using mean-value-based functional reasoning, *Neurocomputing* 9:39–61, 1995, with kind permission of Elsevier Science-NL, Sara Burgerhartstraat 25, 1055 KV Amsterdam, The Netherlands.

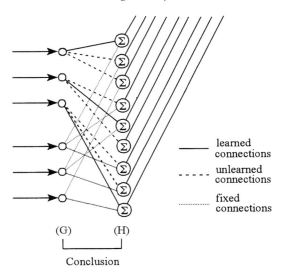

(G)   (H)

└──────────┘

Conclusion

Figure 8   Learning method of parameters in the conclusion for mean-value-based functional reasoning (there are partially fixed connections). Reprinted from K. Watanabe *et al.*, Fuzzy-neural network controllers using mean-value-based functional reasoning, *Neurocomputing* 9:39–61, 1995, with kind permission of Elsevier Science-NL, Sara Burgerhartstraat 25, 1055 KV Amsterdam, The Netherlands.

Thus, the mean-value-based functional reasoning model learns 14 parameters (w_{aj}, w_{bk}), $j, k = 1, \ldots, 7$. Furthermore, if the parameter β is fixed as an averaging parameter with respect to the number of inputs, then only seven parameters w_{aj}, $j = 1, \ldots, 7$, are learned (see Fig. 8). Consequently, the mean-value-based functional approach can drastically reduce the number of parameters that have to be learned in the conclusion.

C. TRAINING

By applying the back-propagation algorithm [32–34], one can train the connection weights of the FGNN controller, and consequently identify the control rules and finely tune the membership functions in the antecedent part. In the following, a case based on the specialized learning architecture will be explained, in which the FGNN is trained so that the output deviations of the plant are minimized without using the pattern data generated by an expert. However, note that we merely change the delta quantities in the output layer for a case based on the generalized learning architecture or the feedback error learning architecture.

Consider the hierarchical multilayered neural network consisting of M layers, and denote the input–output relation of any unit by $f(\cdot)$, the input to the jth unit at the kth layer by i_j^k, and the corresponding output from its unit by o_j^k. The weight that connects the jth unit at the kth layer and the lth unit at the $(k+1)$th layer is denoted by $w_{j,l}^{k,k+1}$.

Let Case A denote the case when the input to the kth layer is the output through the function $f(\cdot)$ and the input to the $(k+1)$th layer is calculated by the summation (i.e., $\sum$) operation. Similarly, let Case B be the case when the input to the kth layer is output through the function $f(\cdot)$ and the input to the $(k+1)$th layer is calculated by the product (i.e., $\prod$) operation.

Under the preceding conditions, we have the following input–output relation of a unit:

$$i_l^{k+1} = \sum_j w_{jl}^{k,k+1} o_j^k, \qquad o_l^{k+1} = f(i_l^{k+1}) \tag{19}$$

for Case A, and

$$i_l^{k+1} = \prod_j w_{jl}^{k,k+1} o_j^k, \qquad o_l^{k+1} = f(i_l^{k+1}) \tag{20}$$

for Case B.

For the specialized learning, the following cost function is considered:

$$J = \tfrac{1}{2} \sum_{i=1}^{m} (y_{di} - y_i)^2, \tag{21}$$

which gives the weights $w_{ij}^{k,k+1}$ that minimize J. Here, m denotes the number of plant outputs, y_{di} the ith desired reference, and y_i the ith output of the plant. Then the delta quantities, δ_j^M in the jth unit at the output layer M and δ_j^k in the jth unit at any intermediate layer k, are given by

$$\text{Output layer:} \qquad \delta_j^M = f'(i_j^M) \sum_{i=1}^{m} (y_{di} - y_i) \frac{\partial y_i}{\partial u_j}; \tag{22}$$

Intermediate layer:

$$\delta_j^k = \begin{cases} f'(i_j^k) \sum_l \delta_l^{k+1} w_{jl}^{k,k+1} & \text{for case A,} \\[2em] f'(i_j^k) \sum_l \delta_l^{k+1} w_{jl}^{k,k+1} \left(\prod_{i \neq j} w_{il}^{k,k+1} o_i^k \right) & \text{for case B,} \end{cases} \tag{23}$$

where u_j denotes the jth input to the plant.

Thus, the application of Case B to the delta calculation is used at layers D and F. Here, from the definitions of Eqs. (15) and (16), f' in the first unit at layers D and F is evaluated by

$$f'(i_j^k) = 2\ln(0.5)i_j^k o_j^k \qquad \text{for layer } D, \tag{24}$$

$$f'(i_j^k) = -(o_j^k)^2 \qquad \text{for the first unit of layer } F, \tag{25}$$

and $f'(i_j^k) = 1$ for other linear units. Note also that the Jacobian $\partial y_i / \partial u_j$ in Eq. (21) can be approximated as

$$\frac{\Delta y_i}{\Delta u_j} \simeq \frac{\partial y_i}{\partial u_j} + \sum_{l \neq j} \frac{\partial y_i}{\partial u_l} \frac{\Delta u_l}{\Delta u_j} \tag{26}$$

if the control inputs are coupled, or

$$\frac{\partial y_i(kT)}{\partial u_j(kT)} \simeq \frac{\Delta y_i(kT)}{\Delta u_j(kT)}, \qquad i = j, \tag{27}$$

if the control inputs are decoupled, where $\Delta u_j(\cdot)$ and $\Delta y_i(\cdot)$ are generated from the input and output data at the sampling instant kT, $\Delta = 1 - z^{-1}$, z^{-1} is the one-step delay operator, k is the discrete time, and T is the sampling period. If the plant is originally a discrete-time system with no time delay in the input, then one must evaluate $\partial y_i(kT)/\partial u_j[(k-1)T]$ instead of the previous equation.

The preceding results yield the following update equations for the connection weights:

$$w_{ij}^{k-1,k}(t+1) = w_{ij}^{k-1,k}(t) + \eta \delta_j^k o_i^{k-1} + \xi \Delta w_{ij}^{k-1,k}(t)$$

$$\text{for Case A,} \quad (28)$$

$$w_{ij}^{k-1,k}(t+1) = w_{ij}^{k-1,k}(t) + \eta \delta_j^k o_i^{k-1} \left(\prod_{l \neq i} w_{lj}^{k-1,k} o_l^{k-1} \right) + \xi \Delta w_{ij}^{k-1,k}(t)$$

$$\text{for Case B,} \quad (29)$$

where t denotes the tth update time, η is a small positive constant that means a learning rate, $\Delta w_{ij}^{k-1,k}(t)$ is an increment of the connection weight at the tth step, and ξ is a small positive constant used as a stabilizing factor. Therefore, the connection weights w_c, w_d, w_a, and w_b can be updated by using Eq. (28). Note that Eq. (29) is not required to update any connection weight, because the connection weights associated with Case B are all fixed as unity.

V. ATTITUDE CONTROL APPLICATION EXAMPLE

In this section, the effectiveness of the mean-value-based functional reasoning method is illustrated by designing and simulating an unlearning fuzzy controller for the attitude control problem of a flexible satellite.

A. TWO-INPUT–SINGLE-OUTPUT REASONING

In this subsection, the design method of the conclusion is illustrated by using an example for two-input–single-output reasoning.

Consider the following attitude control problem of a flexible satellite described by [35]:

$$\ddot{\theta}(t) = 1.764u(t), \tag{30}$$

$$\ddot{\phi}(t) = -\omega^2\phi(t) + 4.358u(t), \tag{31}$$

$$y(t) = \theta(t) + 4.358\phi(t), \tag{32}$$

where $\omega^2 = 33.15 \times 10^{-4}$ [rad^2/s^2], $\theta(t)$ is the center body rotation due to rigid body motion, $\phi(t)$ is the center body rotation due to flexural motion, $y(t)$ is the measurement of the attitude, and $u(t)$ is the control input torque produced by reaction jets.

Although this control is a regulator problem, to retain the generality of the problem, it is regarded as a tracking problem with a reference $y_d = 0$. Then the tracking error is defined by $e = y_d - y$ and the corresponding derivative is assumed to be constructed as $\dot{e} = \{e[kT] - e[(k-1)T]\}/T$, where $k = 0, 1, \ldots$ and T is the sampling period of 0.01 [s].

Define each membership function of five labels for e and $\dot{e}$ on the support set $[-L, L] = [-6, 6]$ as shown in Fig. 9. Let the mean values on e and $\dot{e}$, defined on the support set, be respectively e_{cj} and $\dot{e}_{ck}$, $j, k = NB, NM, \ldots, PB$. In addition, introducing two new parameters α and β in (18) gives the following

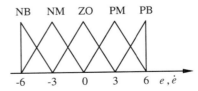

Figure 9   Membership functions of five labels for e and $\dot{e}$.

Table I
Control Rules for Five Labels ($\alpha = 1$, $\beta = 2$)

e	$\dot{e}$				
	NB	NM	ZO	PM	PB
NB	-6	-4.5	-3	-1.5	0
NM	-4.5	-3	-1.5	0	1.5
ZO	-3	-1.5	0	1.5	3
PM	-1.5	0	1.5	3	4.5
PB	0	1.5	3	4.5	6

Reprinted with the permission of the Society of Instrument and Control Engineers, Japan.

representation:

$$\sigma_i = [\alpha/\beta \quad 1/\beta]\begin{bmatrix} e_{cj} \\ \dot{e}_{ck} \end{bmatrix}, \qquad |\sigma_i| \leq L. \tag{33}$$

Note here that the upper and lower limits of the conclusion are constrained as $|u| \leq L$.

Case 1. Table I shows the control rule for $\alpha = 1$ and $\beta = 2$. If we pick up the rule 1 of $e = NB$, $\dot{e} = NB$ as an example, it is seen that the deviation data are assumed to be distributed on the sliding line $2\sigma_1 = -12$, that is, $\sigma_1 = -6$ (see Fig. 10). For the regulator problem of $y_d = 0$, this is equivalent to the output

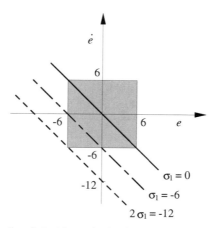

Figure 10 Switching line for rule 1 with $\alpha = 1$, $\beta = 2$. Reprinted with the permission of the Society of Instrument and Control Engineers, Japan.

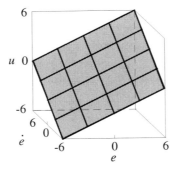

Figure 11   Conclusion constants for $\alpha = 1$, $\beta = 2$. Reprinted with the permission of the Society of Instrument and Control Engineers, Japan.

data being distributed on the sliding line $\dot{y} + y = 6$. Therefore, it is found that the control is determined so as to obtain the output data distribution on an ideal sliding line $\dot{y} + y = 0$ by moving $\dot{y} + y = 6$ to the origin. Figure 11 depicts the constant values in the conclusion for this case. It should be noted that, when defining the deviation error as $e = y - y_d$, we must use the relation such that $k_i = -|\sigma_i|$ as in (12).

Case 2.   Table II shows the control rule for the case when the conclusion is a sliding line faster than that of Case 1 by setting $\alpha = 2$ and $\beta = 2$. The interpretation of the sliding line is shown in Fig. 12 for rule 1 consisting of $e = NB$ and $\dot{e} = NB$. The corresponding constant values in the conclusion are shown in Fig. 13.

In the following, we will show some cases that blend Cases 1 and 2 for any control rule.

Table II
Control Rules for Five Labels ($\alpha = 2$, $\beta = 2$)

		$\dot{e}$		
e	NB	NM	ZO	PM	PB
NB	-6	-6	-6	-4.5	-3
NM	-6	-4.5	-3	-1.5	0
ZO	-3	-1.5	0	1.5	3
PM	0	1.5	3	4.5	6
PB	3	4.5	6	6	6

Reprinted with the permission of the Society of Instrument and Control Engineers, Japan.

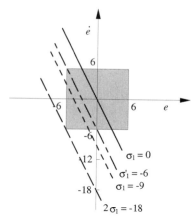

Figure 12   Switching line for rule 1 with $\alpha = 2$, $\beta = 2$. Reprinted with the permission of the Society of Instrument and Control Engineers, Japan.

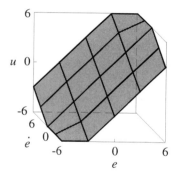

Figure 13   Conclusion constants for $\alpha = 2$, $\beta = 2$. Reprinted with the permission of the Society of Instrument and Control Engineers, Japan.

Table III

Control Rules for Five Labels ($\alpha = 2$, $\beta = 2$ if $\dot{e} = NB$ or PB; otherwise $\alpha = 1$, $\beta = 2$)

e		$\dot{e}$			
	NB	NM	ZO	PM	PB
NB	-6	-4.5	-3	-1.5	-3
NM	-6	-3	-1.5	0	0
ZO	-3	-1.5	0	1.5	3
PM	0	0	1.5	3	6
PB	3	1.5	3	4.5	6

Reprinted with the permission of the Society of Instrument and Control Engineers, Japan.

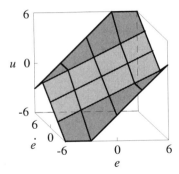

Figure 14   Conclusion constants for $\alpha = 2$, $\beta = 2$ if $\dot{e} = NB$ or PB. Reprinted with the permission of the Society of Instrument and Control Engineers, Japan.

Case 3.   For the case when $|\dot{e}|$ is very large, to set a sliding line faster than that of Case 1, we determine the parameters α and β such that

$$\begin{cases} \alpha = 2, \beta = 2, & \text{if } \dot{e} = NB \text{ or } PB, \\ \alpha = 1, \ \beta = 2, & \text{otherwise.} \end{cases}$$

The corresponding control rule is tabulated in Table III and the constant values in the conclusion are shown in Fig. 14.

Case 4.   For the case when $|e|$ is relatively small, to set a sliding line faster than that of Case 1, we determine the parameters α and β such that

$$\begin{cases} \alpha = 2, \ \beta = 2, & \text{if } e = NM \text{ or } ZO \text{ or } PM, \\ \alpha = 1, \ \beta = 2, & \text{otherwise.} \end{cases}$$

Table IV

Control Rules for Five Labels ($\alpha = 2$, $\beta = 2$ if $e = NM$ or ZO or PM; otherwise $\alpha = 1$, $\beta = 2$)

			$\dot{e}$		
e	NB	NM	ZO	PM	PB
NB	−6	−4.5	−3	−1.5	0
NM	−6	−4.5	−3	−1.5	0
ZO	−3	−1.5	0	1.5	3
PM	0	1.5	3	4.5	6
PB	0	1.5	3	4.5	6

Reprinted with the permission of the Society of Instrument and Control Engineers, Japan.

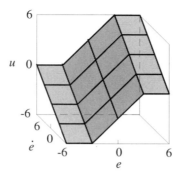

Figure 15 Conclusion constants for $\alpha = 2$, $\beta = 2$ if $e = NM$ or ZO or PM. Reprinted with the permission of the Society of Instrument and Control Engineers, Japan.

The corresponding control rule is tabulated in Table IV and the constant values in the conclusion are shown in Fig. 15.

Case 5. As a mix of Cases 3 and 4, we determine the parameters α and β such that

$$\begin{cases} \alpha = 2, \ \beta = 2, & \text{if } \dot{e} = NB \text{ or } PB, \text{ or } e = NM \text{ or } ZO \text{ or } PM, \\ \alpha = 1, \ \beta = 2, & \text{otherwise.} \end{cases}$$

The corresponding control rule is tabulated in Table V and the constant values in the conclusion are shown in Fig. 16.

Table V

Control Rules for Five Labels ($\alpha = 2$, $\beta = 2$ if $\dot{e} = NB$ or PB or $e = NM$ or ZO or PM; otherwise $\alpha = 1$, $\beta = 2$)

e	$\dot{e}$				
	NB	NM	ZO	PM	PB
NB	−6	−4.5	−3	−1.5	−3
NM	−6	−4.5	−3	−1.5	0
ZO	−3	−1.5	0	1.5	3
PM	0	1.5	3	4.5	6
PB	3	1.5	3	4.5	6

Reprinted with the permission of the Society of Instrument and Control Engineers, Japan.

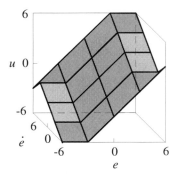

Figure 16   Conclusion constants for $\alpha = 2$, $\beta = 2$ if $\dot{e} = NB$ or PB or $e = NM$ or ZO or PM. Reprinted with the permission of the Society of Instrument and Control Engineers, Japan.

Figure 17 shows the attitude control results when the control rules determined by Cases 1–5 were applied to the control object. Note here that an adaptive input scaling method with the initial scalers 10^4 was used [31] and the output scaler was fixed to the value 1.6. It is seen from this figure that the result of Case 4 is good for the case when the undershoot is not allowed, whereas the result of Case 5 is good for the case when a little undershoot is allowed.

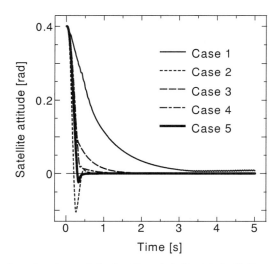

Figure 17   Control results using control rules: Cases 1–5. Reprinted with the permission of the Society of Instrument and Control Engineers, Japan.

B. THREE-INPUT–SINGLE-OUTPUT REASONING

In this subsection, we further consider a case in which the accelerative error information is taken into account, in addition to the deviation error and its derivative information. The accelerative error is assumed to be constructed by $\ddot{e} \triangleq \{\dot{e}[kT] - \dot{e}[(k-1)T]\}/T$. Introducing the parameters α, β, and γ, Eq. (14) can be rewritten as

$$\sigma_i = [\alpha/\gamma \quad \beta/\gamma \quad 1/\gamma] \begin{bmatrix} e_{cj} \\ \dot{e}_{ck} \\ \ddot{e}_{c\ell} \end{bmatrix}, \qquad |\sigma_i| \leq L. \qquad (34)$$

Here, the membership functions for e and $\dot{e}$ are defined as shown in Fig. 6, but those for $\ddot{e}$ are assumed to consist of three labels as shown in Fig. 18, where their mean values are denoted by $\ddot{e}_{c\ell}$, $\ell = NB, ZO, PB$.

1. Design of Switching Plane as an Overdamped or Critically Damped Response

Using three inputs e, $\dot{e}$, $\ddot{e}$ and defining the following sliding plane:

$$\ddot{e} + \beta\dot{e} + \alpha e = 0, \qquad (35)$$

we determine the parameters α and β such that the characteristic equation associated with (35) has two real and unequal roots, or two real and equal roots, which are, respectively, overdamped and critically damped responses. Hereafter, it is assumed that $\gamma = 3$.

Case 6. When allocating the characteristic roots for (35) as multiple roots of -1, it follows that $\alpha = 1$ and $\beta = 2$, and the corresponding control rules are tabulated in Table VI.

Case 7. When allocating the characteristic roots for (35) as two distinct roots of -1 and -2, it follows that $\alpha = 2$ and $\beta = 3$, and the corresponding control rules are tabulated in Table VII.

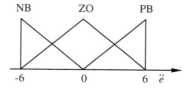

Figure 18 Membership functions of three labels for $\ddot{e}$.

Table VI

Control Rules for Five Labels
($\alpha = 1$, $\beta = 2$, $\gamma = 3$)

e	$\dot{e}$				
	NB	NM	ZO	PM	PB
(a) $\ddot{e} = NB$					
NB	-6	-6	-4	-2	0
NM	-6	-5	-3	-1	1
ZO	-6	-4	-2	0	2
PM	-5	-3	-1	1	3
PB	-4	-2	0	2	4
(b) $\ddot{e} = ZO$					
NB	-6	-4	-2	0	2
NM	-5	-3	-1	1	3
ZO	-4	-2	0	2	4
PM	-3	-1	1	3	5
PB	2	0	2	4	6
(c) $\ddot{e} = PB$					
NB	-4	-2	0	2	4
NM	-3	-1	1	3	5
ZO	-2	0	2	4	6
PM	-1	1	3	5	6
PB	0	2	4	6	6

Table VII

Control Rules for Five Labels
($\alpha = 2$, $\beta = 3$, $\gamma = 3$)

e	$\dot{e}$				
	NB	NM	ZO	PM	PB
(a) $\ddot{e} = NB$					
NB	-6	-6	-4	-3	0
NM	-6	-6	-4	-1	2
ZO	-6	-5	-2	1	4
PM	-6	-3	0	3	6
PB	-4	-1	2	5	6
(b) $\ddot{e} = ZO$					
NB	-6	-6	-4	-1	2
NM	-6	-5	-2	1	4
ZO	-6	-3	0	3	6
PM	-4	-1	2	5	6
PB	-2	1	4	6	6
(c) $\ddot{e} = PB$					
NB	-6	-5	-2	1	4
NM	-6	-3	0	3	6
ZO	-4	-1	2	5	6
PM	-2	1	4	6	6
PB	0	3	6	6	6

Case 8. When allocating the characteristic roots for (35) as multiple roots of -2, it follows that $\alpha = 4$ and $\beta = 4$, and the corresponding control rules are tabulated in Table VIII.

Case 9. When allocating the characteristic roots for (35) as multiple roots of -3, it follows that $\alpha = 9$ and $\beta = 6$, and the corresponding control rules are tabulated in Table IX.

Figure 19 shows the attitude control results when the control rules determined by Cases 6–9 were applied to the control object. It is seen from this figure that the faster control result with no oscillations is obtained from Case 6 to Case 9.

2. Design of Switching Plane as an Underdamped Response

In this subsection, we will determine the parameters α and β such that the roots $(-\beta \pm \sqrt{\beta^2 - 4\alpha})/2$ of the characteristic equation for (35) are complex numbers, in which case we have the so-called underdamped response.

Table VIII

Control Rules for Five Labels
$(\alpha = 4,\ \beta = 4,\ \gamma = 3)$

e	$\dot{e}$				
	NB	NM	ZO	PM	PB
(a) $\ddot{e} = NB$					
NB	-6	-6	-6	-6	-2
NM	-6	-6	-6	-2	2
ZO	-6	-6	-2	2	6
PM	-6	-2	2	6	6
PB	-2	2	6	6	6
(b) $\ddot{e} = ZO$					
NB	-6	-6	-6	-4	0
NM	-6	-6	-4	0	4
ZO	-6	-4	0	4	6
PM	-4	0	4	6	6
PB	0	4	6	6	6
(c) $\ddot{e} = PB$					
NB	-6	-6	-6	-2	2
NM	-6	-6	-2	2	6
ZO	-6	-2	2	6	6
PM	-2	2	6	6	6
PB	2	6	6	6	6

Table IX

Control Rules for Five Labels
$(\alpha = 9,\ \beta = 6,\ \gamma = 3)$

e	$\dot{e}$				
	NB	NM	ZO	PM	PB
(a) $\ddot{e} = NB$					
NB	-6	-6	-6	-6	-6
NM	-6	-6	-6	-5	1
ZO	-6	-6	-2	4	6
PM	-5	1	6	6	6
PB	4	6	6	6	6
(b) $\ddot{e} = ZO$					
NB	-6	-6	-6	-6	-6
NM	-6	-6	-6	-3	3
ZO	-6	-6	0	6	6
PM	-3	3	6	6	6
PB	6	6	6	6	6
(c) $\ddot{e} = PB$					
NB	-6	-6	-6	-6	-4
NM	-6	-6	-6	-1	5
ZO	-6	-4	2	6	6
PM	-1	5	6	6	6
PB	6	6	6	6	6

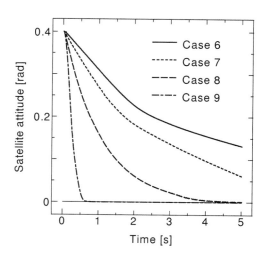

Figure 19   Control results using control rules: Cases 6–9. Reprinted with the permission of the Society of Instrument and Control Engineers, Japan.

Case 10. When setting the imaginary part of the complex conjugate roots as $\sqrt{3}$ with a fixed $\beta = 1$, it follows that $\alpha = 1$. The corresponding control rules are tabulated in Table X. Note that in the following $\beta = 1$ is used for all cases.

Case 11. Setting the imaginary part of the complex conjugate roots as $\sqrt{5}$ gives $\alpha = 1.5$. The corresponding control rules are tabulated in Table XI.

Case 12. Setting the imaginary part of the complex conjugate roots as $\sqrt{7}$ yields $\alpha = 2$. The corresponding control rules are tabulated in Table XII.

Case 13. Setting the imaginary part of the complex conjugate roots as $\sqrt{11}$ gives $\alpha = 3$. The corresponding control rules are tabulated in Table XIII.

Figure 20 depicts the attitude control results when the control rules determined by Cases 10–13 were applied to the control object. It is seen from this figure

Table X

Control Rules for Five Labels
($\alpha = 1$, $\beta = 1$, $\gamma = 3$)

e	NB	NM	ZO	PM	PB
		(a) $\ddot{e} = NB$			
NB	-6	-5	-4	-3	-2
NM	-5	-4	-3	-2	1
ZO	-4	-3	-2	-1	0
PM	-3	-2	-1	0	1
PB	-2	-1	0	1	2
		(b) $\ddot{e} = ZO$			
NB	-4	-3	-2	-1	0
NM	-3	-2	-1	0	1
ZO	-2	-1	0	1	2
PM	-1	0	1	2	3
PB	0	1	2	3	4
		(c) $\ddot{e} = PB$			
NB	-2	-1	0	0	2
NM	-1	0	1	2	3
ZO	0	1	2	3	4
PM	1	2	3	4	5
PB	2	3	4	5	6

Table XI

Control Rules for Five Labels
($\alpha = 1.5$, $\beta = 1$, $\gamma = 3$)

e	NB	NM	ZO	PM	PB
		(a) $\ddot{e} = NB$			
NB	-6	-6	-5	-4	-3
NM	-5.5	-4.5	-3.5	-2.5	-1.5
ZO	-4	-3	-2	-1	0
PM	-2.5	-1.5	-0.5	0.5	1.5
PB	-1	0	1	2	3
		(b) $\ddot{e} = ZO$			
NB	-5	-4	-3	-2	-1
NM	-3.5	-2.5	-1.5	-0.5	0.5
ZO	-2	-1	0	1	2
PM	-0.5	0.5	1.5	2.5	3.5
PB	1	2	3	4	5
		(c) $\ddot{e} = PB$			
NB	-3	-2	-1	0	1
NM	-1.5	-0.5	0.5	1.5	2.5
ZO	0	1	2	3	4
PM	1.5	2.5	3.5	4.5	5.5
PB	3	4	5	6	6

Table XII

Control Rules for Five Labels
$(\alpha = 2,\ \beta = 1,\ \gamma = 3)$

e	$\dot e$ NB	NM	ZO	PM	PB
	(a) $\ddot e = NB$				
NB	-6	-6	-6	-5	-4
NM	-6	-5	-4	-3	-2
ZO	-4	-3	-2	-1	0
PM	-2	-1	0	1	2
PB	0	1	2	3	4
	(b) $\ddot e = ZO$				
NB	-6	-5	-4	-3	-2
NM	-4	-3	-2	-1	0
ZO	-2	-1	0	1	2
PM	0	1	2	3	4
PB	2	3	4	5	6
	(c) $\ddot e = PB$				
NB	-4	-3	-2	-1	0
NM	-2	-1	0	1	2
ZO	0	1	2	3	4
PM	2	3	4	5	6
PB	4	5	6	6	6

Table XIII

Control Rules for Five Labels
$(\alpha = 3,\ \beta = 1,\ \gamma = 3)$

e	$\dot e$ NB	NM	ZO	PM	PB
	(a) $\ddot e = NB$				
NB	-6	-6	-6	-6	-6
NM	-6	-6	-5	-4	-3
ZO	-4	-3	-2	-1	0
PM	-1	0	1	2	3
PB	2	3	4	5	6
	(b) $\ddot e = ZO$				
NB	-6	-6	-6	-5	-4
NM	-5	-4	-3	-2	-1
ZO	-2	-1	0	1	2
PM	1	2	3	4	5
PB	4	5	6	6	6
	(c) $\ddot e = PB$				
NB	-6	-5	-4	-3	-2
NM	-3	-2	-1	0	1
ZO	0	1	2	3	4
PM	3	4	5	6	6
PB	6	6	6	6	6

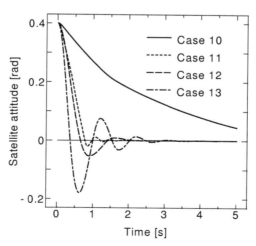

Figure 20 Control results using control rules: Cases 10–13. Reprinted with the permission of the Society of Instrument and Control Engineers, Japan.

that the larger the imaginary part allocated, the larger is the oscillating control response obtained, as expected.

VI. MOBILE ROBOT EXAMPLE

In this section, an FGNN controller based on mean-value-based functional reasoning is applied to the tracking control problem of a mobile robot with two independent driving wheels.

A. MODEL OF A MOBILE ROBOT

Let the mobile robot be rigidly moving on a plane as shown in Fig. 21. The absolute coordinate system $O - XY$ is assumed to be fixed on the plane. Then the dynamic behavior of the robot is described by the following equations of motion [36]:

$$I_v \ddot{\phi} = D_r l - D_l l, \tag{36}$$

$$M \dot{v} = D_r + D_l. \tag{37}$$

For the right and left wheels, the dynamics of the driving system is described by

$$I_w \ddot{\theta}_i + c \dot{\theta}_i = k u_i - r D_i, \qquad i = r, l, \tag{38}$$

where the parameters and variables are defined as follows: I_v, the moment of inertia around the center of gravity (c.g.) of the robot; M, the mass of the robot; D_l, D_r, the left and right driving forces; l, the distance between the left or the

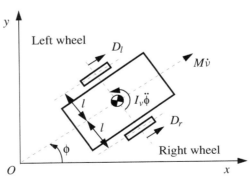

Figure 21   Mobile robot model.

right wheel and the c.g. of the robot; ϕ, the azimuth of the robot; v, the velocity of the robot; I_w, the moment of inertia of the wheel; c, the viscous friction factor; k, the driving gain factor; r, the radius of the wheel; θ_i, the rotational angle of the wheel; and u_i, the driving input.

On the other hand, the geometrical relationships among the variables ϕ, v, θ_i are given by

$$r\dot{\theta}_r = v + l\dot{\phi}, \tag{39}$$
$$r\dot{\theta}_l = v - l\dot{\phi}. \tag{40}$$

From these equations, defining the state variable for the robot as $\mathbf{x} = [v \quad \phi \quad \dot{\phi}]^T$, the manipulated variable as $\mathbf{u} = [u_r \quad u_l]^T$, and the output variable as $\mathbf{y} = [v \quad \phi]^T$, one obtains the following state equations:

$$\dot{\mathbf{x}} = A\mathbf{x} + B\mathbf{u}, \tag{41}$$
$$\mathbf{y} = C\mathbf{x}, \tag{42}$$

where

$$A = \begin{bmatrix} a_1 & 0 & 0 \\ 0 & 0 & 1 \\ 0 & 0 & a_2 \end{bmatrix}, \qquad B = \begin{bmatrix} b_1 & b_1 \\ 0 & 0 \\ b_2 & -b_2 \end{bmatrix}, \qquad C = \begin{bmatrix} 1 & 0 & 0 \\ 0 & 1 & 0 \end{bmatrix},$$

$$a_1 = -2c/(Mr^2 + 2I_w), \qquad a_2 = -2cl^2/(I_v r^2 + 2I_w l^2),$$

$$b_1 = kr/(Mr^2 + 2I_w), \qquad b_2 = krl/(I_v r^2 + 2I_w l^2).$$

B. SIMULATION EXAMPLES

Figure 22 shows the block diagram of the path control system of the mobile robot. This system consists of two FGNNs; one is for processing the information of the velocity error $e_v = v_d - v$ and its rate $\dot{e}_v$; and the other is for processing the information of the azimuth error $e_\phi = \phi_d - \phi$ and its rate $\dot{e}_\phi$. Here, v_d and ϕ_d denote the reference velocity and reference azimuth, respectively. The system also contains a net which combines two consequent torques u_v and u_ϕ generated from two FGNNs and determines the right and left driving torques, that is, u_r and u_l.

In the simulations, we apply the FGNN using the mean-value-based functional reasoning described in Section II.C. To simulate the mobile robot model, the fourth-order Runge–Kutta–Gill method was used with an integration step of 1 [ms]. It is also assumed that the control sampling period is 50 [ms]. The

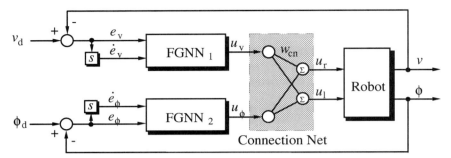

Figure 22   Fuzzy Gaussian neural network controller for a mobile robot with two independent drive wheels. Reprinted from K. Watanabe *et al.*, Fuzzy-neural network controllers using mean-value-based functional reasoning, *Neurocomputing* 9:39–61, 1995, with kind permission of Elsevier Science-NL, Sara Burgerhartstraat 25, 1055 KV Amsterdam, The Netherlands.

physical parameters of the mobile robot are as follows:

$$I_v = 10 \text{ [kgm}^2], \quad M = 200 \text{ [kg]}, \quad l = 0.3 \text{ [m]},$$
$$I_w = 0.005 \text{ [kgm}^2], \quad c = 0.05 \text{ [kgm}^2/\text{s]}, \quad r = 0.1 \text{ [m]}, \quad k = 5.$$

A circular trajectory with a radius of 1.5 [m] is considered, in which the reference velocity v_d is 0.25 [m/s] and the initial value of the state variable is given as $\mathbf{x} = [0 \quad 0 \quad 0]^T$.

We used the 49 control rules in which the seven membership functions shown in Fig. 23 were applied to each input variable. The center values of the seven

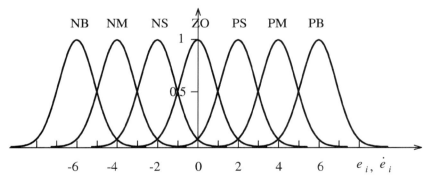

Figure 23   Gaussian membership functions with seven labels. Reprinted from K. Watanabe *et al.*, Fuzzy-neural network controllers using mean-value-based functional reasoning, *Neurocomputing* 9:39–61, 1995, with kind permission of Elsevier Science-NL, Sara Burgerhartstraat 25, 1055 KV Amsterdam, The Netherlands.

membership functions, w_c, were $-6, -4, -2, 0, 2, 4, 6$, and the reciprocal values of the deviation w_d were all unity so as to equally allocate all membership functions on the support set $[-6, 6]$. Note that all initial scalers of the adaptive input scaling method were set as 10^4. Note also that the weights of the connection net, w_{cn}, are usually fixed as $1.0, 1.0, 1.0, -1.0$ as discussed in [11]. However, here we set them as $1.5, 1.5, 5.0, -5.0$, which means that the output torque u_v from FGNN$_1$ was scaled as $1.5u_v$ and the output torque u_ϕ from FGNN$_2$ was also scaled as $5.0u_\phi$.

1. Effect of Input Scaling

In this simulation, the connection weights w_a and w_b for each FGNN were learned, under the assumption that the other connection weights in the FGNNs were not learned; that is, the learning rates of w_c and w_d were all fixed to 0.

The control results of the velocity and azimuth for the case where $\alpha = 3$ and $\beta = 2$, that is, when the initial parameters for w_a and w_b are 1.5 and 0.5, respectively, are given in Figs. 24 and 25, where the learning rates of w_a and

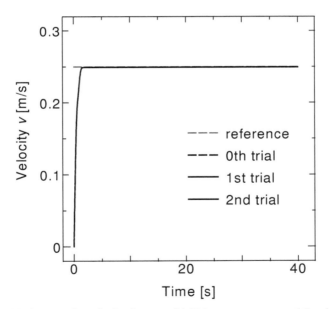

Figure 24 Velocity control results for the case of initial parameters $w_a = 1.5$ and $w_b = 0.5$. Reprinted from K. Watanabe *et al.*, Fuzzy-neural network controllers using mean-value-based functional reasoning, *Neurocomputing* 9:39–61, 1995, with kind permission of Elsevier Science-NL, Sara Burgerhartstraat 25, 1055 KV Amsterdam, The Netherlands.

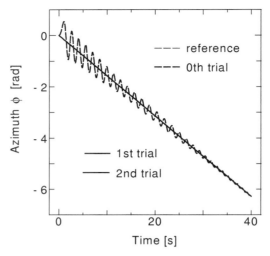

Figure 25 Azimuth control results for the case of initial parameters $w_a = 1.5$ and $w_b = 0.5$. Reprinted from K. Watanabe *et al.*, Fuzzy-neural network controllers using mean-value-based functional reasoning, *Neurocomputing* 9:39–61, 1995, with kind permission of Elsevier Science-NL, Sara Burgerhartstraat 25, 1055 KV Amsterdam, The Netherlands.

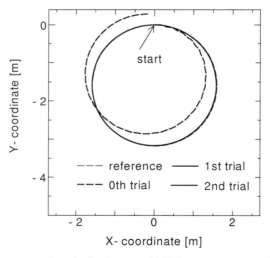

Figure 26 Trajectory control results for the case of initial parameters $w_a = 1.5$ and $w_b = 0.5$. Reprinted from K. Watanabe *et al.*, Fuzzy-neural network controllers using mean-value-based functional reasoning, *Neurocomputing* 9:39–61, 1995, with kind permission of Elsevier Science-NL, Sara Burgerhartstraat 25, 1055 KV Amsterdam, The Netherlands.

w_b were separately determined for the velocity and azimuth: $\eta_a^1 = 0.0005$ and $\eta_b^1 = 0.001$ for the FGNN associated with the velocity, and $\eta_a^2 = 0.001$ and $\eta_b^2 = 0.0005$ for the FGNN associated with the azimuth. It is seen from these figures that, after the first trial, a very fast response is obtained for the velocity and azimuth of the robot. The corresponding circular path in the (x, y) coordinate is also depicted in Fig. 26.

It is also remarked that, in this case, the learning of the parameters w_a and w_b does not necessarily contribute to the control of the velocity and azimuth of the

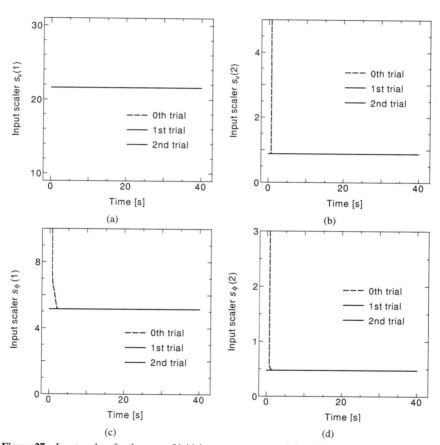

Figure 27 Input scalers for the case of initial parameters $w_a = 1.5$ and $w_b = 0.5$. Reprinted from K. Watanabe *et al.*, Fuzzy-neural network controllers using mean-value-based functional reasoning, *Neurocomputing* 9:39–61, 1995, with kind permission of Elsevier Science-NL, Sara Burgerhartstraat 25, 1055 KV Amsterdam, The Netherlands.

robot, because the initial parameters for w_a and w_b are set suitably. This was confirmed by the fact that setting all learning rates to 0 for w_a and w_b in both FGNNs gave the same results as before. As seen from Fig. 27a–d, the main contribution to the control of the trajectory of the robot is only the adjustment of the input scaling for e_ϕ and $\dot{e}_\phi$.

2. Effect of the Learning of Parameters in the Conclusion

In this case, the parameters in the conclusion were modified using the values $\alpha = 3.2$ and $\beta = 2$; that is, the initial parameters for w_a and w_b were 1.6 and 0.5, respectively. In addition, the learning rates of w_a and w_b were also changed as $\eta_a^1 = \eta_b^1 = 0.005$ for the FGNN associated with the velocity, and $\eta_a^2 = \eta_b^2 = 0.001$ for the FGNN associated with the azimuth.

The corresponding control results are shown in Figs. 28–30, together with their input scaling adjustments shown in Fig. 31. It is seen from these figures that satisfactory trajectory and azimuth are obtained after the second trial. As observed from Fig. 28, the velocity response is still improved up to the sixth trial.

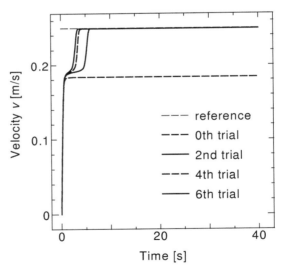

Figure 28   Velocity control results for the case of initial parameters $w_a = 1.6$ and $w_b = 0.5$, where $\eta_a^1 = \eta_b^1 = 0.005$ and $\eta_a^2 = \eta_b^2 = 0.001$. Reprinted from K. Watanabe *et al.*, Fuzzy-neural network controllers using mean-value-based functional reasoning, *Neurocomputing* 9:39–61, 1995, with kind permission of Elsevier Science-NL, Sara Burgerhartstraat 25, 1055 KV Amsterdam, The Netherlands.

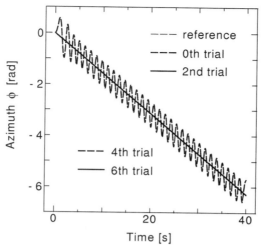

Figure 29 Azimuth control results for the case of initial parameters $w_a = 1.6$ and $w_b = 0.5$, where $\eta_a^1 = \eta_b^1 = 0.005$ and $\eta_a^2 = \eta_b^2 = 0.001$. Reprinted from K. Watanabe *et al.*, Fuzzy-neural network controllers using mean-value-based functional reasoning, *Neurocomputing* 9:39–61, 1995, with kind permission of Elsevier Science-NL, Sara Burgerhartstraat 25, 1055 KV Amsterdam, The Netherlands.

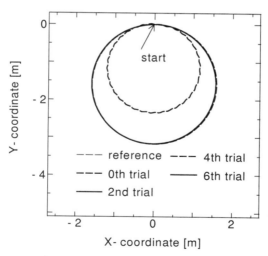

Figure 30 Trajectory control results for the case of initial parameters $w_a = 1.6$ and $w_b = 0.5$, where $\eta_a^1 = \eta_b^1 = 0.005$ and $\eta_a^2 = \eta_b^2 = 0.001$. Reprinted from K. Watanabe *et al.*, Fuzzy-neural network controllers using mean-value-based functional reasoning, *Neurocomputing* 9:39–61, 1995, with kind permission of Elsevier Science-NL, Sara Burgerhartstraat 25, 1055 KV Amsterdam, The Netherlands.

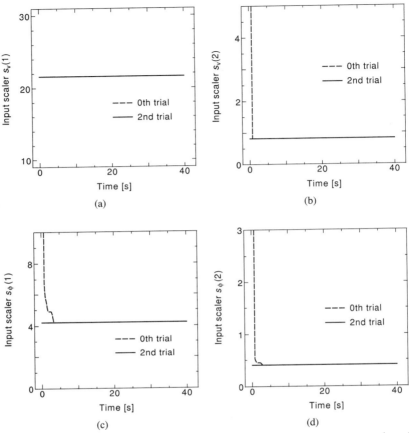

Figure 31 Input scalers for the case of initial parameters $w_a = 1.6$ and $w_b = 0.5$, where $\eta_a^1 = \eta_b^1 = 0.005$ and $\eta_a^2 = \eta_b^2 = 0.001$.

To check the effect of the learning of the parameter w_b on the control performance, the results for the case where the learning rate of w_b was fixed to 0 are depicted in Figs. 32–34. From these figures, it is seen that the control of the velocity is inferior to that of the case where both w_a and w_b are learned simultaneously. Therefore, to improve the velocity response, the learning rate of w_a was change to $\eta_a^1 = 0.05$ for the FGNN associated with the velocity. The corresponding results are shown in Figs. 35–37. From Fig. 35, it is understood that a very fast velocity response is obtained, even though the parameter w_b is not learned.

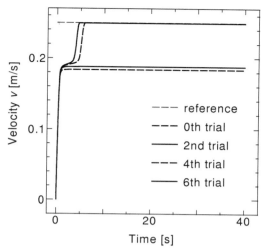

Figure 32 Velocity control results for the case of initial parameters $w_a = 1.6$ and $w_b = 0.5$, where $\eta_a^1 = 0.005$ and w_b was fixed. Reprinted from K. Watanabe *et al.*, Fuzzy-neural network controllers using mean-value-based functional reasoning, *Neurocomputing* 9:39–61, 1995, with kind permission of Elsevier Science-NL, Sara Burgerhartstraat 25, 1055 KV Amsterdam, The Netherlands.

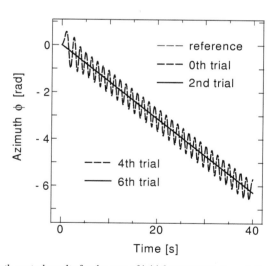

Figure 33 Azimuth control results for the case of initial parameters $w_a = 1.6$ and $w_b = 0.5$, where $\eta_a^1 = 0.005$ and w_b was fixed. Reprinted from K. Watanabe *et al.*, Fuzzy-neural network controllers using mean-value-based functional reasoning, *Neurocomputing* 9:39–61, 1995, with kind permission of Elsevier Science-NL, Sara Burgerhartstraat 25, 1055 KV Amsterdam, The Netherlands.

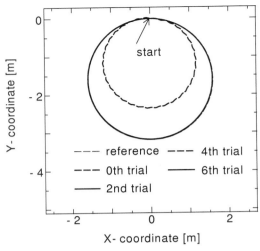

Figure 34 Trajectory control results for the case of initial parameters $w_a = 1.6$ and $w_b = 0.5$, where $\eta_a^1 = 0.005$ and w_b was fixed. Reprinted from K. Watanabe *et al.*, Fuzzy-neural network controllers using mean-value-based functional reasoning, *Neurocomputing* 9:39–61, 1995, with kind permission of Elsevier Science-NL, Sara Burgerhartstraat 25, 1055 KV Amsterdam, The Netherlands.

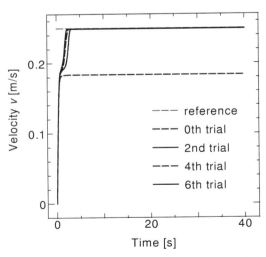

Figure 35 Velocity control results for the case of initial parameters $w_a = 1.6$ and $w_b = 0.5$, where $\eta_a^1 = 0.05$ and w_b was fixed. Reprinted from K. Watanabe *et al.*, Fuzzy-neural network controllers using mean-value-based functional reasoning, *Neurocomputing* 9:39–61, 1995, with kind permission of Elsevier Science-NL, Sara Burgerhartstraat 25, 1055 KV Amsterdam, The Netherlands.

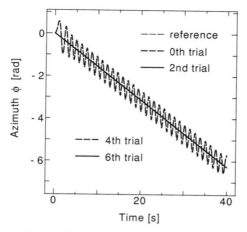

Figure 36 Azimuth control results for the case of initial parameters $w_a = 1.6$ and $w_b = 0.5$, where $\eta_a^1 = 0.05$ and w_b was fixed.

VII. CONCLUSIONS

We have presented a mean-value-based functional reasoning scheme, in addition to the usual input-data-based functional reasoning and the simplified reasoning schemes, in which the conclusion consists of a function of mean values on each membership function in the antecedent. It was shown that the constant

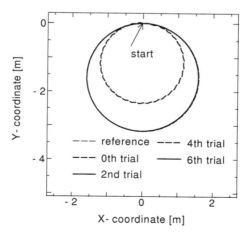

Figure 37 Trajectory control results for the case of initial parameters $w_a = 1.6$ and $w_b = 0.5$, where $\eta_a^1 = 0.05$ and w_b was fixed.

parameters in the conclusion of these functional reasoning schemes can be rationally designed through the use of VSS control theory. Furthermore, some fuzzy neural network controllers were developed by using these functional reasoning schemes. It was proved that the fuzzy neural network which uses the mean-value-based functional reasoning scheme allows the number of learning parameters in the conclusion to be reduced drastically, compared to those of the input-data-based functional reasoning and the usual simplified reasoning schemes.

Recently, a stochastic fuzzy control approach that includes the present result as a special case has been proposed [37–41]. This approach is based on using the so-called multiple model adaptive control [42], and is closely related to the conventional model-based control. The proposed stochastic fuzzy controller can be designed as a conventional stochastic control consisting of a static fuzzy observer part and a feedback gain part [39]. Also, this type of fuzzy controller can assure the bounded stability of the controlled system through a robust control approach [43], instead of assuring an ideally asymptotic stability [9–11]. For their practical application to robotics and mechatronics, the reader is referred to [44–46].

REFERENCES

[1] M. Sugeno. *Fuzzy Control*, pp. 67–136. Nikkan-kogyo-shinbun-sha, Tokyo, 1988 (in Japanese).
[2] T. Takagi and M. Sugeno. Fuzzy identification of systems and its applications to modeling and control. *IEEE Trans. Systems Man Cybernet.* 15:116–132, 1985.
[3] M. Mizumoto. Simple fuzzy theories. *Computrol* 28:32–45, 1989 (in Japanese).
[4] M. Mizumoto. Fuzzy reasoning methods for fuzzy control. *J. Soc. Instrument Control Engineers* 28:959–963, 1989 (in Japanese).
[5] E. H. Mamdani. Application of fuzzy algorithms for control of simple dynamic plant. *Proc. IEE* 121:1585–1588, 1974.
[6] E. H. Mamdani. Advances in the linguistic synthesis of fuzzy controller. *Internat. J. Man Machine Studies* 8:669–679, 1976.
[7] H. Ichihashi and H. Tanaka. PID-Fuzzy Hybrid Controller. In *Proceedings of the Fourth Fuzzy System Symposium* (Tokyo), pp. 97–102, 1988 (in Japanese).
[8] K. Tanaka. *Advanced Fuzzy Control*. Kyoritsu-syuppan, Tokyo, 1994 (in Japanese).
[9] K. Tanaka and M. Sano. A robust stabilization problem of fuzzy control systems and its application to backing up control of a truck-trailer. *IEEE Trans. Fuzzy Systems* 2:119–134, 1994.
[10] K. Tanaka, T. Ikeda, and H. O. Wang. Robust stabilization of a class of uncertain nonlinear systems via fuzzy control: quadratic stabilization, H^∞ control theory, and linear matrix inequalities. *IEEE Trans. Fuzzy Systems* 4:1–13, 1996.
[11] H. O. Wang, K. Tanaka, and M. F. Griffin. An approach to fuzzy control of nonlinear systems: stability and design issues. *IEEE Trans. Fuzzy Systems* 4:14–23, 1996.
[12] N. Matsunaga and S. Kawaji. Fuzzy control of VSS type and its robustness. *J. Japan Soc. Fuzzy Theory Systems* 4:1147–1155, 1992 (in Japanese).
[13] C.-C. Kung and S.-C. Lin. Fuzzy controller design: a sliding mode approach. In *Fuzzy Reasoning in Information, Decision and Control Systems* (S. G. Tzafestas and A. N. Venetsanopoulos, Eds.), pp. 277–306. Kluwer Academic, Dordrecht, 1994.

[14] J. C. Wu and T. S. Liu. A sliding-mode approach to fuzzy control design. *IEEE Trans. Control Systems Technol.* 4:141–151, 1996.

[15] S. Horikawa, T. Furuhashi, and Y. Uchikawa. On fuzzy modeling using fuzzy neural networks with the back-propagation algorithm. *IEEE Trans. Neural Networks* 3:801–806, 1992.

[16] K. Watanabe and J. Tang. Learning controller based on fuzzy Gaussian neural network. In *Proceedings of the Second Intelligent System Symposium* (Nagoya), pp. 255–260, 1992 (in Japanese).

[17] K. Watanabe and J. Tang. Control of a robot vehicle using fuzzy Gaussian neural network. In *Proceedings of the Second Intelligent System Symposium* (Nagoya), pp. 261–266, 1992 (in Japanese).

[18] J.-S. Roger Jang and C.-T. Sun. Functional equivalence between radial basis function networks and fuzzy inference systems. *IEEE Trans. Neural Networks* 4:156–159, 1993.

[19] J. Nie and D. A. Linkens. Learning control using fuzzified self-organizing radial basis function network. *IEEE Trans. Fuzzy Systems* 1:280–287, 1993.

[20] K. Watanabe, J. Tang, M. Nakamura, S. Koga, and T. Fukuda. Mobile robot control using fuzzy Gaussian neural networks. In *Proceedings of the 1993 IEEE/RSJ International Conference on Intelligent Robots and Systems* (Yokohama), Vol. 2, pp. 919–925, 1993.

[21] I. Hayashi and M. Umano. Perspectives and trends of fuzzy-neural networks. *J. Japan Soc. Fuzzy Theory Systems* 5:178–190, 1993.

[22] H. Ichihashi, T. Miyoshi, and K. Nagasaka. Computed tomography by neuro-fuzzy inversion. In *Proceedings of the International Joint Conference on Neural Networks* (Nagoya), Vol. 1, pp. 709–712, 1993.

[23] T. Watanabe and H. Ichihashi. Fuzzy control of a robotic manipulator by the feedback error learning. *Trans. Inst. Systems Control Inform. Engineers* 3:212–217, 1990 (in Japanese).

[24] K. Watanabe, J. Tang, M. Nakamura, S. Koga, and T. Fukuda. A fuzzy-Gaussian neural network and its application to a mobile robot control. *IEEE Trans. Control Systems Technol.* 4:193–199, 1996.

[25] K. Watanabe, K. Hara, and S. G. Tzafestas. Fuzzy controller design using the mean-value-based functional reasoning. In *Proceedings of the International Joint Conference on Neural Networks* (Nagoya), Vol. 3, pp. 2983–2986, 1993.

[26] K. Watanabe. Fuzzy controller design using the mean-value-based functional reasoning. *Trans. Soc. Instrument Control Engineers* 31:1106–1113, 1995 (in Japanese).

[27] R. A. DeCarlo, S. H. Zak, and G. P. Matthews. Variable structure control of nonlinear multivariable systems: a tutorial. *Proc. IEEE* 76:212–232, 1988.

[28] K. Watanabe, K. Hara, S. Koga, and S. G. Tzafestas. Mean-value-based functional reasoning and its realization as a fuzzy-neural-network controller. In *Proceedings of the First Asian Control Conference* (Tokyo), Vol. 3, pp. 435–438, 1994.

[29] K. Watanabe, K. Hara, S. Koga, and S. G. Tzafestas. Fuzzy-neural network controllers using the mean-value-based functional reasoning. *Neurocomputing* 9:39–61, 1995.

[30] M. Teshnehlab and K. Watanabe. A fuzzy neural network controller based on Gaussian potential functions. In *Proceedings of the Third International Conference on Fuzzy Logic, Neural Nets and Soft Computing* (Iizuka), pp. 193–196, 1994.

[31] K. Watanabe and S. G. Tzafestas. Fuzzy logic controller as a compensator in the problem of tracking control of manipulators. In *Proceedings IFToMM-jc International Symposium on Theory of Machines and Mechanisms* (Nagoya), pp. 98–103, 1992.

[32] D. E. Rumelhart and J. L. McClelland, and the PDP Research Group, Eds., *Parallel Distributed Processing: Explorations in the Microstructures of Cognition,* Vol. 1: *Foundations.* MIT Press, Cambridge, MA, 1986.

[33] R. Hecht-Nielsen. *Neurocomputing.* Addison–Wesley, New York, 1987.

[34] K. Watanabe and S. G. Tzafestas. Learning algorithms for neural networks with the Kalman filters. *J. Intell. Robotic Systems* 3:305–319, 1990.

[35] S. Daley and K. F. Gill. A justification for the wider use of fuzzy logic control algorithms. *Proc. Inst. Mechanical Engineers* 199-C1: 43–49, 1985.

[36] M. Saito and T. Tsumura. Collision avoidance among multiple mobile robots—a local approach based on non-linear programming. *Trans. Inst. Systems Control Inform. Engineers* 3:252–260, 1990 (in Japanese).

[37] K. Watanabe. Stochastic fuzzy control (1st report, theoretical derivation). *Trans. Japan Soc. Mechanical Engineers C* 62:1005–1012, 1996.

[38] K. Watanabe. Stochastic fuzzy control (2nd report, relationships among a priori probabilities, fuzzy sets and control rules). *Trans. Japan Soc. Mechanical Engineers C* 62:1013–1018, 1996.

[39] K. Watanabe and A. Nomiyama. Stochastic fuzzy control (3rd report, application to trajectory tracking control of a mobile robot). *Trans. Japan Soc. Mechanical Engineers C* 62:1019–1025, 1996.

[40] K. Watanabe. Stochastic fuzzy control, I: Theoretical derivation. In *Proceedings FUZZ–IEEE/IFES '95* (Yokohama), Vol. 2, pp. 547–554, 1995.

[41] K. Watanabe and A. Nomiyama, and J. Tang. Stochastic fuzzy control, II: relationships among a priori probabilities, fuzzy sets, and control rules. In *Proceedings of the Fourth International Conference on Soft Computing* (Iizuka), Vol. 1, pp. 359–362, 1996.

[42] K. Watanabe. *Adaptive Estimation and Control*. Prentice–Hall, Hemel Hempstead, 1992.

[43] K. Watanabe, A. Nomiyama, and J. Tang. A design of stochastic fuzzy controller using a robust state feedback stabilization. In *Proceedings of the Fourth International Conference on Soft Computing* (Iizuka), Vol. 1, pp. 378–383, 1996.

[44] J. Tang, A. Nomiyama, and K. Watanabe. Stochastic fuzzy control law for path tracking in mobile robot. In *Proceedings of the 1996 Japan–U.S.A. Symposium on Flexible Automation* (Boston), Vol. 1, pp. 615–622, 1996.

[45] J. Tang, A. Nomiyama, and K. Watanabe. Stochastic fuzzy control for an autonomous mobile robot. In *Proceedings of the 1996 IEEE International Conference on Systems, Man and Cybernetics* (Beijing), Vol. 1, pp. 316–321, 1996.

[46] K. Watanabe and K. Noda. Position control of prismatic link using a stochastic fuzzy controller with a robust servo structure. In *Proceedings of the 1996 IEEE International Conference on Systems, Man and Cybernetics* (Beijing), Vol. 1, pp. 304–309, 1996.

Fuzzy Neural Network Systems in Model Reference Control Systems

Yie-Chien Chen

Department of Control Engineering
National Chiao-Tung University
Hsinchu, Taiwan

Ching-Cheng Teng

Department of Control Engineering
National Chiao-Tung University
Hsinchu, Taiwan

I. INTRODUCTION

In this chapter, we propose a model reference control system that uses fuzzy neural networks (FNNs). The proposed model reference control system belongs to indirect adaptive control. The controlled plant is identified by the fuzzy neural network identifier (FNNI), which approximates the system and provides the sensitivity of the plant for the fuzzy neural network controller (FNNC). This is a real adaptation system that can learn to control a complex system and adapt to a wide range of variations in plant parameters. Unlike most other adaptive learning neural controllers [1–8], the FNNC presented in this chapter is based not only on the theory of neural network computing but also on that of fuzzy logic [9].

Though the proposed control scheme is a slight modification of those in [6, 10], we believe that our structure is more reasonable for a fuzzy logic control system. Because the place for the reference model (RM) in the proposed system is specially considered, the FNNC is designed such that the actual output of the system will track the desired output of the reference model. Moreover, we can simply take the error (between the actual output and the desired output) and the change in this error as the input for the FNNC [2].

II. FUZZY NEURAL NETWORK

In this section, we study the fuzzy inference system first. Later, such a system is implemented by using the FNN which is a four-layered fuzzy neural network. Because the generalized fuzzy neural network (GFNN) is the basis of the FNN, the FNN will inherit the general properties from the GFNN. To capture the important concept of the FNN, the construction, learning process, and corresponding operations of the FNN will be described in the following subsections.

A. FUZZY INFERENCE SYSTEM

The main goal of fuzzy inference systems is to model human decision making within the conceptual framework of fuzzy logic and approximate reasoning. As is well known, a fuzzy inference system consists of four important parts: the fuzzification interface, knowledge base, decision-making unit, and defuzzification interface [11]. A fuzzy inference system is a model having the format of a fuzzy controller, which has been the most developed area of fuzzy set theory in engineering [12].

1. Generalized Modus Ponens

In this subsection, the operations of a fuzzy inference system are discussed based on the generalized modus ponens (GMP) [13]. A general fuzzy inference system with n inputs and p outputs can be described in the following format:

$$\frac{\text{Premise: } x \text{ is } A'}{\begin{array}{l}\text{Implication 1: If } x \text{ is } A^1, \text{ then } y \text{ is } B^1 \text{ else}\\ \text{Implication 2: If } x \text{ is } A^2, \text{ then } y \text{ is } B^2 \text{ else}\end{array}}$$

$$\vdots$$

$$\frac{\text{Implication } m: \text{If } x \text{ is } A^m, \text{ then } y \text{ is } B^m}{\text{Conclusion: } y \text{ is } B'}$$

where $x_i \in X_i$, $y_i \in Y_i$, and X_i and Y_i are the universe of discourse of the corresponding inputs and outputs, respectively. The n-array variable $x = [x_1, \ldots, x_n]$ denotes the input vector and the p-array variable $y = [y_1, \ldots, y_p]$ denotes the output vector. Vectors $\mathbf{A}^i = [A_1^i, A_2^i, \ldots, A_n^i]$ and $\mathbf{B}^i = [B_1^i, B_2^i, \ldots, B_n^i]$ are vectors of linguistic values referring to the fuzzy variables x and y, respectively. Vector $\mathbf{A}' = [A_1', \ldots, A_n']$ is the input observation vector and $\mathbf{B}' = [B_1', \ldots, B_p']$ is the output observation vector [14]. In a fuzzy inference system, A_i' is the result of applying fuzzification for numerical input x_i. This is the first step for rule reasoning. This means that a fuzzy inference system can be used for any nonfuzzy application.

2. Rule Inference

According to the compositional rule of inference [15], B_i' can be obtained by taking the sup-$*$ composition of fuzzy set $\mathbf{A}'$ and fuzzy relation $\mathbf{A}^j \to B_i^j$:

$$B_i' = (A_1' \text{ and } A_2' \text{ and } \cdots A_n') \circ (A_1^j \text{ and } A_2^j \text{ and } \cdots A_n^j), \qquad (1)$$

where $\circ$ denotes the sup-$*$ composition operation. "$*$" is the t-norm operator. The sup-min and sup-product composition are often used. The fuzzy relation used here is fuzzy implication. Note that Eq. (1) calculates only the jth individual consequence for B_i'. Now, let us consider the whole set of rules in the generalized modus ponens. The overall output fuzzy set B_i' can be obtained by taking the union of all the individual conclusions [13], that is,

$$B_i' = (A_1' \text{ and } A_2' \text{ and } \cdots A_n') \circ \bigcup_{j=1,\ldots,m} \left[(A_1^j \text{ and } A_1^j \text{ and } \cdots A_1^j) \to B_i^j \right]. \qquad (2)$$

If we take Larsen's product fuzzy implication [13] and the sup-product composition on Eq. (2), then we obtain the membership function of output fuzzy set B_i' in the following equation:

$$\mu_{B_i'}(y_i) = \bigvee_x \left[\prod_{l=i}^n \mu_{A_l'}(x) \cdot \bigvee_{j=l}^m \prod_{l=1}^n \mu_{A_l^j}(x) \mu_{B_i^j}(y_i) \right], \qquad (3)$$

where $\vee$ denotes the pairwise maximum operator. For simplicity, we take fuzzy singletons [15] on A_l' for $l = 1, \ldots, n$, that is,

$$\mu_{A_l'}(x) = \begin{cases} 1, & \text{if } x = x_1, \\ 0, & \text{otherwise.} \end{cases} \qquad (4)$$

Substituting Eq. (4) into Eq. (3), we obtain the following equation:

$$\mu_{B_i'}(y_i) = \bigvee_{j=l}^m \left(\prod_{l=1}^n \mu_{A_l^j}(x) \right) \mu_{B_i^j}(y_i). \qquad (5)$$

Because we want to use numerical output in most of the applications, Eq. (5) must be transformed to a numerical output by taking defuzzification. There are various methods of the defuzzification. Here we use the center of area (COA) method. Then the numerical output y_i inferred from the fuzzy logical rules can be determined from the output fuzzy set B_i' as follows:

$$y_i = \frac{\int_{Y_i} \mu_{B_i'}(y_i) y_i \, dy_i}{\int_{Y_i} \mu_{B_i'}(y_i) \, dy_i} \qquad \text{for } i = 1, \ldots, p. \tag{6}$$

3. Simplified Fuzzy Inference System

Many researchers have dealt with the modification of fuzzy inference systems by using different types of fuzzy logical rules, for example, [11, 16–18]. The motivation for modifying the fuzzy inference systems is as follows:

1. The pairwise maximum operation in Eq. (3) causes extreme difficulty in parameterizing the fuzzy logical rules. Also, it makes conventional estimation methods inapplicable.
2. The integral in Eq. (6) requires numerical analysis methods in computer simulations.

We can see that the modification of fuzzy inference systems is practically needed. However, the modified fuzzy inference system must still be a universal approximator [19]. A universal approximator means that, given a function $F: R^n \to R$ which is continuous, there exists a fuzzy inference system f such that f can approximate F uniformly on a compact subset of R^n to any degree of accuracy.

By taking the fuzzy singletons to represent the output fuzzy sets B_i^j, that is,

$$\mu_{B_i^j}(y_i) = \begin{cases} 1, & \text{if } y_i = \beta_i^j, \\ 0, & \text{if } y_i = \beta_i^j, \end{cases}$$

where the β_i^j are the fuzzy singletons, then a discrete form to calculate the ith numerical output y_i is obtained as follows:

$$y_i = \frac{\sum_{j=1}^{m} \beta_i^j \left(\prod_{l=1}^{n} \mu_{A_l^j}(x_l) \right)}{\sum_{j=1}^{m} \left(\prod_{l=1}^{n} \mu_{A_l^j}(x_l) \right)}. \tag{7}$$

The simplified fuzzy inference system in Eq. (7) has been proved to be a universal approximator by Jou [12] and Wang [19]. Because Eq. (7) is only an algebraical form, the disadvantages mentioned previously have disappeared.

We rewrite the generalized modus ponens based on the simplified fuzzy inference system as follows:

Premise: x is $\mathbf{A'}$

Implication 1: If x is A^1, then y is $[\beta_1^1, \beta_2^1, \ldots, \beta_p^1]$ else

Implication 2: If x is A^2, then y is $[\beta_1^2, \beta_2^2, \ldots, \beta_p^2]$ else

$$\vdots$$

Implication m: If x is A^m, then y is $[\beta_1^m, \beta_2^m, \ldots, \beta_p^m]$

Conclusion: y is $[\beta_1', \beta_2', \ldots, \beta_p']$

Each fuzzy if-then rule has the format

If x_1 is A_1^j and $\cdots$ and x_n is A_n^j, then y_1 is β_1^j, $\ldots$, y_p is β_p^j.

We can see that this fuzzy rule representation is exactly the same as Sugeno's fuzzy rules [17]. This indicates that if we use max-product inference to Sugeno's type of fuzzy rules, then we will get the same equation as Eq. (7).

4. Fuzzy Inference System and Neural Network

We know that neural networks have the capability of highly parallel distributed processing and learning from experience. An automatic structure for a fuzzy inference system utilizing the learning capability of a neural network is reasonable. In this subsection, we will construct a four-layered neural network structure to implement the fuzzy inference system as stated in Eq. (7). The construction of the neural network is restricted by the following conditions:

1. The fuzzy inference system can be directly pointed out in the neural network.
2. Every node at each layer has the physical meaning according to the fuzzy inference system.
3. The overall operations are equal to Eq. (7).

Figure 1 shows a GFNN. This fuzzy neural network consists of four layers. Nodes at layer 1 are input nodes which represent input linguistic variables. Nodes at layer 2 are membership nodes which act like membership functions. The membership node is responsible for mapping an input linguistic variable into a possibility distribution [13] for the variable being equal to it. The rule node resides in layer 3. Thus, all the connections between membership nodes and rule nodes indicate the if part or premise of fuzzy rules. The last layer node is the output node. The connections between rule nodes and output nodes indicate the then part or consequence of fuzzy rules.

In Fig. 1, we use the feedforward arrow to represent the whole connection links between the term nodes and the rule nodes. The arrow is used to indicate the connections of the antecedent part (if part) of the fuzzy rules. All of the connections

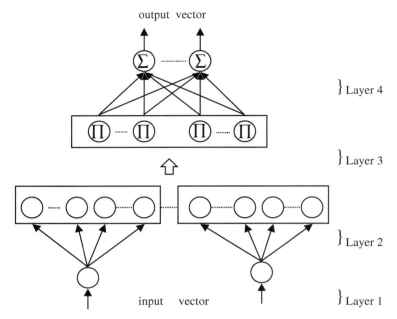

Figure 1   Generalized fuzzy neural network (GFNN).

of the fuzzy neural network must be predetermined because we want to develop an automatic fuzzy inference system. Obviously, there are many ways to set up a fuzzy neural network. In this chapter, we introduce an FNN based on the GFNN. It will be stated in the next section.

B. STRUCTURE OF THE FUZZY NEURAL NETWORK

The structure diagram of the proposed FNN is shown in Fig. 2. The specialty of the proposed FNN lies in the conditions for setting up the connections between layer 2 and layer 3. Its construction is directly based on the fuzzy rules without adjustment. For example, if we encounter the jth fuzzy rule described as follows:

$$\text{If } x_1 \text{ is } A_1^j \text{ and } x_2 \text{ is } A_2^j \cdots \text{ and } x_n \text{ is } A_n^j, \text{ then } y \text{ is } \beta_j,$$

then a connection structure based on these fuzzy rules is illustrated in Fig. 3. This forms the jth component of the FNN. For generality, we must consider m fuzzy rules which can be considered independently like dealing with the jth fuzzy rules. The complete fuzzy neural network is illustrated in Fig. 2.

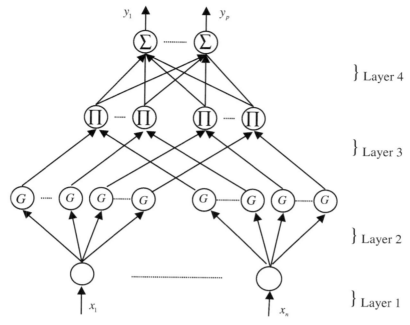

Figure 2   Structure diagram of the FNN.

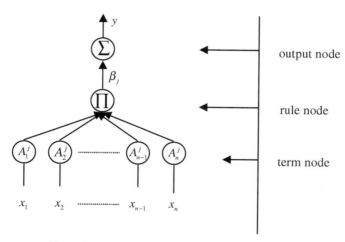

Figure 3   Construction of the jth component of the FNN.

However, we must emphasize that the FNN does not ensure that $A_i^j \neq A_i^k$ for $j \neq k$. The advantages of the FNN are as follows:

1. The structure of the FNN allows us to construct the fuzzy system rule by rule. In other words, we can implement each fuzzy rule without considering the other fuzzy rules.
2. If the prior knowledge of an expert is available, then we can directly add some nodes (rules nodes and term nodes) on the FNN.
3. We do not take an ordinary fuzzy partition of the input space; thus, the number of rules will not increase exponentially with the number of inputs.
4. Elimination of the redundant nodes (rule nodes and term nodes) are also rule by rule. This means that if we eliminate a rule node, then the associated terms are also removed from the FNN.

The disadvantage of the FNN lies in the requirements of a large amount of term nodes. As we see in Fig. 2, although some term sets are almost the same, we still require $m \times n$ term nodes at layer 2 for n inputs and m fuzzy rules.

C. LAYERED OPERATION OF THE FUZZY NEURAL NETWORK

We will consider the proposed FNN as a special type of neural network. Here, special type means both the special connections and the node operations. In the FNN, every layer and every node have the practical meaning because the FNN is constructed directly based on both fuzzy rules and fuzzy inference.

With the four-layer structure of the FNN, we will define the basic function of a node [20]; each node performs two actions using two different functions. The first function is the aggregation function $g^k(\cdot)$ which provides the input for the node, that is,

$$\text{Net input} = g^k(x^k; W^k), \tag{8}$$

where the superscript indicates the layer number, x^k denotes the input vector, and W^k denotes the connection weights vector. This notation will also be used in the following equations. The second function is the nonlinear activation function $f^k(\cdot)$ which gives the output an activation value as a function of its net input, that is,

$$\text{Output} = O_i^k = f^k(g^k), \tag{9}$$

where O_i^k is the ith output of the kth layer. Next, we will indicate the signal propagation, the basic function, and the practical meaning for every node at each layer.

Layer 1 (Input Layer)

The nodes at this layer are used to directly transmit the numerical inputs to the next layer. The output of the ith input node (O_i^1) is equal to the numerical input (x_i^1), that is,

$$g_j^1(x_i^1; W_{ij}^1) = W_{ij}^1 \cdot x_i^1, \tag{10}$$

$$O_j^1 = f_j^1(g_j^1) = g_j^1(x_i^1; W_{ij}^1), \tag{11}$$

where the weights at layer 1 are assumed to be unity, so no weight is adjusted here.

Layer 2 (Linguistic Term Layer)

At this layer, every node performs a membership function. The Gaussian function, a particular example of a radially symmetric function [21], is used as the membership function. The jth term set of the ith input maps the input x_i^2 into the membership degree, that is,

$$g_{ij}^2(x_i^2; W_{ij}^2) = g_{ij}^2(x_i^2; m_{ij}; \sigma_{ij}) = -\frac{(x_i^2 - m_{ij})^2}{\sigma_{ij}^2}, \tag{12}$$

$$O_{ij}^2 = f_{ij}^2(g_{ij}^2) = \exp\left[-\frac{(x_i^2 - m_{ij})^2}{\sigma_{ij}^2}\right], \tag{13}$$

where m_{ij} and σ_{ij} denote the mean (center) and variance (width) with respect to A_i^j. The adjusted weights at layer 2 are m_{ij}'s and σ_{ij}'s.

Layer 3 (Rule Layer)

This layer implements the related links for the term nodes and rule nodes. In other words, the antecedent matching will be determined here. The node at layer 3 performs the product operation. The net input and output of the jth rule node are

$$g_{ij}^3(x_{ij}^3; W_{ij}^3) = \prod_{i=1}^{n} W_{ij}^3 \cdot x_{ij}^3, \tag{14}$$

$$O_j^3 = f_j^3(g_j^3) = g_j^3(x_{ij}^3; W_{ij}^3), \tag{15}$$

where W_{ij}^3 is the connection weight between the jth term node of the ith input and the jth rule node. There is no weight adjusted here, that is, $W_{ij}^3 = 1$, $\forall i, j$.

Layer 4 (Output Layer)

This layer performs the defuzzification to get numerical outputs. The W_{ij}^4 connection weight between the ith rule node and the jth output node represents the consequence fuzzy singletons. If we use COA defuzzification, the node operation is

$$g_j^4(x_i^4; W_{ij}^4) = \sum_{i=1}^{m} W_{ij}^4 \cdot x_i^4, \tag{16}$$

$$O_j^4 = \frac{f_j^4(g_j^4)}{\sum_{i=1}^{m} x_i^4} = \frac{g_i^4(x_i^4; W_{ij}^4)}{\sum_{i=1}^{m} x_i^4}. \tag{17}$$

Equation (17) is fully based on Eq. (7), so the FNN with COA defuzzification will be a universal approximator. However, we need to note that the adopted FNN here is modified to be nonnormalized, that is, the operation in layer 4 is simply modified as

$$O_j^4 = g_i^4(x_i^4; W_{ij}^4) = \sum_{i=1}^{m} W_{ij}^4 \cdot x_i^4. \tag{18}$$

A nonnormalized FNN exhibits the desired performance for the identification and control of nonlinear systems. Moreover, there are two advantages of an FNN without a normalization process:

1. A faster training rate than the one which is normalized.
2. A much simpler form of the input–output sensitivity equations than in a normalized FNN.

In the next section, we will show that such a nonnormalized FNN can approximate any real continuous function.

D. Supervised Learning

The adjusted parameters in the FNN can be divided into two categories based on the if (premise) part and then (consequence) part of the fuzzy rules. In the premise part, we are asked to fine tune the mean and variance of the Gaussian functions, whereas, in the consequence part, the adjusted parameters are the consequence weights. Once the FNN has been initialized, a gradient-descent-based back-propagation (BP) algorithm [22–24] is employed to adjust the parameters of the FNN by using the training patterns. The main goal of supervised learning is to minimize the error function:

$$E = \tfrac{1}{2}(d(k) - y(k))^2, \tag{19}$$

where $y(k)$ is the output of the FNN and $d(k)$ is the desired output for the ith input pattern. If θ_{ij} is the adjusted parameter, then the learning rule used is

$$\theta_{ij}(k+1) = \theta_{ij}(k) - \eta \cdot \frac{\partial E}{\partial \theta_{ij}} + \alpha \cdot \Delta\theta_{ij}(k), \tag{20}$$

$$\Delta\theta_{ij}(k) = \theta_{ij}(k) - \theta_{ij}(k-1), \tag{21}$$

where η is the learning rate and α, between 0 and 1, is the momentum parameter (a value of 0.9 is often chosen for α).

To derive the learning law based on the back-propagation algorithm, we shall derive the computation of the $\partial E/\partial W_{ij}$ layer. We start this procedure from layer 4 because the error is back-propagated from this layer.

Layer 4

At this layer, the adjusted weights are W_{ij}^4. Using Eqs. (20) and (21), the adaptive rule of W_{ij}^4 is derived as follows:

$$
\begin{aligned}
-\frac{\partial E}{\partial W_{ij}^4} &= -\frac{\partial E}{\partial O_j^4} \cdot \frac{\partial O_j^4}{\partial W_{ij}^4} \\
&= (d_j^4 - O_j^4) \cdot x_i^4 \\
&= \delta_j^4 \cdot x_i^4,
\end{aligned}
\tag{22}
$$

where

$$\delta_j^4 = (d_j^4 - O_j^4),$$

and δ_j^4 is the error signal with respect to the jth output node. Hence, the consequence weights are updated by

$$W_{ij}^4(k+1) = W_{ij}^4 + \eta_W \cdot \delta_j^4(k) \cdot x_i^4(k) + \alpha_W \cdot \Delta W_{ij}^4(k), \tag{23}$$

where η_W and α_W are the learning rate and the momentum parameter for adjusting the parameter W_{ij}^4, respectively.

Layer 3

Only the error signal δ_i^3 needs to be computed and propagated because there is no weight adjustment at this layer. The error signal δ_i^3 is derived as follows:

$$\delta_i^3 = -\frac{\partial E}{\partial g_i^3} = -\sum_{j=1}^{p} \frac{\partial E}{\partial O_j^4} \cdot \frac{\partial O_j^4}{\partial O_i^3} \cdot \frac{\partial O_i^3}{\partial g_i^3} = \sum_{j=1}^{p} \delta_j^4 \cdot W_{ij}^4, \tag{24}$$

where p is the number of output nodes.

Layer 2

The adjusted parameters are m_{ij} and σ_{ij} at this layer. Using Eqs. (20) and (21), we can derive the adaptive rule of m_{ij} as follows:

$$
\begin{aligned}
-\frac{\partial E}{\partial m_{ij}} &= -\frac{\partial E}{\partial g_j^3} \cdot \frac{\partial g_j^3}{\partial O_{ij}^2} \cdot \frac{\partial O_{ij}^2}{\partial m_{ij}} \\
&= \delta_{ij}^3 \cdot \left(\prod_{i=1}^{n} O_{ij}^2 \right) \cdot \frac{2(x_{ij}^2 - m_{ij})}{\sigma_{ij}^2} \\
&= \delta_{ij}^2 \cdot \frac{2(x_{ij}^2 - m_{ij})}{\sigma_{ij}^2},
\end{aligned}
\tag{25}
$$

where

$$
\delta_{ij}^2 = \delta_j^3 \cdot \prod_{i=1}^{n} O_{ij}^2,
\tag{26}
$$

and δ_{ij}^2 is the error signal with respect to the jth term set of the ith input. Similarly, the adaptive rule of σ_{ij} is derived as follows:

$$
\begin{aligned}
-\frac{\partial E}{\partial \sigma_{ij}} &= -\frac{\partial E}{\partial g_j^3} \cdot \frac{\partial g_j^3}{\partial O_{ij}^2} \cdot \frac{\partial O_{ij}^2}{\partial \sigma_{ij}} \\
&= \delta_j^3 \cdot \left(\prod_{i=1}^{n} O_{ij}^2 \right) \cdot \frac{2(x_{ij}^2 - m_{ij})^2}{\sigma_{ij}^3} \\
&= \delta_{ij}^2 \cdot \frac{2(x_{ij}^2 - m_{ij})^2}{\sigma_{ij}^3}.
\end{aligned}
\tag{27}
$$

Thus, the update rules for m_{ij} and σ_{ij} are

$$
m_{ij}(k+1) = m_{ij}(k) + \eta_m \cdot \sigma_{ij}^2 \cdot \frac{2(x_{ij}^2 - m_{ij})}{\sigma_{ij}^2} + \alpha_m \cdot \Delta m_{ij}(k),
\tag{28}
$$

$$
\sigma_{ij}(k+1) = \sigma_{ij}(k) + \eta_\sigma \cdot \sigma_{ij}^2 \cdot \frac{2(x_{ij}^2 - m_{ij})^2}{\sigma_{ij}^3} + \alpha_\sigma \cdot \Delta \sigma_{ij}(k),
\tag{29}
$$

where η_m, η_σ and σ_m, σ_σ are the learning rates and the momentum parameters for adjusting the parameters m_{ij} and σ_{ij}, respectively.

E. INITIALIZATION OF THE FUZZY NEURAL NETWORK

The parameters of the FNN have clear physical meanings. This is one of the differences between the FNN and a typical back-propagation neural network (BPNN) [25, 26]. To initialize the connection weights of the BPNN, random values are frequently used because the relation between the weights and input–output data of the BPNN is unknown. In contrast to the BPNN, the parameters of the FNN have a clear relationship with the input–output data. Thus, the initial FNN can be constructed to a good approximation of an unknown function based on input–output data. Now, we will briefly describe an on-line initialization.

On-Line Initialization

In the on-line initialization method, the initialization takes place immediately after each training pattern has been presented. Let m be the default fuzzy rule number. Let X_i denote the universe of discourse of the input x_i and let a_i, b_i be the lower and upper bounds of X_i, that is, if $x_i \in X_i$, then $x_i \in [a_i, b_i]$. Suppose, at instant k, $1 \le k \le m - 2$, a training pattern $(x_1(k), \ldots, x_n(k); \ y(k))$ is presented. We can directly set the parameters

$$\begin{cases} \beta^k = y(k) & \text{and} & m_{ik} = x_i(k), & 1 \le i \le n \text{ for } 1 \le k \le m - 2, \\ \beta^k = 1 & \text{and} & m_{ik} = a_i, & 1 \le i \le n \text{ for } k = m - 1, \\ \beta^k = 1 & \text{and} & m_{ik} = b_i, & 1 \le i \le n \text{ for } k = m. \end{cases} \quad (30)$$

In this way, when $m - 2$ training patterns are presented, we can obtain m consequence weights (β^k, $k = 1, \ldots, m$) and the centers for the input fuzzy sets (A_i^k, $k = 1, \ldots, m$).

The remaining problem is how to determine the corresponding width (σ_{ik}) for A_i^k; this is also the main problem in the on-line initialization method. Though we can match the first $m - 2$ training pairs quite well by choosing σ_{ik} to be sufficiently small, we will have large approximation errors for other input–output pairs [19]. Therefore, the reasonable choice of σ_{ik} should make the input membership functions cover the input range in a good way. Moreover, the method in [19] results in a fixed value of σ_{ik} once the m training pairs are fed into the fuzzy neural network. We expect to obtain a more flexible result to satisfy our requirements.

In the fuzzy neural network systems [11, 16, 27], the initial parameter values can be easily set in such a way that the membership functions are equally spaced along the operating range of each input variable. Then these membership functions will satisfy ϵ-*completeness* [13, 15], which means that, given a value x of one of the inputs in the operating range, we can always find a linguistic label A

such that $\mu_A(x) \geq \epsilon$. In this manner, the fuzzy inference system can provide a smooth transition and sufficient overlapping from one linguistic label to another. Note especially that if the ϵ-completeness condition is not satisfied, there may be no fuzzy rules fired when the input data are fed into the fuzzy neural network. Thus, we want to present a flexible method to properly choose σ_{ik} such that the input membership functions can satisfy ϵ-completeness.

Before going further to show the choice and characteristic of σ_{ik}, we want to introduce the following notation. We note that the following notation is based on a fixed k or A_i^k, $1 \leq k \leq m$:

A_i^R: the closet fuzzy set of A_i^k on the right side of A_i^k,
A_i^L: the closet fuzzy set of A_i^k on the left side of A_i^k,
m_{iR}: the corresponding center of A_i^R,
m_{iL}: the corresponding center of A_i^L. The special choice for σ_{ik} is

$$\sigma_{ik} = \frac{\max\{|m_{ik} - m_{iR}|, |m_{ik} - m_{iL}|\}}{\sqrt{|\ln \lambda_i|}}, \tag{31}$$

where λ_i is the overlapping factor, $0 < \lambda_i < 1$. We now show that, by choosing σ_{ik} this way, the membership functions of the linguistic labels A_i^j, $j = 1, \ldots, m$, will cover X_i with a good property.

THEOREM 1.   *The fuzzy set* $\mathbf{A}_i = (A_i^1, A_i^2, \ldots, A_i^m)$, *where each linguistic label* A_i^j *has a Gaussian membership function constructed by the preceding initial* m_{ik} *[see Eq. (30)] and* σ_{ik} *[see Eq. (31)], will satisfy. That is,*

for all $x_i \in X_i$, *there exists* $k \in 1, 2, \ldots, m$ *such that* $\mu_{A_i^k}(x_i) \geq \epsilon = \lambda_i$,

where λ_i, $0 < \lambda_i < 1$, *is the overlapping factor.*

Proof.   Because $x_i \in X_i$, there must exist $k \in 1, 2, \ldots, m$, such that $m_{ik} \leq x_i \leq m_{iR}$ or $m_{iL} \leq x_i \leq m_{ik}$. We will prove this theorem under several different cases as follows:

1. If $m_{ik} \leq x_i \leq m_{iR}$ and $|m_{ik} - m_{iR}| \geq |m_{ik} - m_{iL}|$, then we have

$$\sigma_{ik} = \frac{|m_{ik} - m_{iR}|}{\sqrt{|\ln \lambda_i|}}.$$

By using the Gaussian membership function, we can obtain

$$\mu_{A_i^k}(x_i) \geq \mu_{A_i^k}(m_{iR})$$

$$= \exp\left[-|\ln \lambda_i| \cdot \left(\frac{m_{ik} - m_{iL}}{m_{ik} - m_{iR}}\right)^2\right]$$

$$= \lambda_i.$$

2. If $m_{iL} \leq x_i \leq m_{ik}$ and $|m_{ik} - m_{iR}| \geq |m_{ik} - m_{iL}|$, then σ_{ik} is the same as shown in case 1. Thus, we have

$$\mu_{A_i^k}(x_i) \geq \mu_{A_i^k}(m_{iL})$$

$$= \exp\left[-|\ln \lambda_i| \cdot \left(\frac{m_{iL} - m_{ik}}{m_{ik} - m_{iR}}\right)^2 \right]$$

$$\geq \exp\left[-|\ln \lambda_i| \cdot \left(\frac{m_{iR} - m_{ik}}{m_{ik} - m_{iR}}\right)^2 \right]$$

$$= \lambda_i.$$

The proof for the other cases induced by $|m_{ik} - m_{iR}| \leq |m_{ik} - m_{iL}|$ is very similar to cases 1 and 2. This completes the proof. ∎

Although we can incorporate prior expert information to choose a better initial parameter of the FNN, we finally gave up this attempt, because we believe that the proposed on-line initialization method is efficient and sufficient in practical applications. In fact, based on our simulation results in Section V, this is indeed true.

The on-line initialization method can be summarized as follows:

Step 1.

For $k = 1, 2, \ldots, m$ and $i = 1, 2, \ldots, n$, let

$$\begin{cases} \beta^k = y(k) & \text{and} & m_{ik} = x_i(k), & 1 \leq i \leq n \text{ for } 1 \leq k \leq m - 2, \\ \beta^k = 1 & \text{and} & m_{ik} = a_i, & 1 \leq i \leq n \text{ for } k = m - 1, \\ \beta^k = 1 & \text{and} & m_{ik} = b_i, & 1 \leq i \leq n \text{ for } k = m. \end{cases}$$

Step 2.

Let

$$\sigma_{ik} = \frac{\max\{|m_{ik} - m_{iR}|, |m_{ik} - m_{iL}|\}}{\sqrt{|\ln \lambda_i|}}.$$

III. MAPPING CAPABILITY OF THE FUZZY NEURAL NETWORK

In this section, we will show that the FNN can be used effectively for any real continuous function approximation. That is, an FNN with an arbitrarily large number of fuzzy logical rules can approximate any continuous function in $C(R^n)$ over a compact subset of R^n. It is described in the following theorem.

UNIVERSAL APPROXIMATION THEOREM. *For any given real function h:*
$R^n \rightarrow R^m$, *continuous on a compact set* $K \subset R^n$ *and arbitrary* $\epsilon > 0$, *there
exists an FNN system f such that* $\| f(x) - h(x) \| < \epsilon$. *Here* $\| \cdot \|$ *can be referred
to any norm.*

This theorem will be proved by using the Stone–Weierstrass theorem. We begin
with a single-output case and extend it to a multiple-output case later.

A. PROOF OF SINGLE-OUTPUT CASE

The structure diagram of the proposed FNN is shown in Fig. 2. The single
output of the FNN can be expressed as

$$y(x) = \sum_{j=1}^{m} \beta_j \cdot \phi_j(x), \qquad (32)$$

where

$$\phi_j(x) = \prod_{i=1}^{n} \mu_{A_i}^j(x_i) = \prod_{i=1}^{n} \exp\left[-\frac{(x_i - m_{ij})^2}{\sigma_{ij}^2} \right]$$

is a function of the input $x = (x_1, x_2, \ldots, x_n)$ and the link weight β_j is the
output action strength. Let Φ be of the form: $\prod_{i=1}^{n} \exp(-((x_i - b)/a)^2)$, where
$a, b \in R$. Let F^n be the family of the function $y: R^n \rightarrow R$ in the form of

$$y(x) = \sum_{j=1}^{m} \beta_j \cdot \phi_j, \quad \text{for } \beta_j \in R, \ \{\phi_j\} \in \Phi, \ x \in R^n, \ j = 1, 2, \ldots, m. \quad (33)$$

To prove the universal approximation, the following definitions [28] are necessary. A family A of real-valued functions defined on a set K is an algebra if A
is closed under addition, multiplication, and scalar multiplication. For example,
the set of all polynomials is an algebra. A family A is uniformly closed if A has
the property that $f \in A$ whenever $f_n \in A$, $n = 1, 2, \ldots$, and $f_n \rightarrow f$ uniformly on K. The uniform closure of A, denoted by B, is the set of all functions
which are limits of uniformly convergent sequences of members of A. By Weierstrass' famous theorem, it is known that the set of continuous functions on $[a, b]$
is the uniform closure of the set of polynomials on $[a, b]$. A separates points on
a set K if for every x, y in K, $x \neq y$, there exists a function f in A such that
$f(x) \neq f(y)$; A vanishes at no point of K if for each x in K there exists f in A
such that $f(x) \neq 0$.

STONE–WEIERSTRASS THEOREM [28]. *Let A be a set of real continuous
functions on a compact set K. If* (1) *A is an algebra;* (2) *A separates points on K;*

(3) *A vanishes at no point of K; then the uniform closure of A consists of all real continuous functions on K.*

To prove our main result, we will begin with the following lemmas:

LEMMA 1.   *Let F^n be defined as in Eq. (33). Then F^n is an algebra.*

Proof.   Let g_1, $g_2 \in F^n$, and $g_1 = \sum_{p=1}^{s} \alpha_p \cdot p$, $g_2 = \sum_{q=1}^{t} \gamma_q \cdot b_q$, where α_p, $\gamma_q \in R$, a_p, $b_q \in \Phi$, $p = 1, 2, \ldots, s$, $q = 1, 2, \ldots, t$, with

$$a_p = \prod_{i=1}^{n} \exp\left(- \left(\frac{x_i - m_{ip}}{\sigma_{ip}} \right)^2 \right), \qquad b_q = \prod_{i=1}^{n} \exp\left(- \left(\frac{x_i - v_{iq}}{u_{iq}} \right)^2 \right).$$

1. Since

$$\begin{aligned}
g_1 + g_2 &= \sum_{p=1}^{s} \alpha_p \cdot a_p + \sum_{q=1}^{t} \gamma_q \cdot b_q \\
&= (\alpha_1 \cdot a_1 + \alpha_2 \cdot a_2 + \cdots + \alpha_s \cdot a_s) \\
&\quad + (\gamma_1 \cdot b_1 + \gamma_2 \cdot b_2 + \cdots + \gamma_t \cdot b_t) \\
&= \sum_{k=1}^{s+t} r_k \cdot \theta_k
\end{aligned}$$

thus, $r_k = \alpha_k$ if $k \le s$, and $r_k = \gamma_{k-s}$ if $k > s$. So $r_k \in R$ (α_k, $\gamma_k \in R$). Furthermore, $\theta_k = a_k$ if $k \ge s$, and $\theta_k = b_{k-s}$ if $k > s$, so $\theta_k \in \Phi$ ($a_k, b_{k-s} \in \Phi$). That is, $g_1 + g_2 \in F^n$. This proves that F^n is closed under addition.

2. Let $d \in R$ be a scalar. Then we have $d \cdot g_1 = \sum_{p=1}^{s} (d \cdot \alpha_p) a_p \equiv \sum_{p=1}^{s} \beta_p \cdot a_p$, where $\beta_p \in R$. That is, $d \cdot g_1 \in F^n$. This proves that F^n is closed under scalar multiplication.

3. Let

$$\begin{aligned}
g_1 \cdot g_2 &= \left(\sum_{p=1}^{s} \alpha_p \cdot a_p \right) \cdot \left(\sum_{q=1}^{t} \gamma_q \cdot b_q \right) \\
&= \sum_{k=1}^{st} \varphi_k \cdot c_k,
\end{aligned} \qquad (34)$$

where

$$\varphi_k = \alpha_p \cdot \gamma_q,$$
$$c_k = a_p \cdot b_q$$

and

$$p = [(k-1)/t] + 1,$$
$$q = ((k-1) \bmod t) + 1.$$

Thus

$$c_k = \prod_{i=1}^{n} \exp\left(-\left(\frac{x_i - m_{ip}}{\sigma_{ip}}\right)^2\right) \cdot \prod_{i=1}^{n} \exp\left(-\left(\frac{x_i - v_{iq}}{u_{iq}}\right)^2\right)$$
$$= \prod_{i=1}^{n} \exp\left(-\left(\frac{x_i - m_{iq}}{\sigma_{ip}}\right)^2 - \left(\frac{x_i - v_{iq}}{u_{iq}}\right)^2\right). \tag{35}$$

After some computations, Eq. (34) can be written as

$$\eta_k \cdot \prod_{i=1}^{n} \exp\left(-\left(\frac{x_i - \omega_k}{\lambda_k}\right)^2\right) = \eta_k \cdot \zeta_k, \tag{36}$$

where

$$\eta_k = \prod_{i=1}^{n} \exp\left(-\frac{(m_{ip} - v_{iq})^2}{u_{iq}^2 + \sigma_{ip}^2}\right),$$

$$\omega_k = \frac{u_{ip}^2 \cdot m_{ip} + \sigma_{ip}^2 \cdot v_{iq}}{u_{iq}^2 + \sigma_{ap}^2},$$

$$\lambda_k = \frac{\sigma_{ip} \cdot u_{iq}}{\sqrt{u_{iq}^2 + \sigma_{ip}^2}},$$

and $\zeta_k \in \Phi$. Substituting into Eq. (33), we obtain

$$\sum_{k=1}^{st} \varphi_k \cdot c_k = \sum_{k=1}^{st} \varphi_k \cdot \eta_k \cdot \zeta_k = \sum_{k=1}^{st} \rho_k \cdot \zeta_k.$$

That is, $g_1, g_2 \in F^n$. This proves that F^n is closed under multiplication.

By cases 1–3 we conclude that F^n is an algebra. ∎

LEMMA 2.   *F^n separates points on K.*

Proof.   We prove this by constructing a function f. That is, we specify the number of fuzzy sets defined in K, the parameters of the Gaussian membership functions, and the number of fuzzy rules, such that the resulting f [in the form of (33)] has the property that $f(\underline{x}^0) \neq f(\underline{y}^0)$ for arbitrary $\underline{x}^0, \underline{y}^0 \in K$ with $\underline{x}^0 \neq \underline{y}^0$.

Let $\underline{x}^0 = (x_1^0, x_2^0, \ldots, x_n^0)$ and $\underline{y}^0 = (y_1^0, y_2^0, \cdots, y_n^0)$. We choose two fuzzy rules for the fuzzy rule base, and let the Gaussian membership functions be

$$\mu_{A_i^1}(x_i) = \exp\left(-\frac{(x_i - x_i^0)^2}{2}\right),$$

$$\mu_{A_i^2}(x_i) = \exp\left(-\frac{(x_i - y_i^0)^2}{2}\right).$$

Then f can be expressed as

$$f(x) = \beta_1 \cdot \prod_{i=1}^n \exp\left(-\frac{(x_i - x_i^0)^2}{2}\right) + \beta_2 \cdot \prod_{i=1}^n \exp\left(-\frac{(x_i - y_i^0)^2}{2}\right),$$

where β_1, β_2 are the link weights. With this f, we have

$$f(\underline{x}^0) = \beta_1 + \beta_2 \cdot \prod_{i=1}^n \exp\left(-\frac{(x_i - y_i^0)^2}{2}\right),$$

$$f(\underline{y}^0) = \beta_1 \cdot \prod_{i=1}^n \exp\left(-\frac{(x_i - x_i^0)^2}{2}\right) + \beta_2.$$

Because $\underline{x}^0 \neq \underline{y}^0$, there must be some i such that $x_i^0 \neq y_i^0$. Hence, we have $\prod_{i=1}^n \exp(-(x_i - y_i^0)^2/2) \neq 1$. If we choose $\beta_1 = 1$ and $\beta_2 = 0$, then it is easy to find that $f(\underline{x}^0) = \beta_1 \neq f(\underline{y}^0)$. ∎

LEMMA 3. *F^n vanishes at no point of K.*

Proof. From Eq. (33), if we choose $\beta_j > 0$, $j = 1, 2, \ldots, m$, then $y > 0$ for any $x \in K$. That is, any $y \in F^n$ with $\beta_j > 0$ can serve as the required f. ∎

Therefore, that the FNN having only a single output is a universal approximator is a direct consequence of the Stone–Weierstrass theorem and Lemmas 1–3. ∎

From the previous proof, we can conclude that, given a real function $h: R^n \rightarrow R$, continuous on K, and $\epsilon > 0$, there exists an FNN system $y \in B$, where B is the uniform closure of F^n, such that $|y(x) - h(x)| < \epsilon$ for every x in K. That is, an FNN system with an arbitrarily large number of fuzzy logical rules can approximate any real continuous function in $C(R^n)$ over a compact subset of R^n.

B. EXTENSION TO MULTIPLE-OUTPUT CASE

In the following, we will extend the previous results to the FNN of multiple outputs. Let us consider the following example.

Suppose f_1 and f_2 are functions of x, where $x = (x_1, x_2, \ldots, x_n)$. Furthermore, assume function f_1 can be approximated with m rules, whose structure is shown in Fig. 4a and function f_2 can be approximated with p rules, whose structure is shown in Fig. 4b.

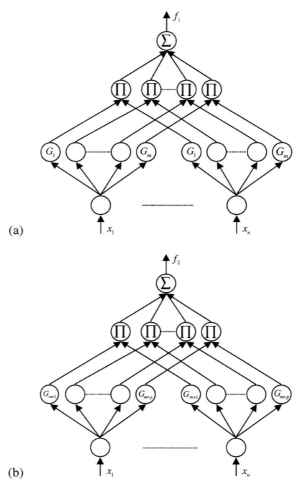

(a)

(b)

Figure 4 Structure diagrams of f_1 and f_2.

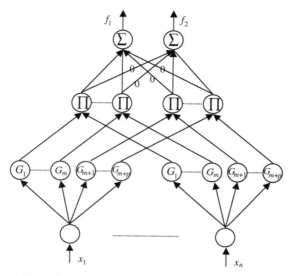

Figure 5 Structure diagram of a multiple-output system.

It is easy for us to combine these two individual FNNs, each with a single output, into a new FNN having two outputs. The new structure is shown in Fig. 5. The new structure has $m+p$ rules; the first m rules are constructed from f_1 and the last p rules are constructed from f_2. With this assignment, the consequence weights of the last p rules associated with the first output (f_1) are set to zero, so are the consequence weights of the first m rules associated with the second output (f_2).

Based on the previous discussion, because the FNN of a single output can perform the universal approximation, there must exist an FNN of multiple outputs with an arbitrarily large number of fuzzy logical rules that can perform the universal approximation on each output. This completes the proof of the universal approximation theorem.

IV. MODEL REFERENCE CONTROL SYSTEM USING A FUZZY NEURAL NETWORK[1]

Figure 6 shows the proposed model reference control system using a fuzzy neural network. The control scheme must perform two major tasks: (1) system identification and (2) plant control. The former is achieved by using the fuzzy

[1]Parts of this section are reprinted from *Fuzzy Sets and Systems* 73:291–312, 1995 with kind permission of Elsevier Science-NL, Sara Burgerhartstraat 25, 1055 KV Amsterdam, The Netherlands.

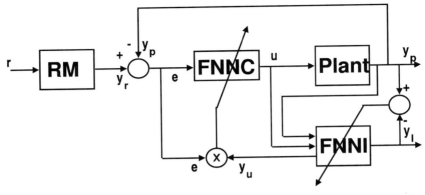

Figure 6 Model reference control system using a fuzzy neural network.

neural network identifier (FNNI) to estimate the dynamics of the controlled plant. The latter is achieved by using the fuzzy neural network controller (FNNC) to generate the control signals. The control action generated by the FNNC is updated by observing the controlled results through the FNNI.

A. OVERALL STRUCTURE OF THE SYSTEM

Fuzzy Neural Network Identifier

The objective of the FNNI is to mimic the dynamic characteristics of the controlled plant. Training of the FNNI is similar to plant identification except that the plant identification here is done automatically by a fuzzy neural network which is capable of modeling nonlinear plants [9]. The FNNI is trained by the preceding algorithm to predict the state vector of the plant y_I, with the actual value of the state of the plant y_p used as the desired response. The training process stops when the error signal between y_I and y_p is small enough. If changes in the system parameters or the environment occur, the FNNI is triggered on again to begin relearning.

Fuzzy Neural Network Controller

The fuzzy neural network here serves as a controller. The FNNC is expected to approximate an optimal control surface. The surface is encoded in the form of fuzzy rules, which are represented by the interconnection weights embedded in the FNNC. Thus, the weights can be modified to established different control rules. As time goes on and the system accumulates more experience, it learns to control the plant more effectively. A controlled plant is identified by the FNNI, which provides the sensitivity of the plant to the FNNC. The plant sensitivity is used in Eq. (41).

Reference Model

The reference model specifies the desired output performance of the control system. The controller is designed such that the actual output of the system will track the desired output of the reference model. This goal can be achieved by minimizing $e = (y_r - y_p)$.

Note that our structure is different from that in [6], in which the reference model is placed on the left side. We believe that our structure is more reasonable for a fuzzy system.

B. TRAINING THE FUZZY NEURAL NETWORK IDENTIFIER AND THE FUZZY NEURAL NETWORK CONTROLLER

Let the cost function, E_I, for training pattern k be proportional to the sum of the square of the difference between the plant output $y(k)$ and the actual output $y_I(k)$ of FNNI, and let E_I be defined by

$$E_I = \tfrac{1}{2}[y(k) - y_I(k)]^2. \tag{37}$$

Then the gradient of error in Eq. (37) with respect to an arbitrary weighting vector $W_I \in R^n$ is as follows:

$$\frac{\partial E_I}{\partial W_I} = e_I(k)\frac{\partial e_I(k)}{\partial W_I} = -e_I(k)\frac{\partial y_I(k)}{\partial W_I}$$
$$= -e_I(k)\frac{\partial O_I(k)}{\partial W_I}, \tag{38}$$

where $e_I(k) = y(k) - y_I(k)$ is the error between the plant and the FNNI response. $O_I(k)$ is the actual output of the identifier (FNNI).

The weight can be adjusted by the following formula:

$$W_I(k+1) = W_I(k) + \Delta W_I(k) = W_I(k) + \eta_I\left(-\frac{\partial E_I}{\partial W_I}\right), \tag{39}$$

where η_I is a learning rate.

Similarly, let the cost function, E_C, for training pattern k be proportional to the sum of the square of the difference between the desired output $y_r(k)$ of the reference model and the plant output $y(k)$, and let E_C be defined by

$$E_C = \tfrac{1}{2}[y_r(k) - y(k)]^2. \tag{40}$$

Then the gradient of error in Eq.(40) with respect to an arbitrary weighting vector $W_C \in R^n$ is as follows:

$$
\begin{aligned}
\frac{\partial E_C}{\partial W_C} &= e_C(k)\frac{\partial e_C(k)}{\partial W_C} = -e_C(k)\frac{\partial y(k)}{\partial W_C} \\
&= -e_C(k)\frac{\partial y(k)}{\partial u(k)} \cdot \frac{\partial u(k)}{\partial W_C} = -e_C(k)y_u(k) \cdot \frac{\partial O_C(k)}{\partial W_C},
\end{aligned} \tag{41}
$$

where $e_C(k) = y_r(k) - y(k)$ is the error between the actual plant and desired reference output, $O_C(k)$ is the output of the controller (FNNC), and $S = y_u(k) = \partial y(k)/\partial u(k)$ is called the plant sensitivity.

The weight can be adjusted by the following formula:

$$W_C(k+1) = W_C(k) + \Delta W_C(k) = W_C(k) + \eta_C\left(-\frac{\partial E_C}{\partial W_C}\right), \tag{42}$$

where η_C is a learning rate.

The plant sensitivity can be computed as follows:

$$
\begin{aligned}
\frac{\partial y_j}{\partial u_i} &= \sum_{a=1}^{R_I} W_{aj} \cdot \left\{ \sum_{k=1}^{N_{m_i}} \frac{\partial O_{ka}^{(3)}}{\partial O_{ik}^{(2)}} \cdot \frac{\partial O_{ik}^{(2)}}{\partial u_i} \right\} \\
&= \sum_{a=1}^{R_I} W_{aj} \cdot \left\{ \prod_{b \neq i} O_{bL}^{(2)} \cdot \frac{\partial O_{ik}^{(2)}}{\partial u_i} \right\} \\
&= \sum_{a=1}^{R_I} W_{aj} \cdot \left\{ O_{ka}^{(3)} \cdot (-2) \cdot \frac{(u_i - m_{ik})}{(\delta_{ik})^2} \right\},
\end{aligned} \tag{43}
$$

where m_{ik} and σ_{ik} are, respectively, the mean (or center) and the variance (or width) of the Gaussian function in the kth term of the ith input linguistic variable u_i. The superscript denotes the layer number. The link weight W_{aj} is the output action strength of the jth output associated with the ath rule. N_{m_i} is the number of fuzzy sets of the ith input linguistic variable u_i. R_I is the number of rules in the FNNI. Some convergence theorems for selecting appropriate learning rates have been proved in [29]. The interested reader is referred to [29].

V. SIMULATION RESULTS

In this section, we test the model reference control system with an example. The number of inputs for the FNNC is denoted by n_C and that of the FNNI by n_I. R_C and R_I denote the number of rules in the FNNC and FNNI. P_C and P_I are the inputs to the FNNC and FNNI.

Example: A Nonlinear Unstable Plant

In this case the plant is described by the differential equation [8]

$$\dot{y} = \tfrac{1}{2}y^2 - y + u.$$

The reference model is 1. The objective is to control the nonlinear plant such that the desired value is 2.0. This problem is a stability regulation problem.

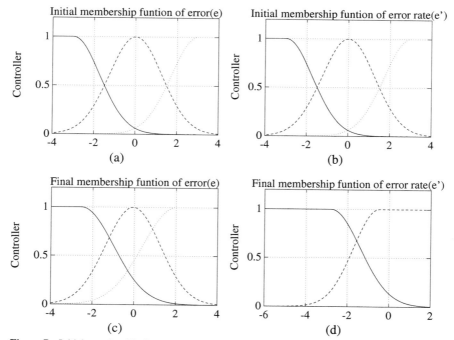

Figure 7 Initial membership functions of controller: (a) error and (b) error rate. Final membership functions of controller: (c) error and (d) error rate.

Assume that for the FNNC, $P_C = \{e(t), \dot{e}(t)\}$; each input variable has three fuzzy partition sets. We have $R_C = 3 \times 3 = 9$ rules, and $n_C = 9 + 2 \times (3+3) = 21$. For the FNNI, $P_I = \{u(t), y(t)\}$; each input variable has three fuzzy partition sets. We have $R_I = 3 \times 3 = 9$ rules, and $n_I = 9 + 2 \times (3+3) = 21$. See Figs. 7a and b and 8a and b.

Each cycle takes 6 seconds. After 10 cycles, the plant can be controlled very effectively, with a step size of 0.015 seconds.

In the simulation, we find a way to cancel the redundant rules. In each cycle, if the value of a particular consequence link rule is smaller than $1/R_C = 1/9$ for the FNNC or $1/R_I = 1/9$ for the FNNI, then we eliminate that rule. In the final simulation result, we find that the FNNC has five rules and the FNNI has four rules. See Figs. 7c and d and 8c and d and Tables I and II. The final result is shown in Fig. 9.

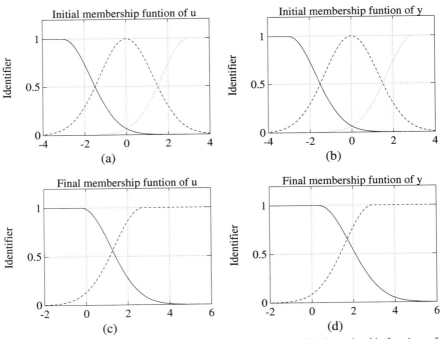

Figure 8 Initial membership functions of identifier: (a) u and (b) y. Final membership functions of identifier: (c) u and (d) y.

Table I
Learned Rule Weight Matrix for the FNNC

		$\dot{e}$	
e	NM	ZE	PM
NM	0.000	−2.880	0.000
ZE	1.165	0.042	0.000
PM	0.012	1.161	0.000

Table II
Learned Rule Weight Matrix for the FNNI

		u	
y	NM	ZE	PM
NM	0.000	0.000	0.000
ZE	0.000	0.305	−0.608
PM	0.000	2.059	2.477

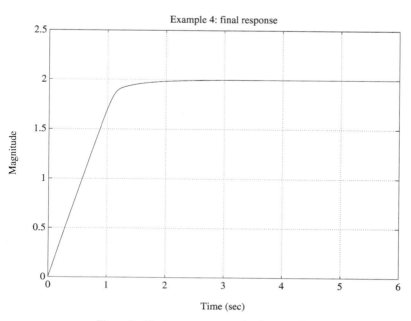

Figure 9   Final system response for the example.

VI. CONCLUSIONS

We have investigated a fuzzy neural network structure which can be successfully applied to a model reference control system. First, we study a simple fuzzy-logic-based neural network system, in which the knowledge of rules is explicitly expressed in the weights of the neural network and inferences are executed efficiently at a high rate. Then the capability of the universal approximation of the fuzzy neural network is proved in detail.

In model reference control, two fuzzy neural networks are used. One is for a controller (FNNC); the other is for an identifier (FNNI). The system has been tested for its on-line adaptive ability, robustness, and interpolation ability. The simulation result showed that combining fuzzy logic and neural network computing appears to be a feasible way of dealing with real-time application.

REFERENCES

[1] C. W. Anderson. Learning to control an inverted pendulum using neural networks. *IEEE Control Systems Mag.* 9:31–37, 1989.
[2] A. G. Barto, R. S. Sutton, and C. W. Anderson. Neuron like adaptive elements that can solve difficult learning control problems. *IEEE Trans. Systems Man Cybernet.* 13:834–846, 1983.
[3] A. Guez and J. Selinsky. A trainable neuromorphic controller. *J. Robotics Systems* 5:363–388, 1988.
[4] C. C. Ku and K. Y. Lee. Diagonal recurrent neural networks for dynamic systems control. *IEEE Trans. Neural Networks* 6:144–156, 1995.
[5] Y. Li, A. K. C. Wang, and F. Yang. Optimal neural network control. In *IFAC–INCOM*, pp. 41–46, 1992.
[6] K. S. Narendra and K. Parthasarathy. Identification and control of dynamical systems using neural networks. *IEEE Trans. Neural Networks* 1:4–27, 1990.
[7] G. J. Wang and D. K. Miu. Unsupervised adaptation neural-network control. In *Proceedings of the IEEE International Joint Conference on Neural Networks*, Vol. 3, pp. 421–428, 1990.
[8] H. Ying, W. Siler, and J. J. Buckley. Fuzzy control theory: a nonlinear case. *Automatica* 26:513–520, 1990.
[9] Y. C. Chien. Adaptive fuzzy logic controller using neural networks. Masters Thesis, National Chiao-Tung University, 1992.
[10] Y. C. Chien, Y. C. Chen, and C. C. Teng. Model reference adaptive fuzzy logic controller design using fuzzy neural network. In *Proceedings of the First Asian Fuzzy Systems Symposium*, (Singapore), pp. 334–340, 1993.
[11] J. S. Jang. ANFIS: adaptive-network-based fuzzy inference system. *IEEE Trans. Systems Man Cybernet.* 23:665–684, 1993.
[12] C. C. Jou. On the mapping capability of fuzzy inference system. In *Proceedings of the International Joint Conference on Neural Networks*, (Baltimore), pp. 708–713, 1992.
[13] C. C. Lee. Fuzzy logical in controller system: fuzzy logical controller, II. *IEEE Trans. Systems Man Cybernet.* 20:419–435, 1990.
[14] A. Lotfi and A. C. Tsoi. Adaptive membership function for fuzzy inference systems. In *Proceedings of the First Asian Fuzzy Systems Symposium*, (Singapore), pp. 628–635, 1993.

[15] C. C. Lee. Fuzzy logical in controller system: fuzzy logical controller, I. *IEEE Trans. Systems Man Cybernet.* 20:404–418, 1990.

[16] S. Horikawa, T. Furuhashi, and Y. Uchikawa. On fuzzy modeling using fuzzy neural networks with back-propagation algorithm. *IEEE Trans. Neural Networks* 3:801–806, 1992.

[17] M. Sugeno and T. Yasukawa. A fuzzy-logical-based approach to qualitative modeling. *IEEE Trans. Fuzzy Systems* 1:7–31, 1993.

[18] M. Sugeno and G. T. Kang. Structure identification of fuzzy model. *Fuzzy sets systems* 28:15–33, 1988.

[19] L. X. Wang. *Adaptive Fuzzy Systems and Control: Design and Stability Analysis.* Prentice–Hall, Englewood Cliffs, NJ, 1994.

[20] C. T. Lin and C. S. G. Lee. Neural-network-based fuzzy logical control and decision system. *IEEE Trans. Comput.* 40:1320–1336, 1991.

[21] J. Moody and C. J. Darken. Fast learning in networks of locally-turned procession units. *Neural Comput.* 1:281–294, 1989.

[22] D. E. Rumelhart and J. L. McClelland, Eds. *Parallel Distributed Processing*, Vol. I. MIT Press, Cambridge, MA, 1986.

[23] P. J. Webos. Back propagation through time: what it does and how to do it. *Proc. IEEE* 78:1550–1560, 1990.

[24] R. Hecht-Nielsen. Theory of the backup-propagation neural network. In *Proceedings of the IEEE International Joint Conference on Neural Networks*, Vol. 1, pp. 593–605, 1989.

[25] Y. F. Wang, J. B. Cruz, and J. H. Mulligan. Multiple training for back-propagation neural networks for use in associative memories. *Neural Networks* 6:1169–1175, 1993.

[26] B. Kosko. *Neural Networks and Fuzzy Systems.* Prentice–Hall, Englewood Cliffs, NJ, 1992.

[27] C. W. Xu and Y. Z. Lu. Fuzzy modeling identification and self-learning for dynamical systems. *IEEE Trans. Systems Man Cybernet.* 17:683–689, 1987.

[28] W. Rudin. *Principles of Mathematical Analysis*, 3rd ed. McGraw–Hill, New York, 1976.

[29] Y. C. Chen and C. C. Teng. A model reference control structure using a fuzzy neural network. *Fuzzy Sets Systems* 73:291–312, 1995.

Wavelets in Identification

A. Juditsky
Institut de Recherche en
Informatique et Systemes
Aleatoires (IRISA)
Campus Universitaire de
Beaulieu
35042 Rennes Cedex
France

Q. Zhang
Institut de Recherche en
Informatique et Systemes
Aleatoires (IRISA)
Campus Universitaire de
Beaulieu
35042 Rennes Cedex
France

B. Delyon
Institut de Recherche en
Informatique et Systemes
Aleatoires (IRISA)
Campus Universitaire de
Beaulieu
35042 Rennes Cedex
France

P.-Y. Glorennec
Institut de Recherche en
Informatique et Systemes
Aleatoires (IRISA)
Campus Universitaire de
Beaulieu
35042 Rennes Cedex
France

A. Benveniste
Institut de Recherche en
Informatique et Systemes
Aleatoires (IRISA)
Campus Universitaire de
Beaulieu
35042 Rennes Cedex
France

I. INTRODUCTION, MOTIVATIONS, BASIC PROBLEMS

In his inspiring tutorial [1], Ljung quoted the following:

An engineer, who is faced with [characterizing, or predicting, the behavior of his or her plant based on recorded data] *has the following perspective*:

- How can I best use the information in the observed data to calculate a model of the system's properties?
- How can I know if the model is any good, and how can I trust it for simulation and design purposes?
- How shall I manipulate the input signals to obtain as much information as possible about the system?
- What kind of software support is available for doing the tasks?

Later on in the same article, Ljung discusses the question of model nature and structure. By model nature, we have in mind the following classification:

- physical models,
- semiphysical models, also called "gray-box" models,
- black-box models.

This chapter mainly concentrates on the last category, namely black-box models. And, within black-box models, we shall concentrate on the less popular ones in the control community, namely those that are *nonlinear and nonparametric* in nature. Here "nonlinear" means that our model class will not be restricted to linear input–output maps. And "nonparametric" means that our models do have parameters, but in a quantity that is not *a priori* fixed, but fully depends on the data; consequently, convergence issues and quality of fit cannot be assessed in terms of the involved parameters, but rather more globally in terms of the global behavior. "Nonlinear and nonparametric" thus will be our general perspective throughout this chapter. Although this setting may appear quite technical, more familiar and even some exotic ones will also be covered, such as *neural networks* [2–4], *wavelets* [5, 6], *and fuzzy models* [7]. A typical form of the kind of model class that we shall consider is the popular single hidden layer neural network for static systems:

$$f_n(x) = \sum_{i=1}^{n} c_i \sigma(a_i^T x + t_i) + c_0, \tag{1}$$

where σ is the sigmoid function, $x \in \mathbf{R}^d$ is the input, n is the number of neurons, and the (c_i, a_i, t_i)'s are the adjustable parameters. This is clearly nonlinear in x, and the size n of the network is to be tuned on the data. In addition, in this case, the model is also nonlinear in the parameters.

Such models have gained increasing interest, as reflected, for instance, in the articles [2–4]. This is due to their ability to encompass truly nonlinear behaviors, including those involved in classification and, more generally, decision procedures. Referring to Ljung's practical problem setting given previously, the following practical questions must be investigated when using nonlinear nonparametric models such as (1):

- *How can good nonlinear nonparametric models extrapolate or predict behaviors outside the range of data used for their identification, fitting, tuning, or training?*[1] Predicting behaviors is one of the main purposes of system identification. It is not usual to ask such a question about linear system identification, because good linear model fitting generally also provides good prediction for truly

---

[1] These are more or less equivalent words used by different communities; we shall use any one of these indifferently.

linear plants. However, this is of primary concern in our case, because nonlinear systems are in essence not easily predictable outside the range of available observations. This question is also related to that of the appropriate choice of inputs for identification.

• *How can nonlinear nonparametric models be used for system monitoring and diagnostics?* Such models are, in principle, good candidates for system monitoring, because they are able to describe systems behaviors at *all* operating points simultaneously, thus preventing confusion between changes in operating point and changes in systems behavior. However, it is not clear how changes could be interpreted using such models, that is, how diagnostics could be performed.

Then the user is faced with a second question, namely how identification should be performed:

• *How should data be used to fit a nonlinear nonparametric model?* Though different situations can occur, we shall mainly investigate the classical situation in which noisy input–output measurements are available.

• *How can one take advantage of any kind of prior knowledge for some partial tuning or pretuning of the model?* Such a coarsely tuned model is sometimes sufficient, and sometimes used as an initial guess for system identification. Also, prior information can be critical for diagnostics. Again, linear systems engineering can serve as a guide for us: response times, resonant modes, delay, and others are typical qualitative information that engineers may have from experience about their plants, and they know how to reflect this prior knowledge into linear models. For nonlinear nonparametric models, no obvious alternative seems to exist: what kind of prior knowledge is relevant for such models, and how does one express it? Thus it seems that the engineer must entirely rely on fitting from data, without taking advantage of some prior knowledge he or she may have; we shall see that fuzzy models and their rules may be good candidates to express such prior knowledge.

• *What kind of software support is available for doing the tasks?*

Moving one step further toward a mathematical formulation of our problems, we may translate some of these questions into the following more technical ones:

• *How does one assess the quality of approximation?* Given a true system f and an approximation $\hat{f}$ of it, how does one measure the quality of approximation? No parametric distance can be used. And because nonlinear systems are considered, the usual operator norms from linear system theory cannot be considered. In the second part of this section, based on a few examples, we shall introduce the distance measures we shall use throughout this chapter. These will mainly be L_p-type norms involving $f - \hat{f}$ and possibly some derivatives of it. Note that using such distance measures involves some kind of prior knowledge, namely the assumption that the system under consideration belongs to the considered space, and this is a smoothness prior information.

• *How does one measure the quality of fit from noisy data?* This is really assessing the quality of system identification. We shall naturally use figures of merit of the form $\mathbf{E}\| f - \hat{f}_N \|$, where $\| \cdot \|$ denotes a norm such as discussed previously, $\hat{f}_N$ is the estimate of f based on an N-sample record, and $\mathbf{E}$ is the expectation with respect to all kinds of uncertainties (input, output, and noise).

• *What plays the role of "Cramer–Rao bounds," and what does it mean for an estimator to be "optimal"?* Such criteria are important in assessing the relative performance of estimators, especially because of the very large variety of models and identification procedures proposed so far.

• *How efficient are identification algorithms really in terms of computational cost and quality of conditioning?* Because our model classes often are nonlinear in the parameters, tuning procedures may be of prohibitive cost and may further be ill-behaved (cf. the well-known "backpropagation" algorithm for neural network training).

• *What kind of coarse or qualitative property can be asserted about the models we consider,* apart from smoothness prior information such as discussed previously?

These are some of the issues that we shall discuss throughout this chapter. The chapter is organized as follows. The remainder of this section is devoted to the two applications we selected for a more detailed discussion. Then we discuss some basic mathematical problems relevant to our nonparametric setting, and justify, by the way, the use of some specific distance measures between systems and their estimates.

In Section II, the classical background of nonparametric estimation is visited. First, so-called "linear" estimators (be careful that systems and models are nevertheless nonlinear) are presented and discussed: kernel, piecewise-polynomial, and projection estimators are typical instances. Then the issue of selecting the "model order" is discussed and generalized cross validation is introduced. In a second subsection, convergence rates and performance criteria are analyzed, and it is shown that classical linear estimators perform poorly for systems with sparse singularities—such nonlinear systems frequently occur in practice. Some existing nonlinear estimation techniques which provide spatial adaptation are briefly discussed in the last subsection; these include sigmoid-based neural networks and an interesting alternative proposed by Leo Breiman, namely the "hinging hyperplanes" which are, in fact, piecewise linear models such as used in control by Sontag in the early 1980s [8]. Such nonlinear estimators with spatial adaptation are not supported by satisfactory mathematical analysis, however. This motivates investigating wavelets.

Wavelets are introduced in Section III and their contribution to function approximation theory is briefly reported. In particular, orthonormal bases of wavelets for L^2-type spaces are presented. The importance of *Besov spaces* of functions is emphasized, for modeling smooth systems with sparse singularities.

Besov spaces are closely related to the more usual Sobolev spaces. The optimality of wavelet expansions in Besov spaces of functions is discussed. The central role of Besov spaces for wavelets was pointed out by Yves Meyer.

How wavelets can be best used for estimation is the topic of Section IV. We report on and discuss the simple and elegant method of "wavelet shrinking" as introduced by David Donoho and co-workers.

Building orthonormal bases of wavelets, for even medium-large-dimensional input spaces (say, ≥ 10), becomes prohibitive in terms of memory requirements. Thus an alternative method is proposed in Section V, which is still based on wavelets, but in a different manner. This method is suitable for sparse training data sets, that is, data sets whose cardinality does not grow exponentially with the dimension of the input space.

Now, the question of how to practically express available prior knowledge for nonparametric models is still open. In Section VI, we discuss a proposal toward achieving this, which is based on fuzzy models and their associated rules. An extension of the usual fuzzy models is proposed to capture the multiresolution aspects of wavelet-based estimators.

The experimental results of some of these methods are reported in Section VII.

Finally, both the practical and the mathematical aspects are summarized and discussed in Section VIII.

A. TWO APPLICATION EXAMPLES

1. Modeling a Gas Turbine System: An Example of Identification of a Static Nonlinear System

Here we briefly present the case study of a gas turbine system, as an example of identification of a static nonlinear system. Results and experiments will be reported in Section VII. Gas turbines are power motors, typically used in electrical power generators and aircraft. Usually a gas turbine system is composed of a compressor, one or several combustion chambers, and an expansion turbine. The compressor produces high-pressure air which is then mixed with the fuel. This mixed gas is burned in the combustion chambers to increase its temperature and pressure. The burned gas is then forwarded to the expansion turbine. The pressure of the gas drives the rotor of the expansion turbine, which, in turn, drives the compressor. The residual energy can then be used for producing electricity, and the gas is rejected at the exhaust of the expansion turbine.

One of the purposes of our joint study with European Gas Turbine SA, Belfort, and Alcatel-Alsthom-Recherche, Marcoussis, was to develop a monitoring and diagnostics system for the joint system (combustion chambers, expansion turbine). Monitoring is based on the measured pressure in the compressor, the rotation velocity of the turbine, and measurements from the thermocouples available at the

exhaust of the expansion turbine. Thus no direct observation is available on the
status of the combustion chambers. Hence a semiphysical model has been developed
that predicts the profile of the temperature at the exhaust of the expansion
turbine using the pressure in the compressor, the mean temperature at the exhaust
of the expansion turbine, and the rotation velocity of the turbine [9, 10].
This model consists of two parts: first the unknown temperature profile within the
chambers is modeled as a linear regression involving one parameter per chamber;
then, based on basic thermodynamics, a relation between this profile and the temperature
profile at the exhaust of the expansion turbine is given. Because the gas
flow rotates within the turbine during its expansion, a phase shift between the two
input and output temperature profiles is exhibited. Therefore, some phase shift
parameter appears in the model which makes it strongly nonlinear. This model
is semiphysical and inaccurate because the input temperature profile uses as a
regression function some waveform based on qualitative knowledge, and very
simplified thermodynamics is used for gas diffusion in the expansion turbine.

This semiphysical modeling was for the purpose of monitoring the turbine
system. Despite its inaccurate nature, the model has been successfully used for
developing a monitoring system of the combustion chambers; see [10]. Unfortunately,
this model is not entirely satisfactory for some other purposes, such as the
monitoring of the thermocouples installed at the exhaust of the expansion turbine.
The purpose of this discussion is to compare results from this semiphysical model
with some alternative nonparametric identification method based on wavelets, and
to discuss the two questions of the respective accuracy of fit and explicative power
of these two styles of models.

2. Modeling the Hydraulic Actuator of a Robot Arm: An Example of Identification of a Dynamic Nonlinear System

Now let us consider the modeling of the actuator of a robot arm.[2] It is a hydraulically
driven arm. By controlling the position of a valve, the oil pressure in
the transmission circuit is regulated. The oil pressure drives the motion of the arm.
What we want to model is the relationship between the position of the valve and
the oil pressure, both quantities being measured. In fact, the valve directly regulates
the oil streams injected in the transmission circuit. Hence variation of the oil
pressure depends not only on the position of the valve, but also on the quantity of
the oil accumulated in the transmission circuit, which, in turn, is reflected by the
oil pressure. Clearly this is a dynamic system: variation of its output (oil pressure)
depends on both its input (the position of the valve) and its state (reflected by the

[2]This application has been borrowed from Linköping University, while Q. Zhang was visitor at the
Automatic Control Group.

oil pressure). We tried to model this dynamic system with linear autoregressive exogenous (ARX) models, but the results were not satisfactory. Therefore, we decided to apply some nonlinear nonparametric model and see if we could improve the performance of the modeling.

3. Prediction of Glycemic Variations: An Example of Identification of a Dynamic Nonlinear System with Imprecise and Incomplete Data

Glycemic variations depend on several factors which are not easily quantifiable and, moreover, may vary with time. Diet, physical activity, stress and emotions, and proximity of meal have effects that doctors know how to *qualitatively* assess. For a healthy person, glycemic regulation is ensured via the secretion of insulin by the pancreas. In the case of organic deficiency, for diabetic persons, insulin must be artificially injected. Deciding the amount for injection is very difficult, because morphology, future physical activity, time of meal, glucide richness of meal, present glucose concentration, and results of the previous day have to be taken into account. Moreover, injected insulin acts with delay, and its efficiency is reduced as glucose concentration becomes higher. Lastly, hypoglycemia is almost always followed by hyperglycemia. For optimum glycemic control, it would be better to anticipate before the glucose level rises, as it occurs for endogenic insulin secretion in healthy persons. To summarize, we have to deal with a nonlinear, unstable system, with time delay.

Doctors have devised empirical rules allowing diabetic persons to approximatively compute for themselves the insulin level for injection. For diabetic persons using a pump, the insulin injection rate has two parts: the basic flow rate, denoted $B_a(t)$, and providing about 50% of daily insulin needs, and a variable part, the bolus, denoted $B_o(t)$, which is a flash injection to assimilate a recent meal.

Nevertheless, despite the doctor's experience, it is very difficult to manually obtain a more or less constant glycemic level, in part because a good control should take into account up to six input variables, which is far beyond human control capability. This motivated us to propose a predictive glycemic model, as a basis for automatic injection control. This model uses as a basis the empirical rules of doctors, and takes into account the qualitative nature of the available data. For this proposal, we have several "self-supervision notebooks," that is, daily support to control the context and the treatment of insulin-dependent diabetic patients under pump operation. Thus each day the diabetic writes in his notebook 1/time and actual glycemia, 2/time, importance, and quality of his meal, 3/activity, and 4/insulin injection. The experimental results on this case study are reported in Section VII.C.

B. BASIC MATHEMATICAL PROBLEMS

Here we establish the general framework of the *nonparametric regression* we shall use throughout this chapter, and we justify the use of particular distance measures between a true system and its estimate in the sequel.

Problem 1 (Nonparametric regression). Let (X, Y) be a pair of random variables with values in $\mathcal{X} = \mathbf{R}^d$ and $\mathcal{Y} = \mathbf{R}$, respectively. A function $f: \mathcal{X} \mapsto \mathcal{Y}$ is said to be the *regression function of Y on X* if

$$\mathbf{E}(Y|X) = f(X). \tag{2}$$

A typical case is $Y = f(X) + e$, where e is zero mean and independent of X. For $N \geq 1$, $\hat{f}_N$ shall denote an estimator of f based on the random sample $\mathcal{O}_1^N = \{(X_1, Y_1), \ldots, (X_N, Y_N)\}$ of size N from the distribution of (X, Y), that is, a map

$$\hat{f}_N: \mathcal{O}_1^N \mapsto \hat{f}_N(\mathcal{O}_1^N, \cdot), \tag{3}$$

where, for fixed $\mathcal{O}_1^N$, $x \mapsto \hat{f}_N(\mathcal{O}_1^N, x)$ is an estimate of the regression function $f(x)$. The family of estimators $\hat{f}_N$, $N \geq 1$, is said to be *parametric* if $\hat{f}_N \in F$ for all $N \geq 1$, where F is some set of functions which are defined in terms of a *fixed* number of unknown parameters. Otherwise the family of estimators $\hat{f}_N$, $N \geq 1$, is said to be *nonparametric*.

For the sake of convenience, we shall often refer to X and Y as the *input* and *output*, respectively (although they do not need to be such in actual applications). Our objective in this section is to give a short overview of some basic instances of nonparametric regression. Two typical problems are considered in the statistical literature, namely the

- *nonparametric regression with random design (or sampling)*, where it is assumed that the variables X_i are random, independent, and identically distributed on $[0, 1]^d$ with density $g(x)$, and the
- *nonparametric regression with deterministic design (or sampling)*, where it is assumed that the input variables X_i are nonrandom; the simplest case of deterministic design is the *regular design*, where the inputs X_i form a regular grid (for instance, $f: \mathbf{R} \to \mathbf{R}$ and $X_i = i/N$).

In the remainder of this section, we consider the random design only, although the observations (X_i, Y_i) are allowed to be dependent.

Nonparametric Regression for Static Systems

This is the simplest case. The considered system has the form

$$Y_i = f(X_i) + e_i, \qquad i = 1, \dots, N, \tag{4}$$

where $f(x)$: $\mathbf{R}^d \mapsto \mathbf{R}$, and, for the sake of simplicity, we assume that e_i are independent Gaussian random variables with $\mathbf{E}e_i = 0$ and $\mathbf{E}e_i^2 = \sigma_e^2$.

Adaptive Classification and Density Estimation[3]

The problem of classification (discriminant analysis or statistical pattern recognition) is usually formulated as follows. Let X be a random variable with values in $\mathbf{R}^d$, and let the label Z denote a random variable which takes values in some finite set $\mathcal{Z} = \{z_1, \dots, z_M\}$; the symbol z shall denote a generic element of this finite set. We want to guess the value of Z when X is observed. We consider the case in which the random vector X has probability density $f(x)$ and conditional densities $f(x|z)$ given that $Z = z$; the general case is handled similarly. We call a *solution* any measurable function $g: \mathcal{X} \mapsto \mathcal{Z}$, and $\mathbf{P}(g(X) \neq Z)$ is the corresponding *error probability*. The distribution of the pair (X, Z) is defined by the distribution μ of X and the regression functions

$$\mathbf{p}(z|x) = \mathbf{P}(Z = z | X = x) = \frac{\mathbf{p}(z) f(x|z)}{f(x)}, \qquad x \in \mathbf{R}^d,$$

where Bayes' rule has been used for the second equality, and $\mathbf{p}(z) = \mathbf{P}(Z = z)$. The functions $f(x|z)$ are also called *a posteriori densities*. The solution $g^*(x)$ is called *Bayesian* or *maximum a posteriori* (MAP), if

$$\mathbf{p}(g^*(x)) f(x|g^*(x)) = \max_z \mathbf{p}(z) f(x|z) \qquad \text{a.e. } x. \tag{5}$$

The Bayesian solution g^* minimizes the error probability, that is,

$$\mathcal{L}^* \overset{\Delta}{=} \min_g \mathbf{P}(g(X) \neq Z) = \mathbf{P}(g^*(X) \neq Z), \tag{6}$$

and $\mathcal{L}^*$ is called the *Bayesian error probability*.

In *adaptive classification*, we want to minimize the error probability when the true $\mathbf{p}(z)$ and $f(x|z)$ are unknown and a training sample $\mathcal{O}_1^N = \{(X_1, Z_1), \dots, (X_N, Z_N)\}$ of N independent observations distributed as (X, Z) is available. We assume that the training sample $\mathcal{O}_1^N$ and the test sample (X, Z) are independent. The estimate $g_N(X)$ of Z is now a measurable function of X and $\mathcal{O}_1^N$, and the following *conditional error probability* is a quantity of interest:

$$\mathcal{L}_N = \mathbf{P}(g_N(X) \neq Z | \mathcal{O}_1^N). \tag{7}$$

---

[3] In this section we follow the presentation of [11, Chap. 10].

In particular, we search for a sequence of estimates g_N such that

$$\mathcal{L}_N \to \mathcal{L}^* \qquad \text{almost surely.} \tag{8}$$

Referring to (5), the Bayesian solution can be approximated by the function g_N characterized by

$$\hat{\mathbf{p}}(g_N(x)) \hat{f}(x|g_N(x)) = \max_z \hat{\mathbf{p}}(z) \hat{f}(x|z), \tag{9}$$

where $\hat{f}(\cdot|z)$ are estimates of $f(\cdot|z)$ based on $\mathcal{O}_N$. There is a simple way to measure the conditional error probability $\mathcal{L}_N$ for the adaptive classifiers which satisfy (9): Devroye and Györfi [11] have shown that, if the random vector X is distributed with some density f and g_N is defined via (9), then

$$0 \le \mathcal{L}_N - \mathcal{L}^* \le \sum_{z \in \mathcal{Z}} \int |\mathbf{p}(z) f(x|z) - \hat{\mathbf{p}}(z) \hat{f}(x|z)| \, dx.$$

Different versions of this result were proved in [12–14], among others. This result implies that the classification error can be bound using the L_1-norm[4] of the estimation error of the density $\mathbf{p}(z) f(x|z)$. Thus we have related the problem of adaptive classification to that of estimating the density of a random variable in the L_1-norm. Other advantages of considering the averaged L_1-norm are discussed in [11]. Alternative distance measures for densities are often considered, for example, the averaged L_2-norm (often used, because it seems to be the easiest to estimate) or L_∞-norm.

Nonparametric Regression with Dynamics

Consider the following dynamical system:

$$Y_i = f(\Phi_i) + e_i, \qquad i = 1, \ldots, N,$$

where $Y_i \in \mathbf{R}$ and $\Phi_i \in \mathbf{R}^d$ are observed, and e_i is a white noise as before. We assume that

$$\Phi_i = (Y_{i-1}, \ldots, Y_{i-m}; U_i, \ldots, U_{i-p}), \tag{10}$$

where $U_i \in \mathbf{R}$ denote the inputs ($m + p = d$). For example, if $\Phi_i = (Y_{i-1}, \ldots, Y_{i-d})$, then

$$Y_i = f(Y_{i-1}, \ldots, Y_{i-d}) + e_i. \tag{11}$$

In analogy with the corresponding parametric model, we call this system a *nonparametric autoregression* or a *functional autoregression* of dimension d

---

[4]Recall that for a function $g: \mathbf{R}^d \to \mathbf{R}$ the L_p-norm is defined for $0 < p < \infty$: $\|g\|_p = (\int |g(x)|^p dx)^{1/p}$, and for $p = \infty$: $\|g\|_\infty = \operatorname{ess\,sup}_x |g(x)|$.

[FAR(d)]. As an interesting application, we can consider a simple controlled FAR model for adaptive control:

$$Y_i = f(\Phi_i) + U_i + e_i, \tag{12}$$

where $\Phi_i = (Y_{i-1}, \ldots, Y_{i-m})$, and U_i is the control. The following question can be considered: how does one choose the control (U_i) for the system (12) to track some reference trajectory $y = (y_i)$, or, at least, how does one choose U_i in order to minimize $\mathbf{E}Y_i^2$, or, simply, to stabilize the system (12)? If the function $f(\Phi)$ was known, we could use the control

$$U_i = -f(\Phi_i)$$

to obtain $Y_i = e_i$. Clearly, this is a "minimum variance" control, since $\mathbf{E}Y_i^2 \geq \sigma_e^2 = \mathbf{E}e_i^2$. If f is unknown, a possible solution consists in performing nonparametric "certainty equivalence control": compute an estimate $\hat{f}_N$ of the regression function f based on the observations of the input–output pair $(\Phi_i, Y_i - U_i)$, and then take

$$U_i = -\hat{f}_i(\Phi_i). \tag{13}$$

To analyze the certainty equivalence control (13), let us consider the control cost

$$Q_N = \frac{1}{N}\sum_{i=1}^{N} Y_i^2 = \frac{1}{N}\sum_{i=1}^{N}(f(\Phi_i) - \hat{f}_i(\Phi_i))^2 + \frac{1}{N}\sum_{i=1}^{N} e_i^2.$$

It is easily checked that

$$\mathbf{E}(\hat{f}_i(\Phi_i) - f(\Phi_i))^2 \to 0 \qquad \text{when } i \to \infty \tag{14}$$

implies $\mathbf{E}Q_N \to \sigma_e^2$, and $\hat{f}_i(\Phi_i) - f(\Phi_i) \to 0$ a.e. implies $Q_N \to \sigma_e^2$ a.e. Thus condition (14) is instrumental in analyzing this problem, and we shall informally discuss how it can be guaranteed.

Denote by $\Phi_0^{i-1} = (\Phi_0, \ldots, \Phi_{i-1})^T$ the vector of all available inputs up to time $i-1$, and by $\varphi_0^{i-1} = (\varphi_0, \ldots, \varphi_{i-1})^T$ the corresponding vector of integration variables. Let $\mathbf{P}$ denote the distribution of the vector sequence (Φ_i) when driven by the unknown "true" model (12)–(13), let $\mathbf{P}_{\Phi_0^{i-1}}(\cdot)$ be a distribution of Φ_0^{i-1}, and let $\mathbf{p}_{\Phi_i|\Phi_0^{i-1}}(\cdot)$ be a conditional density of the distribution of Φ_i given Φ_0^{i-1} (we assume that such a density exists). We have

$$\mathbf{E}\left|\hat{f}_i(\Phi_i) - f(\Phi_i)\right|^2 \sim \int \left|\hat{f}_i(x) - f(x)\right|^2 \mathbf{p}_{\Phi_i|\Phi_0^{i-1}}(x)\, dx\, \mathbf{P}_{\Phi_0^{i-1}}\left(d\varphi_0^{i-1}\right).$$

Note that, if the closed-loop system (12)–(13) is stable, one would reasonably take equal weights for the observations $\Phi_0, \ldots, \Phi_i$ in the estimate $\hat{f}_i$. In such a case the estimate $\hat{f}_i(\Phi)$ is asymptotically (as $i \to \infty$) slowly varying, that is, $\hat{f}_i \sim \hat{f}_{i-1}$. Thus we can write informally

$$\mathbf{E}|\hat{f}_i(\Phi_i) - f(\Phi_i)|^2 \sim \int \mathbf{P}_{\Phi_0^{i-1}}(d\varphi_0^{i-1}) \int |\hat{f}_{i-1}(x) - f(x)|^2 \mathbf{P}_{\Phi_i|\Phi_0^{i-1}}(x)\, dx.$$

The latter integral can be bound in several ways. For instance,

$$\int |\hat{f}_{i-1}(x) - f(x)|^2 \mathbf{P}_{\Phi_i|\Phi_0^{i-1}}(x)\, dx$$

$$\leq \sup_x |\hat{f}_{i-1}(x) - f(x)|^2 \int \mathbf{P}_{\Phi_i|\Phi_0^{i-1}}(x)\, dx$$

$$= \sup_x |\hat{f}_{i-1}(x) - f(x)|^2,$$

which yields the bound

$$\mathbf{E}|\hat{f}_i(\Phi_i) - f(\Phi_i)|^2 \leq \mathbf{E}\sup_x |\hat{f}_{i-1}(x) - f(x)|^2 = \mathbf{E}\|\hat{f}_{i-1} - f\|_\infty^2.$$

On the other hand, if the conditional density is bounded, that is, $\mathbf{P}_{\Phi_i|\Phi_0^{i-1}} \leq C_p$, then

$$\int |\hat{f}_{i-1}(x) - f(x)|^2 \mathbf{P}_{\Phi_i|\Phi_0^{i-1}}(x)\, dx \leq C_p \int |\hat{f}_{i-1}(x) - f(x)|^2\, dx$$

$$= C_p \|\hat{f}_{i-1} - f\|_2^2.$$

Thus, as a conclusion, in any case, the crux in analyzing this adaptive minimum variance nonlinear control consists in getting bounds for the error in estimating the unknown function f. Hence, in addition to proving consistency for the estimates, getting such bounds is an important question.

Discussion

This section about basic mathematical issues can be summarized as follows:

1. Nonparametric estimation of regression functions is instrumental in various problems such as adaptive identification, classification, and control.
2. The averaged L_p-norms of estimation error for various p's are natural candidates as a figure of merit. We shall see later that error measures also involving derivatives of f and $\hat{f}$ will be useful, so that smoothness of estimates can also be guaranteed.
3. Having bounds for the estimation error is of paramount importance. This has been illustrated by the adaptive control example. However, we shall

see later that some estimators can exhibit arbitrarily poor performance for some "bad" systems, so that having error bounds is really needed to prevent the user from getting bad results.

II. "CLASSICAL" METHODS OF NONLINEAR SYSTEM IDENTIFICATION

Throughout this section, Problem 1 is considered. We first discuss some estimators that are *linear*, that is, that satisfy $\widehat{f+g} = \hat{f} + \hat{g}$; note that the functions f, g, and their estimates, are generally nonlinear as functions of their input x. Linear estimators build the folklore of nonparametric estimation; kernel estimators and projections on linear subspaces of functions are typical instances we shall describe. We shall then discuss, both practically and theoretically, some severe practical limitations of linear estimators. Roughly speaking, linear estimators are suitable for systems with "uniform smoothness"; systems with sparse singularities (e.g., hard limiters, quantizers, some mechanical systems) are poorly handled. This motivates the search for new nonlinear estimators; neural networks and some related methods are candidates we shall briefly scan.

A. LINEAR NONPARAMETRIC ESTIMATORS

All estimators presented in this subsection are linear ones, that is, they have a common general form

$$\hat{f}_N(x) = \sum_{i=1}^{N} Y_i W_{N,i}(x), \qquad W_{N,i}(x) = W_{N,i}(x, X_1, \ldots, X_N), \qquad (15)$$

where we recall that $\mathcal{O}_1^N = \{(X_1, Y_1), \ldots, (X_N, Y_N)\}$ is the given random sample observation, and the weights $W_{N,i}(x)$ only may differ.

1. Some Linear Nonparametric Estimators

Kernel Estimators for Regression Functions and Densities

Kernel estimators were first proposed by Nadaraya and Watson in 1964 [15, 16]. The Nadaraya–Watson kernel estimator is an interpolation procedure. It is given by

$$\hat{f}_N(x) = \frac{\sum_{i=1}^{N} Y_i K((x - X_i)/h_N)}{\sum_{i=1}^{N} K((x - X_i)/h_N)}, \qquad (16)$$

where (h_N) is a sequence of positive numbers, $h_N \to 0$ as $N \to \infty$, and K is a function on $\mathbf{R}$ satisfying

$$\lim_{|u| \to \infty} |u| |K(u)| = 0, \qquad \int_{-\infty}^{\infty} |K(u)| \, du < \infty,$$

$$\sup_{u \in \mathbf{R}} |K(u)| < \infty, \qquad \int_{-\infty}^{\infty} K(u) = 1. \tag{17}$$

The positive number h_N is called the *bandwidth* and the function K satisfying (17) is called a *kernel*; in fact, h_N is better interpreted as a scaling factor. Clearly, the Nadaraya–Watson estimator is linear, and has the form (15). Typical examples of kernels are $K(u) = (1/2)1_{\{|u| \le 1\}}$ (rectangular window kernel), $K(u) = (1/\sqrt{2\pi}) \exp(-|u|^2/2)$ (Gaussian kernel), etc. Usually K is chosen to be an even function.

The idea of kernel estimation is simple. Let us explain it for the case of the rectangular kernel in one dimension. In this case the estimator (16) is a simple moving average with equal weights: the estimate at point x is the average of observations Y_i corresponding to X_i's belonging to the "window" $[x - h_N, x + h_N]$. If $h_N \to \infty$, then the estimator tends to $N^{-1} \sum_i Y_i$, the average of all observations, and thus for functions f which are far from being constant, the bias becomes large. If h_N is very small (say, smaller than the pairwise distance between sample points X_i), then the estimator reproduces the observations: $\hat{f}_N = Y_i$. In this extremal case the variance of the error becomes high. Thus increasing h_N tends to increase the bias of estimator, while reducing h_N leads to a larger variance. The optimal choice for h_N corresponds to an equal balance between bias and variance.

Also closely related to estimator (16) is the Parzen–Rosenblatt kernel estimator for densities. Let $X_1, \ldots, X_N$ be independent and identically distributed random variables with common density $f(x)$, $x \in \mathbf{R}^d$. The Parzen–Rosenblatt estimator of density $f(x)$ is a suitably smoothed histogram. It is defined as [17, 18]

$$\hat{f}_N(x) = \frac{1}{N h_N^d} \sum_{i=1}^{N} K\left(\frac{x - X_i}{h_N}\right), \tag{18}$$

where d is the state-space dimension of X and K is a kernel as in (17). Kernel estimate (16) can be easily derived from the Parzen–Rosenblatt one. Recall definition (2) of the regression function, take the Parzen–Rosenblatt estimator (18) for the joint density $f(x, y)$ of (X, Y), and denote it by $\hat{f}_N(x, y)$. Then, replacing, in the following formula

$$f(x) = \frac{\int y f(x, y) \, dy}{\int f(x, y) \, dy},$$

$f(x)$ and $f(x, y)$ by their corresponding Parzen–Rosenblatt estimates, yields kernel estimate (16).

We now state a sample of results about the properties of kernel estimates for the d-dimensional case. Assume that it is known *a priori* that f belongs to the ball $C^s(L)$ in the so-called Hölder space: for s and L positive, let $C^s(L)$ be the family of functions $f(x)$, $x \in [0, 1]^d$, defined by[5]

$$C^s(L) = \left\{ f: \left| f^{(k)}(x) - f^{(k)}(x') \right| \le L|x - x'|^{s-k} \text{ for any } x, x' \in [0, 1]^d \right\},$$
$$k = \lfloor s \rfloor. \tag{19}$$

Note that this is a smoothness prior of the kind we discussed in our introduction. If $s \ge 1$ is an integer, then $C^s(L)$ contains continuous functions having Lipschitz $(s - 1)$th derivative. We can now give a result on the rate of convergence of the kernel estimate. We acknowledge Rosenblatt [19] for the first two statements of it, though it probably belongs to the earlier folklore of nonparametric statistics.

THEOREM 1 (Rosenblatt [19]). *Let $\hat{f}_N$ be a kernel estimate with bandwidth h_N such that $h_N \to 0$ and $Nh_N \to \infty$, with kernel K satisfying $\int x^j K(x)\, dx = 0$ for $j = 1, \dots, k$. Here, x^j denotes any product of the form $x_1^{j_1} x_2^{j_2} \cdots x_d^{j_d}$, where $j_1 + j_2 + \cdots + j_d = j$ and $x_1, \dots, x_d$ are the coordinates of x. Assume that the observations X_i are independent and identically distributed on $[0, 1]^d$ with density $g(x) \ge c > 0$, $g \in C^s(L)$, and that the noise satisfies $\mathbf{E}e_i = 0$ and $\mathbf{E}e_i^2 \le \sigma_e^2 < \infty$. Then*

1. *Uniformly over $f \in C^s(L)$ and $x \in [0, 1]^d$, we have the pointwise bound*

$$\mathbf{E}\left| \hat{f}_N(x) - f(x) \right|^2 \le C\left(L^2 h_N^{2s} + \frac{\sigma_e^2}{Nh_N^d} \right). \tag{20}$$

The optimal value of h_N which minimizes the right-hand side of (20) is given by

$$h_N = \left(\frac{\sigma_e^2}{L^2 N} \right)^{1/(2s+d)}. \tag{21}$$

For this value of h_N,

$$\mathbf{E}\left| \hat{f}_N(x) - f(x) \right|^2 \le CL^{2/(d+2s)} \left(\frac{\sigma_e^2}{N} \right)^{2s/(2s+d)}.$$

2. *If we consider instead the global error measure $\mathbf{E}\| \hat{f}_N - f \|_2^2$, using again the same optimal value (21) for h_N, yields the same bound, uniformly over $f \in C^s(L)$.*

---

[5] $\lfloor s \rfloor$ denotes the maximal integer $k < s$.

Comments

1. As expected from the preceding informal discussion concerning the rectangular kernel, the bound for the estimation error variance given on the right-hand side of (20) is decomposed into *bias* and *variance* terms. And, as expected, the optimal choice of h_N in (21) exactly balances these two terms.

2. Note that we have both pointwise and global bounds, which reflects the local nature of kernel estimates.

3. The properties of the Parzen–Rosenblatt algorithm of density estimation are identical when the unknown density f satisfies $f \in C^s(L)$. Note that, because supp $f \subseteq [0, 1]^d$, the L_1-norm of the error (restricted to the $[0, 1]^d$) is dominated by the L_2-norm. So we get from the second statement of the theorem

$$\mathbf{E}\|\hat{f}_N - f\|_1^2 \leq CL^{2/(d+2s)}\left(\frac{\sigma_e^2}{N}\right)^{2s/(2s+1)},$$

provided h_N is chosen as in (21).

4. Often the following recursive version of the kernel estimator is considered [20, 21]:

$$\hat{f}_n(x) = \begin{cases} \Gamma_n^{-1}(x)\left(\sum_{i=0}^n Y_i h_i^{-d} K\left(\frac{x-X_i}{h_i}\right)\right) & \text{if } \Gamma_n(x) \neq 0, \\ 0 & \text{if } \Gamma_n(x) = 0, \end{cases}$$

$$\Gamma_n(x) = \sum_{i=0}^n h_i^{-d} K\left(\frac{x - X_i}{h_i}\right),$$

or

$$\hat{f}_n(x) = \hat{f}_{n-1}(x) + \Gamma_n^{-1}(x)\left(Y_n - h_n^{-d} K\left(\frac{x - X_n}{h_i}\right)\hat{f}_{n-1}\right),$$

$$\Gamma_n(x) = \Gamma_{n-1}(x) + h_n^{-d} K\left(\frac{x - X_n}{h_n}\right). \tag{22}$$

In this form the algorithm resembles very much the recursive least squares algorithm for estimating the parameters of linear models. When the bandwidth is such that $h_i = hi^{-\alpha}$ for some $0 < \alpha < 1$, the properties of the algorithm (22) in the static regression problem are essentially the same as those of the "off-line version" (16). In [20–22] this algorithm was used to identify stable nonparametric autoregression models of the form (11), and the convergence of this estimator was proved. Furthermore, the same algorithm was used to provide the estimates of $\hat{f}_n$ in the closed-loop system (12)–(13), and the stability of such an adaptive control scheme was proved—[21] and [22] consider essentially the one-dimensional case, and in [20] the general multidimensional case is studied.

Piecewise-Polynomial Estimators

Another nonparametric regression estimator which is commonly used is the piecewise-polynomial one. The idea is the same as for the kernel estimator, though the averaging is made over *bins* (i.e., small cubes) of fixed size δ_N rather than in the h_N-neighborhood of the current point x. It is also closely related to *radial-basis function* (RBF) *networks* with rigid location for the radial functions; see [2, 23]. The simplest example of this method is the piecewise-constant estimator or *regressogram*. The value of the estimate $\hat{f}_N$ in each bin equals the average of observations Y_i such that corresponding X_i belong to the bin. For the sake of clarity, we consider the one-dimensional case.

The piecewise-polynomial estimator can be formally defined in terms of the following optimization problem. Let $\delta_N \to 0$ be a positive sequence, and assume that $\delta_N^{-1} = M$ is an integer. Define $u_l = l\delta_N$, $l = 0, \ldots, M$, and divide the interval $[0, 1]$ into M cubes (bins) of the form $U_1 = [0, u_1)$, $U_2 = [u_1, u_2), \ldots, U_M = [u_{M-1}, 1]$, so each bin has length δ_N. Set $F(x) = (1, x, x^2/2, \ldots, x^k/k!)^T$ and, for each bin U_l, $l = 1, \ldots, M$, solve for $\theta \in \mathbf{R}^{k+1}$ in the least squares sense the system of equations

$$Y_i = \theta^T F\left(\frac{X_i - u_{l-1}}{\delta_N}\right), \qquad X_i \in U_l, \tag{23}$$

and denote by $\hat{\theta}_{N,l}$ the corresponding solution. Then the piecewise-polynomial estimate $\hat{f}_N$ of order k in each bin U_l is expressed as

$$\hat{f}_N(x) = \hat{\theta}_{N,l}^T F\left(\frac{x - u_{l-1}}{\delta_N}\right), \qquad x \in U_l. \tag{24}$$

The value δ_N is called the *binwidth*. As for the bandwidth h_N of the kernel estimate, the binwidth tunes the smoothness: larger δ_N leads to a higher bias, and smaller δ_N results in a higher variance. In order for the least-squares problem in (24) to be nondegenerate, we require that the number of points X_i in each bin be larger than $k + 1$.

Stone [24] has proved a result similar to Theorem 1 for this type of estimate [see (19) for the definition of the Hölder space $C^s(L)$]. We state this result in the general d-dimensional case. Assume that the observations X_i satisfy the assumptions of Theorem 1. Let $\hat{f}_N$ be a piecewise-polynomial estimate of order $k = \lfloor s \rfloor$, with binwidth $\delta_N \to 0$ and $N\delta_N \to \infty$ as $N \to \infty$. Then *statement 1 of Theorem 1 holds with binwidth δ_N substituted for the bandwidth h_N.*

Comments

1. Note that, unlike kernel estimates, piecewise-polynomial estimates compute projections on the fixed set of functions $F((x - u_{l-1})/\delta_N)$, $x \in U_l$ (the *l*th bin). The same remark holds for the projection estimate to follow.

2. As can be seen, piecewise-polynomial and kernel estimates have the same asymptotic accuracy when $N \to \infty$.

3. If f is a smooth function (i.e., $s \geq 1$), the optimal number of bins is $n_\delta \sim \delta_M^{-1}$ which is much less than the number of observations ($n_\delta \sim N^{1/3}$ for $s = 1$). This number is equivalent to the memory size required to implement the algorithm: to reconstruct the estimate, $k = \lfloor s \rfloor$ coefficients are necessary. Thus, if N is large, this algorithm offers a significant advantage, in terms of memory requirements, over kernel estimates in which all measurements should be kept to reconstruct $f(x)$. Also, computing (23)–(24) is of lower computational burden than computing (16). These two points make the piecewise-polynomial estimate more attractive.

4. Unfortunately, there is no reasonable recursive version of the estimate $\hat{f}_n$. Although one can use the recursive least squares algorithm to compute linear regression coefficients $\hat{\theta}_{N,l}$ in (24), the derivations quickly become messy, because the number M of bins depends on N, and so does the number of equations in the algorithm.

Projection Estimates

Another class of function estimates was introduced by Cencov [25], who called them *projection estimates*. The idea consists of expanding the unknown function into its "empirical" Fourier series. Consider the set $\mathcal{W}_2^s(L)$ of functions $f(x)$, $x \in [0, 1]^d$, defined as follows. Each f can be represented by its Fourier series

$$f(x) = \sum_{|j|=1}^{\infty} c_j \Phi_j(x), \tag{25}$$

where $j = (j_1, \ldots, j_d)$ is a multi-index, $x = (x^1, \ldots, x^d)^T$,

$$\Phi_j(x) = \varphi_{j_1}(x^1) \times \cdots \times \varphi_{j_1}(x^d),$$

$\varphi_1 \equiv 1$, $\varphi_{2k}(x) = \sqrt{2} \sin(2\pi k x)$, and $\varphi_{2k+1}(x) = \sqrt{2} \cos(2\pi k x)$, $k = 1, \ldots$. Suppose that the following condition is satisfied:

$$\sum_{j=1}^{\infty} |c_j|^2 (1 + |j|^{2s}) < L^2. \tag{26}$$

In fact, we have $\sum_{j=1}^{\infty} |c_j|^2 (1 + |j|^{2s}) \leq C \|f\|_{s,2}^2$, where $\|f\|_{s,2}$ is the norm of the Sobolev space $\mathcal{W}_2^s$ of functions with all derivatives up to order s being square integrable. Note that this is again a smoothness prior. We assume that *input X is* *uniformly distributed.*[6] We construct the estimate $\hat{f}_N$ as follows:

$$\hat{f}_N(x) = \sum_{j=1}^{m} \hat{c}_j^N \Phi_j(x), \tag{27}$$

where m is the "model order," and the empirical estimates $\hat{c}_j^N$ of Fourier coefficients

$$\hat{c}_j^N = \frac{1}{N} \sum_{i=1}^{N} Y_i \Phi_j(X_i) \tag{28}$$

are substituted for the true ones c_j, $j = 1, \ldots, m$. Note that the assumption that X is uniformly distributed has been used. Note also that the estimate (27)–(28) is linear [cf. (15)] with weights given by

$$W_{N,i}(x) = \sum_{j=1}^{m} \frac{1}{N} \Phi_j(x) \Phi_j(X_i).$$

Cencov [25] has proved the following counterpart of statement 1 of Theorem 1: Let $\hat{f}_N$ be a projection estimate. Then, uniformly over $f \in \mathcal{W}_2^s(L)$ and $x \in [0, 1]^d$,

$$\mathbf{E}\|\hat{f}_N(x) - f(x)\|_2^2 \leq C\left(L^2 m^{-2s} + \frac{\sigma_e^2 m^d}{N}\right). \tag{29}$$

The optimal order m of the model is

$$m = \left\lfloor \left(\frac{L^2 N}{\sigma_e^2}\right)^{1/(2s+d)} \right\rfloor. \tag{30}$$

It balances bias and variance error estimates, and yields the bound

$$\mathbf{E}\|\hat{f}_N(x) - f(x)\|_2^2 \leq CL^{2/(d+2s)}\left(\frac{\sigma_e^2}{N}\right)^{2s/(2s+d)}. \tag{31}$$

The following result, due to Ibragimov and Khas'minskij [26], provides a global uniform bound. Take

$$m = \left\lfloor \left(\frac{N}{\ln N}\right)^{1/(2s+d)} \right\rfloor$$

---

[6]See Section IV for a thorough discussion of this assumption.

for the model order [note that this is slightly different from (30)]. Then, uniformly over $f \in C^s(L)$ [the class $C^s(L)$ is defined in (19)], it holds that

$$\mathbf{E}\| \hat{f}_N - f \|_\infty^2 \leq O\left(\frac{\ln N}{N}\right)^{2s/(2s+d)} \tag{32}$$

Comments

1. Projection estimates have the same rate of convergence (up to a constant) as kernel or piecewise-polynomial ones.

2. The bound (29) for the quadratic error of the algorithms appears rather naturally if we consider the following argument: when we approximate $f \in W_2^s$ using m terms of its Fourier decomposition, the approximation error is $O(m^{-2s/d})$. Furthermore, the stochastic error in each term is of order $O(N^{-1})$. This simple calculus can be repeated for any nonparametric estimate. Obviously, it is beyond our capabilities to reduce the stochastic component of the error. On the contrary, the bias part depends on the method we choose to approximate the function (piecewise-polynomial, trigonometric series, etc.), and this choice of approximant is of primary importance.

3. From the computational point of view, projection estimates are more attractive than piecewise-polynomial estimates, because they use an orthonormal basis of functions (the Fourier basis), which dramatically simplifies the computation of the least squares estimates $\hat{c}_j$ of the Fourier coefficients c_j; cf. (28).

2. Practical Implementation of the Algorithms: Adaptation and Tuning of Their Various Design Parameters, Generalized Cross Validation

As we have seen, the convergence of the estimates strongly depends on the choice of the bandwidth h_N for the kernel estimator, the model order m for the projection estimator, and the binwidth δ_N (or, equivalently, the "model order" $M = \delta^{-1}$) for the piecewise-polynomial estimator. *These design parameters depend on the parameters of the smoothness class $C^s(L)$ or $W_2^s(L)$, which are a priori unknown*—see definition (19) of this class and the use of parameters (s, L) in Theorem 1 and corresponding results for the other estimators. Even if some information about the smoothness parameter s is available, the knowledge of the value L is of importance when the data sample is of bounded length. Let us illustrate this with the following example, where input x is a scalar. Consider the problem of estimating a function $f(x)$ in additive white noise e, with $\sigma_e^2 = 1$. Assume that f has support $[0, 1]$, that all its derivatives are continuous, and that $f(1/2) = 1$, $f(0) = f(1) = 0$. Note that in this case, typically, $\sup_x |f^{(s)}(x)| \approx s^s$; that is, higher-order derivatives become very large in uniform bound. In this case the bounds in Theorem 1 are of order $a_N(s) = (s/N)^{2s/(2s+1)}$

when the parameter is selected for the smoothness s. Assume that the size of the observation sample is $N = 10000$, then $a_N(2) = 0.0110$, $a_N(3) = 0.0095$, but we already have $a_N(4) = 0.0122$ [the value of s which minimizes a_N is $s \approx 3.4814$ with $a_N(s) \approx 0.00946$]. This illustrates the fact that the tightest bound is not obtained by taking the largest possible s, but rather by selecting the most favorable pair (s, L), which is obviously much more difficult.

Given that we only have in practice samples of finite size N, we shall not try to estimate the most favorable pair (s, L), but we shall proceed differently. The model order (or bandwidth, or binwidth, depending on the different estimates) shall be estimated from data using a procedure usually referred to as the *generalized cross validation* (GCV) test. GCV procedures were studied for kernel (see, e.g., [27, 28]), spline (e.g., [29, 30]), and projection estimates (cf. [31, 32]). Let us consider, for instance, the procedure for the projection estimates.[7] To make the model order explicit in formula (27), we shall write $\hat{f}_{m,N}$ instead of $\hat{f}_N$. Set $S^2_{m,N} = N^{-1} \sum_{i=1}^N \|Y_i - \hat{f}_{m,N}(X_i)\|^2$. As for the prediction error variance estimate in parametric prediction error methods, $S^2_{m,N}$ is a *biased* estimate of the error. Thus one cannot minimize $S^2_{m,N}$ with respect to m directly: the result of such a brute-force procedure would give a function $\hat{f}_{m_N,N}(x)$ which perfectly fits the noisy data; this is known as "overfitting" in the neural network literature. The solution rather consists in introducing a penalty which is proportional to the model order m; that is, we search for m_N such that

$$m_N = \arg \min_{m \leq N} \left(S^2_{m,N} + \frac{2\sigma_e^2 m}{N} \right). \tag{33}$$

This technique is clearly equivalent to the celebrated Mallows–Akaike criterion [33, 34]. The following result, due to Polyak and Tsybakov [31], shows the consistency of this procedure. Assume that the Fourier coefficients of f in expansion (25) satisfy $|c_j| \leq \varepsilon_j$, $\sum_{j=1}^{\infty} \varepsilon_j < \infty$, $(j\varepsilon_j)$ is nonincreasing, and σ_e^2 is known. Set $V_{m,N} = \|\hat{f}_{m,N} - f\|_2^2$. Then for the estimate (27), (28), and (33), it holds that

$$\frac{V_{m_N,N}}{\min_m V_{m,N}} \to 1 \qquad \text{a.e. as } N \to \infty.$$

B. Performance Analysis of the Nonparametric Estimators

The performance analysis of nonparametric estimation algorithms and/or identification procedures is much more difficult than for parametric estimation. In fact, the following specific issues are important:

---

[7]In fact, a similar result holds for the spline or piecewise-polynomial ones.

1. What plays the role of the Cramer–Rao bound and Fisher information matrix in our case? Recall that the Cramer–Rao bound reveals the best performance one can expect in identifying the unknown parameter θ from sample data arising from some parametrized distribution $\mathbf{p}_\theta$, $\theta \in \Theta$, where Θ is the domain over which the unknown parameter θ ranges. In the nonparametric case, lower bounds for the best achievable performance are provided by *minimax risk functions*. We shall introduce these lower bounds and discuss associated notions of optimality.

2. For lower bounds, the class of systems on which the best achievable performance is considered, is another important issue. For nonparametric representations of linear systems, L_2, L_∞, H_2, H_∞, with their associated norms, are typical spaces to work with. For (even static) nonlinear systems, however, the choice is much wider. How wide should be the class $\mathcal{F}$ of the systems under consideration; what kind of smoothness should be required? Are we interested in the behavior of the estimate at one particular point x of interest, or are we interested in the global behavior of the estimate? Different distance measures should be used in these two different cases.

1. Lower Bounds for Best Achievable Performance

To compare different nonparametric estimators, it is necessary to introduce suitable figures of merit. It seems first reasonable to build on the mean square deviation (or mean absolute deviation) of some seminorm[8] of the error; we denote it by $\| \hat{f}_N - f \|$. The following seminorms are commonly used in nonparameteric regression: $\| f \| = (\int f^p(x)\, dx)^{1/p}$, $0 < p < \infty$ (L_p-norm), $\| f \| = \sup_x |f(x)|$ (uniform norm, $\mathcal{C}$- or L_∞-norm), $\| f \| = |f(x_0)|$ (absolute value at a fixed point x_0). Then we consider the *risk function*

$$R_{a_N}(\hat{f}_N, f) = \mathbf{E}\big[a_N^{-1}\| \hat{f}_N - f \|\big]^2, \tag{34}$$

where a_N is a normalizing positive sequence. Letting a_N decrease as fast as possible so that the risk still remains bounded yields a notion of a convergence rate. Let $\mathcal{F}$ be a set of functions which contains the "true" regression function f. Then the maximal risk $r_{a_N}(\hat{f}_N)$ of estimator $\hat{f}_N$ on $\mathcal{F}$ is defined as follows:

$$r_{a_N}(\hat{f}_N) = \sup_{f \in \mathcal{F}} R_{a_N}(\hat{f}_N, f).$$

If the maximal risk is used as a figure of merit, the optimal estimator $\hat{f}_N^*$ is the one for which the maximal risk is minimized, that is, such that[9]

$$r_{a_N}(\hat{f}_N^*) = \min_{\hat{f}_N} \sup_{f \in \mathcal{F}} R_{a_N}(\hat{f}_N, f).$$

[8] A seminorm is a norm, except it does not satisfy the condition: $\| f \| = 0$ implies $f = 0$.

[9] To properly understand the statement to follow, the reader should pay attention to definition (3) of an estimator.

We call $\hat{f}_N^*$ *the minimax estimator* and the value

$$\min_{\hat{f}_N} \sup_{f \in \mathcal{F}} R_{a_N}(\hat{f}_N, f)$$

the minimax risk on $\mathcal{F}$. The construction of minimax nonparametric regression estimators for different sets $\mathcal{F}$ is a hard problem. Today, it is only solved asymptotically (for large samples) for some special cases (see, e.g., [35–37]). However, letting a_N decrease as fast as possible so that the minimax risk still remains bounded yields a notion of a best achievable convergence rate, similar to that of parametric estimation. More precisely, we state the following definition:

DEFINITION 1 (Lower rate and minimax rate of convergence).

1. The positive sequence a_N is a *lower rate of convergence for the set* $\mathcal{F}$ *in the seminorm* $\| \cdot \|$ if

$$\liminf_{N \to \infty} r_{a_N}(\hat{f}_N^*) = \liminf_{N \to \infty} \inf_{\hat{f}_N} \sup_{f \in \mathcal{F}} \mathbf{E}\big[a_N^{-1} \| \hat{f}_N - f \|\big] \geq C_0 \qquad (35)$$

   for some positive C_0. The inequality (35) is a kind of negative statement that says that no estimator of the function f can converge to f faster than a_N. This notion can be refined as follows.

2. The positive sequence a_N is called the *minimax rate of convergence* for the set $\mathcal{F}$ in seminorm $\| \cdot \|$, if it is a lower rate of convergence, and if, in addition, there exists an estimator $\hat{f}_N^*$ achieving this rate, that is, such that

$$\limsup_{N \to \infty} r_{a_N}(\hat{f}_N^*) < \infty.$$

Thus, a coarser, but easier approach consists of assessing the estimators by their convergence rates. In this setting, by definition, optimal estimators reach the lower bound as defined in (35) (recall that the minimax rate is not unique: it is defined to within a constant).

Some Negative Results

We state first a negative result, due to Devroye and Györfi [11, 38], which expresses that no convergence rate exists if no smoothness assumption about the unknown regression function f is stated.[10] Consider the following classes of functions on $\mathbf{R}$:

$\mathcal{F}^*$: the class of all functions f such that $f(x) = 0$ for $x > 1$ or $x < 0$, and $|f(x)| \leq C$ for $x \in [0, 1]$.
$\mathcal{F}_0^*$: the class of all continuous functions $f \in \mathcal{F}^*$.

---

[10]Note that convergence can sometimes be proved without any smoothness assumption [39].

$\mathcal{F}_\infty^*$: the class of all functions $f \in \mathcal{F}^*$ having all continuous derivatives on $[0, 1)$ (be careful that the interval is right open).

Let $\hat{f}_N$ be an arbitrary estimate of f. Then for the classes $\mathcal{F}^*$, $\mathcal{F}_0^*$, and $\mathcal{F}_\infty^*$ defined previously (we denote them generically by $\mathcal{F}$)

$$\sup_{\mathcal{F}} \limsup_{N \to \infty} \mathbf{E}\left[a_N^{-1} \int_0^1 \left| \hat{f}_N(x) - f(x) \right| dx \right] = \infty$$

for any positive sequence $a_N \to 0$.

There is also a similar result for the adaptive classification problem: consider the classification problem of Section I.B and the notation therein. Suppose that there are only two classes, that is, $M = |\mathcal{Z}| = 2$. Let a_N be any positive sequence such that $a_N \to 0$ and $\lambda \in [0, 1/2)$. Let g_N be an arbitrary estimator. Then there exists a distribution of the pair (X, Z), with X uniformly distributed on $[0, 1]$, such that

$$\limsup_{n \to \infty} a_N^{-1}(\mathbf{E}\mathcal{L}_N - \mathcal{L}^*) = \infty,$$

where $\mathcal{L}_N$ is associated with g_N through (7).

Thus, *no convergence rate exists for any of the preceding classes $\mathcal{F}^*$, $\mathcal{F}_0^*$, and $\mathcal{F}_\infty^*$.* In other words, the convergence can be arbitrary slow, depending on the unknown function or density f to be estimated! It is a natural consequence of the fact that the preceding classes $\mathcal{F}^*$, $\mathcal{F}_0^*$, and $\mathcal{F}_\infty^*$ are too rich: they contain functions which are extremely difficult to approximate. *In other words, to obtain any interesting rate of convergence, smoothness conditions should be imposed.*

Some Positive Results

Let us now concentrate on the case of deterministic uniform design; that is, the input data X are uniformly sampled in the considered interval. The following result in the case of regular design can be attributed to [26] (for the random design case, see [24, 40]).

THEOREM 2. *Let us consider the Hölder class $C^s(L)$ on $[0, 1]^d$; see (19) for the definition of $C^s(L)$. Consider*

$$\|g\| = \left(\int |g(x)|^p \, dx \right)^{1/p}, \qquad 0 < p < \infty,$$

or

$$\|g\| = |g(x_0)|.$$

Then $N^{-s/(2s+d)}$ *is a lower rate of convergence for the class* $C^s(L)$ *in the semi-norm* $\|\cdot\|$. *Furthermore,* $(N/\ln N)^{-s/(2s+d)}$ *is a lower rate of convergence for the class* $C^s(L)$ *in the norm* $\|g\| = \sup_{x \in [0,1]} |g(x)|$.

Note that to obtain the correct rate of convergence for the distance at a fixed point x_0, the corresponding Lipschitz property is required at x_0 only. Similar results hold when the class $C^s(L)$ is replaced by the class $W_p^s(L)$, $p \geq 2$, where $W_p^s(L)$ is the set of k-times differentiable functions f on $[0,1]^d$ such that

$$\|f\|_2 \leq 1, \qquad \left\| f^{(k)}(t+h) - f^{(k)}(t) \right\|_p \leq L\|h\|^\alpha,$$

$$0 < \alpha \leq 1, \qquad s = k + \alpha.^{11}$$

Then $N^{-s/(2s+d)}$ is also a lower rate of convergence for this class in the L_p-norm of the error.

2. Discussion

Criticizing the Minimax Paradigm

Let us compare the lower rates of convergence of Theorem 2 and the upper bounds obtained in this section for different estimators. One can see that the estimators considered are optimal on the classes W_2^s and C^s in the sense that they reach the minimax optimal rate of convergence.[12] Despite many impressive technical achievements in the preceding work, the general reaction within the statistics community has not been really enthusiastic. For example, according to Donoho, "... a large number of computer packages appeared over last fifteen years, but the work on the minimax paradigm has relatively little impact on software" [41]. One of the arguments supporting this skepticism about methods based on the minimax paradigm—kernel estimators, spline methods, or orthogonal series—is that they are spatially nonadaptive, whereas real functions exhibit a variety of shapes and spatial inhomogeneities. To illustrate this point, let us look at the following example. Consider the function $f(x) = 1_{\{0 \leq x < a\}}$ for some $0 < a < 1$. The Fourier coefficients of this function are

$$c_0 = a, \qquad c_{2k} = \sqrt{2}\frac{\sin^2(\pi ka)}{\pi k}, \qquad c_{2k+1} = \sqrt{2}\frac{\sin(\pi ka)\cos(\pi ka)}{\pi k}.$$

Hence the condition in (26) is not verified for $s \geq 1/2$. Thus we conclude from (32) that the rate of convergence (31) for the projection estimate (27), (28) will not be better than $N^{-1/2}$. Furthermore, because f does not belong to the Sobolev space W_2^s for $s \geq 1/2$, this rate of convergence is minimax. On the other hand, one

---

[11] Although defined in a different way, this $W_p^s(L)$ space coincides for $p = 2$ with the space introduced in formula (25) and subsequent ones.

[12] The projection estimates are also minimax on W_p^s (see [26, Theorem 4.3]).

naturally expects that a procedure to detect the edges of f can be designed which would have a rate of convergence "close" to N^{-1}. Indeed, the linear methods fit very well functions which are, say, "uniformly smooth" or "uniformly nonsmooth." Facing the problem of estimating a function with sparse singularities, the projection method will infer erroneously that the function is "uniformly smooth," but with a pessimistic smoothness parameter.

The minimax paradigm as discussed before does not seem to provide methods with convergence rates of order N^{-1} for the preceding example. Thus the authors of [41] argue that one should construct methods (heuristically, if necessary) which address the "real problem," namely *spatial adaptation*. This point of view has had considerable influence on software development and daily statistical practice, apparently much more than the minimax paradigm. Interesting spatially adaptive methods include all sorts of neural networks, projection pursuit [42], classification and regression trees (CARTs) [43], multivariate adaptive regression splines (MARS) [44], variable bandwidth kernel methods [45], and others. These methods implicitly or explicitly attempt to adapt the fitting method to the form of the function being estimated, by ideas like recursive dyadic partitioning of the space on which the function is defined (CART and MARS) and adaptively estimating a local bandwidth function (variable kernel methods). Citing again David Donoho, one could say that "the spatial adaptivity camp is, to date, a-theoretical, as opposed to anti-theoretical, motivated by the heuristic plausibility of their methods, and pursuing practical improvements rather than hard theoretical results which might demonstrate specific quantitative advantages of such methods. But, in our experience, the need to adapt spatially is so compelling that the methods have spread far in the last decade, even though the case for such methods is not proven rigorously" [41]. To conclude, a deeper investigation is needed to find the proper framework.

Adequate Answer: Besov Spaces and Wavelets

This short analysis reveals the crux in the route to both practical efficiency and mathematical support of the methods. It consists of *finding a parametrized family of functional classes which*

1. *fits our prior knowledge about the smoothness of the function to be estimated (in particular, that f is smooth everywhere, except at a sparse set of points), and*
2. *has associated with it an estimation technique which is minimax within these classes.*

It was the merit of Donoho and Johnstone [46] to recognize that Besov spaces, which play a central role in Meyer's mathematical theory of wavelets [5], provide an adequate answer. They are perfectly suited to nonlinear systems which have sparse singularities and otherwise are smooth. This material will be the topic of Section IV.

However, before discussing wavelets and their use in identification, we briefly scan some popular nonlinear estimates. They all provide the kind of "spatial adaptation" that we advocated before. Some of them are supported by efficient software. And some of them have become extremely successful and their names are now buzzwords widely known beyond the scientific community.

C. NONLINEAR ESTIMATES

Starting in the early 1980s, a variety of techniques have been proposed in the statistics literature, which exhibit this desirable feature of "spatial adaptivity." Among them are the *projection pursuit algorithm* developed in [42] (a very good review of these results can be found in [47]), *recursive partitioning* [43, 48], and related methods (cf., e.g., [44] with discussion). These methods are derived from some mixture of statistic and heuristic arguments and give impressive results in simulations. Their drawback lies in the almost total absence of any theoretical results on their convergence. We refer the reader to the previous references for additional information.

Surprisingly enough, the artificial intelligence (AI) literature has proposed independently and at the same time different techniques with the same feature of "spatial adaptivity." These include various forms of neural networks [3]; see the other tutorial [49] by Ljung. We shall briefly describe these. In addition, we shall sketch a recent technique due to Breiman [50], which practically combines some advantages of neural networks (in particular, the ability to handle very large dimensional inputs) and of constructive wavelet-based estimators (the availability of very fast training algorithms).

Relationship with Neural Networks: Barron's Result

The following result, which was recently published in [51], is the most accurate theoretical result about neural networks available today. Let $\sigma(x)$ be a sigmoidal function [i.e., a bounded measurable function on the real line for which $\sigma(x) \to 1$ as $x \to \infty$ and $\sigma(x) \to 0$ as $x \to -\infty$]. Consider a compactly supported function f with $\text{supp}(f) \subseteq [0, 1]^d$, and assume that

$$C_f = \int_{\mathbf{R}^d} |\omega| |\hat{f}(\omega)| \, d\omega < \infty, \tag{36}$$

where $\hat{f}(\omega)$ denotes the Fourier transform of f. The main result of [51] can be roughly stated as follows: there exists an approximation f_n of the compactly supported function f, of the form

$$f_n(x) = \sum_{i=1}^{n} c_i \sigma \left(a_i^T x + t_i \right) + c_0 \tag{37}$$

(note that f_n is *not* compactly supported), such that

$$\left\| (f_n - f) 1_{[0,1]^d} \right\|_2 \leq 2\sqrt{d} C_f n^{-1/2}. \tag{38}$$

This result provides an upper bound of the minimum distance (in the L_2-norm) between any f satisfying condition (36) and the class of all neural networks of size not larger than n. In the same article, this upper bound is compared with the best achievable convergence rate for any linear estimator in class (36). It is shown that a lower rate for linear estimators is $n^{-1/d}$, compared with the much better rate $n^{-1/2}$ for neural networks, especially for large dimension d. No result is available which takes advantage of the possible improved smoothness of the unknown system f. An iterative algorithm for the construction of the approximation (37) is also proposed. The true problem of system identification, that is, that of neural network training based on noisy input–output data, is not addressed in this paper. Also, neural networks need the backpropagation procedure for their training, a stochastic gradient procedure which is known to be of prohibitive cost. In turn, neural network training works even for very large dimensional input data.

Breiman's Hinging Hyperplanes

We now briefly discuss a recent technique due to Breiman [50], which practically combines some advantages of neural networks (in particular, the ability to handle very large dimensional inputs) and of constructive wavelet-based estimators (the availability of very fast training algorithms). Breiman's technique is a very elegant and efficient way of identifying piecewise linear models based on data collected from an unknown nonlinear system; see [8] for the use of such models in control. Following [50], we call a *hinge function* a function $y = h(x)$, $x \in \mathbf{R}^d$, which consists of two hyperplanes continuously joined together, that is, an open book; see Fig. 1.

If the two hyperplanes are given as

$$y = \langle \beta^+, x \rangle + \beta_0^+, \qquad y = \langle \beta^-, x \rangle + \beta_0^-,$$

where $\langle \cdot, \cdot \rangle$ denotes a scalar product in Euclidean spaces, then an explicit form for the hinge function is either

$$h(x) = \max(\langle \beta^+, x \rangle + \beta_0^+, \langle \beta^-, x \rangle + \beta_0^-),$$

or

$$h(x) = \min(\langle \beta^+, x \rangle + \beta_0^+, \langle \beta^-, x \rangle + \beta_0^-).$$

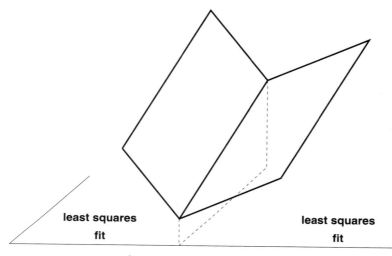

least squares
fit

least squares
fit

Figure 1   Hinge function on $\mathbf{R}^2$. On each side of the corner, the best fit is just performed via linear least squares.

It is proved in [50], using the methods of Barron [51], that there is a constant C such that for any n there are hinge functions $h_1, \ldots, h_n$ such that

$$\left\| f - \sum_{i=1}^{n} h_i 1_{[0,1]^d} \right\|_2 \leq Cn^{-1/2} \tag{39}$$

for any f such that

$$\int_{\mathbf{R}^d} |\omega|^2 |\hat{f}(\omega)| \, d\omega < \infty;$$

that is, Breiman's hinge model is as efficient as neural networks for the L_2-norm. An iterative projection algorithm is proposed to compute the approximation. The interesting point about this iterative approximation technique is that it converges with a magnitude order faster than back propagation does. To understand why this can happen, consider the simplest case where x is of dimension 1, f itself is a hinge function, and we try to fit a single hinge approximant [i.e., $n = 1$ in (39)]. Thus we have to estimate the four unknown parameters $(\beta^{\pm}, \beta_0^{\pm})$. This is done iteratively as follows. First, guess the corner of the hinge (i.e., the x where both arguments in the "max" or "min" are equal); call it $x(0)$. Selecting only those $x > x(0)$ with corresponding y's, a first estimate for, say, (β^+, β_0^+) is obtained by ordinary linear least squares fit, and similarly for $x < x(0)$. Thus we now have a first hinge $\hat{f}(1)$, which yields a new corner $x(1)$, and so on. This converges extremely rapidly. In contrast there is no such fast procedure for a single neuron with adjustable parameters to estimate an unknown single neuron, because the stochastic gradient must be used even in this case. A method based on nested iterations of the preceding kind is proposed in (39) to fit general f's. Reported experimental results show the efficiency of this technique. These experiments show that practically the approximation obtained is much more accurate than is suggested by the estimate in (39). On the other hand, note that a superposition of hinge functions is not smooth, because it is piecewise linear. Also the use of the superposition of hinge functions is especially advocated in (39) for large-dimensional x's. However, as indicated at the beginning of this section, no convergence rate is given for models identified from noisy data [the bound (39) is not a convergence rate for identification, but only a rate of approximation of a given function by some finitely parametrized class of approximants].

III. WAVELETS: WHAT THEY ARE, AND THEIR USE IN APPROXIMATING FUNCTIONS

Warning. Throughout this section, the notation $\hat{\varphi}(\omega)$ denotes the Fourier transform of the function $\varphi(x)$, and *not* the estimator of φ.

A. CONTINUOUS WAVELET TRANSFORM

The continuous wavelet transform and inverse transform of a function f are, respectively, given by Eqs. (41) and (42). These transforms use two functions $\psi(x)$ and $\varphi(x) \in L_2(\mathbf{R}^d)$, both radial (i.e., depending only on $|x|$), known as the *analysis and synthesis wavelets*:

THEOREM 3. *Let ψ and φ be radial functions satisfying*

$$\forall \omega \in \mathbf{R}^d: \int_0^\infty a^{-1}\hat{\varphi}(a\omega)\hat{\psi}(a\omega)\,da = 1, \tag{40}$$

where we recall that $\hat{\varphi}(\omega)$ denotes the Fourier transform of the function $\varphi(x)$. Then for any function $f \in L_2(\mathbf{R}^d)$, the following formulas define an isometry between $L_2(\mathbf{R}^d)$ and a subspace of $L_2(\mathbf{R}^d \times \mathbf{R}_+)$ [6]:

$$u(a, t) = a^{d-1/2} \int f(x)\psi(a(x-t))\,dx, \tag{41}$$

$$f(x) = \int u(a, t)\varphi(a(x-t))a^{d-1/2}\,da\,dt. \tag{42}$$

Here, $a \in \mathbf{R}^+$ and $t \in \mathbf{R}^d$ are, respectively, the dilation and translation factors. Note that the integral (40) does not depend on $\omega \neq 0$ because the functions ψ and φ are radial. For this integral to be properly defined, it is sufficient that, for example, $\hat{\varphi}(\omega)\hat{\psi}(\omega) = O(|\omega|)$; this happens if $\varphi(x)$ and $(1 + |x|)\psi(x)$ are in $L_1(\mathbf{R}^d) \cap L_2(\mathbf{R}^d)$ and ψ has zero integral. Once the integral (40) is well defined and finite, a simple normalization leads to a pair (φ, ψ) which satisfies the assumption.

Examples

One can verify that the following pairs ψ, φ satisfy the assumption:

$$\psi(x) = \sqrt{2}(d - |x|^2)\exp(-|x|^2/2), \qquad \varphi(x) = \sqrt{2}\exp(-|x|^2/2),$$

$$\psi(x) = \varphi(x) = \frac{1}{\sqrt{2}}(d - |x|^2)\exp(-|x|^2/2)$$

and, in the one-dimensional case:

$$\psi(x) = -\text{sign}(x)1_{\{|x|<1\}}, \qquad \varphi(x) = \frac{1 - |x|}{3}1_{\{|x|<1\}},$$

$$\psi(x) = -1_{\{-1 \leq x < -1/2\}} + 1_{\{-1/2 \leq x < 1/2\}} - 1_{\{1/2 \leq x < 1\}},$$

$$\varphi(x) = \lambda^{-1}\exp(-|x|^2/2),$$

with $\lambda = -0.03527343656\ldots$ and $1_{\{A\}}$ is the indicator function of the set A. The choice of possible pairs ψ, φ is very large. In particular, pairs (ψ, φ), with ψ nonsmooth but φ smooth, are allowed.

Time–Frequency Localization

Even this simple construction provides a very interesting property: roughly speaking, the behavior of the function $u(a, t)$, when the scaling factor a is fixed, measures the smoothness of f in the neighborhood of point t. This focusing effect is called "time-frequency localization" (see the discussion in [6, Chap. 2]). It is not provided by the Fourier transform [the behavior of the Fourier transform $\hat{f}(\omega)$ reflects the *global* smoothness of f]. Unfortunately, these localization properties of the continuous wavelet transform cannot be used for estimation, because there is no associated algorithm to compute this transform. For practical purposes, the reconstruction formula (42) has to be discretized:

$$f(x) = \sum_i u_i \varphi(a_i x - t_i); \tag{43}$$

this point will be discussed in Section III.B and in the following sections.

B. DISCRETE WAVELET TRANSFORM: ORTHONORMAL BASES OF WAVELETS AND EXTENSIONS

Multiresolution analysis introduced by Mallat and further developed by Daubechies provides orthonormal bases of $L_2(\mathbf{R})$ of the form $\psi_{j,k}(x) = \{2^{j/2}\psi(2^j x - k): j, k \in \mathbf{Z}\}$; that is, each element of the basis is a translated and dilated version of a single *wavelet* ψ. For a function $f \in L_2(\mathbf{R})$, the inner product $\langle f, \psi_{j,k} \rangle$ performs zooming on f over an $O(2^{-j})$ width interval centered at point $2^{-j}k$. Thus *large j corresponds to checking the function f at fine scales.* This implies that a local singularity of a function f will affect only a small part of its coefficients in this wavelet basis. This is the main difference with the Fourier basis: a local singularity of f would affect the whole Fourier representation.

1. Definition and Construction of Orthogonal Wavelet Bases

To begin, we first discuss the scalar case, that is, that of functions defined on $\mathbf{R}$. Otherwise explicitly stated, all results in this subsection are borrowed from monograph [6].

DEFINITION 2 (Multiresolution analysis).   A multiresolution analysis (MA) consists of a function φ, $\|\varphi\|_2 = 1$, and a sequence $(V_j)_{j \in \mathbf{Z}}$ of spaces defined by

$$\varphi_{jk} = 2^{j/2}\varphi(2^j x - k), \qquad j, k \in \mathbf{Z},$$
$$V_j = \text{Span}\{\varphi_{jk}, \ k \in \mathbf{Z}\},$$

with the properties:

(MA0) $(\varphi_{0k})_{k \in \mathbf{Z}}$ is an orthonormal family;
(MA1) $\bigcap_{j \in \mathbf{Z}} V_j = \{0\}$;
(MA2) $\overline{\bigcup_{j \in \mathbf{Z}} V_j} = L_2(\mathbf{R})$;
(MA3) $V_j \subset V_{j+1}$.

Property (MA3) is equivalent to the existence of a square integrable sequence (h_k) such that

$$\varphi(x) = \sqrt{2} \sum h_k \varphi(2x - k). \qquad (44)$$

We call such a function the *scale function* (also known as the *father wavelet* [5]). Theorem 4 to follow is the basis of the theory; it shows how, starting from a multiresolution analysis and its scale function φ, we can construct very simply an orthonormal basis of $L_2(\mathbf{R})$.

THEOREM 4. *Assume that conditions* (MA0)–(MA3) *are satisfied. Set*[13]

$$\psi(x) = \sqrt{2} \sum g_k \varphi(2x - k), \qquad g_k = (-1)^{k+1} \bar{h}_{1-k},$$
$$\psi_{jk} = 2^{j/2} \psi(2^j x - k),$$
$$W_j = \text{Span}(\psi_{jk}, \ k \in \mathbf{Z}). \tag{45}$$

Then

1. $V_{j+1} = V_j \oplus W_j$ *and* $\{\psi_{jk}: \ j, k \in \mathbf{Z}\}$ *is an orthonormal basis in* $L_2(\mathbf{R})$;
2. $L_2(\mathbf{R}) = V_0 \oplus W_0 \oplus W_1 \oplus \cdots$ *and* $\{\varphi_{0k}, \psi_{jk}: \ j \geqslant 0, k \in \mathbf{Z}\}$ *is an orthonormal basis in* $L_2(\mathbf{R})$.

The function $\psi(x)$ *defined in* (45) *is often referred to as the "mother wavelet."*

Multiresolution analysis and orthonormal wavelets are depicted in Fig. 2. Then Theorem 5 gives the basic tool for building scale functions.

THEOREM 5. *Let* $m_0(\omega)$ *be a trigonometric polynomial*

$$m_0(\omega) = \frac{1}{\sqrt{2}} \sum_{k=K}^{L} h_k e^{-ik\omega}$$

such that

(QMF1) $m_0(0) = 1$;
(QMF2) $m_0(\omega) \neq 0$ *if* $\omega \in [-\pi/2, \pi/2]$;
(QMF3) $|m_0(\omega)|^2 + |m_0(\omega + \pi)|^2 = 1$.

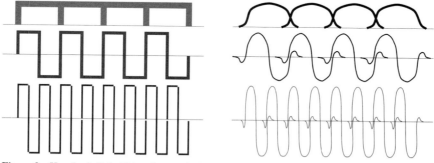

Figure 2 Haar basis (left side) and a wavelet basis (right side). The first row shows the scale function φ and the subsequent rows show wavelets ψ at two successive scales.

---

[13] $\bar{h}$ denotes the complex conjugate of h.

Then the function φ, *with Fourier transform given by*

$$\hat{\varphi}(\omega) = \prod_{j=1}^{\infty} m_0(2^{-j}\omega),$$

satisfies assumptions (MA0)–(MA3) *and* supp(φ) $\subset [K, L]$.

Examples of polynomials satisfying assumptions (QMF1)–(QMF3) are given in [6] and the smoothness properties of φ and ψ are studied. Links with multirate digital signal processing and quadrature mirror filter (QMF) banks are discussed in [52]; see the next subsection.

We now move on to discuss the multidimensional case. There exist two main types of constructions of the wavelet basis with dilation factor 2 in $\mathbf{R}^d$ [6, 10.1]. A first guess simply consists of taking tensor product functions generated by d one-dimensional bases:

$$\Psi_{j_1,k_1,\dots,j_d,k_d}(x) = \psi_{j_1,k_1}(x_1) \times \cdots \times \psi_{j_d,k_d}(x_d). \tag{46}$$

This construction has the drawback of mixing different resolution levels j_i. Alternatively, if such a mixing is not desired, we proceed as follows. Introduce the scale function

$$\Phi(x) = \varphi(x_1) \times \cdots \times \varphi(x_d) \tag{47}$$

and the $2^d - 1$ mother wavelets $\Psi^{(l)}(x)$, $i = 1, \dots, 2^d - 1$, obtained by substituting in (47) some $\varphi(x_j)$'s by $\psi(x_j)$'s. Then the following family is an orthonormal basis of $L_2(\mathbf{R}^d)$:

$$\left\{ \Phi_{0k}(x), \Psi_{jk}^{(1)}(x), \dots, \Psi_{jk}^{(2^d-1)}(x) \right\}, \quad j \in \mathbf{N}_0, \ k = (k_1, \dots, k_d) \in \mathbf{Z}^d, \tag{48}$$

where $\mathbf{N}_0 = \mathbf{N} \cup 0$, and

$$\Phi_{jk}(x) = 2^{jd/2}\Phi(2^j x_1 - k_1, \dots, 2^j x_d - k_d),$$
$$\Psi_{jk}^{(l)}(x) = 2^{jd/2}\Psi^{(l)}(2^j x_1 - k_1, \dots, 2^j x_d - k_d).$$

Note. As formula (48) shows, constructing and storing orthonormal wavelet bases become prohibitively costly for large-dimensional d. This is the main limitation for using the otherwise very efficient techniques which rely on orthonormal wavelet bases (and their generalizations).

2. Orthogonal Wavelet Bases and Quadrature Mirror Filters

For the sake of simplicity, we only discuss the one-dimensional case. Equations (44) and (45) imply that,[14] for $f \in L_2(\mathbf{R})$,

$$\alpha_{jk} = \langle f, \varphi_{jk} \rangle, \qquad \beta_{jk} = \langle f, \psi_{jk} \rangle \qquad (49)$$

satisfy[15]

$$\alpha_{jk} = \sum_l \bar{h}_{l-2k} \alpha_{j+1,l}, \qquad (50)$$

$$\beta_{jk} = \sum_l \bar{g}_{l-2k} \alpha_{j+1,l}. \qquad (51)$$

Introduce the polynomial filters

$$H(z) = \sum_k h_k z^{-k}, \qquad G(z) = \sum_k g_k z^{-k}, \qquad (52)$$

where the coefficients h_k, g_k are as in (44) and (45). Also denote by $\downarrow^{(2)}$ the decimation of a signal by a factor of 2:

$$\downarrow^{(2)} (x_n) = (x_{2n}).$$

Thus, if we consider α_{jk} as a signal indexed by k and denote it by α_j, relations (51) translate into

$$\alpha_j = \downarrow^{(2)} H \alpha_{j+1}, \qquad \beta_j = \downarrow^{(2)} G \alpha_{j+1}$$

and property (QMF3) expresses that the pair (H, G) is QMF [52, 53]. Equations (50) and (51) are used to compute recursively from fine scales to coarse scales the orthonormal wavelet decomposition. Assume that, in addition, the scale function φ is selected so that the computation of the inner product $\langle f, \varphi_{jk} \rangle$ in (49) is performed efficiently for some scale j. Then *formulas (49)–(51) together build a highly efficient procedure for computing the wavelet decomposition of f.* As pointed out at the end of the preceding subsection, orthonormal wavelet bases become prohibitively costly to store for large-dimensional d, however. Scale functions φ are proposed in [6], with vanishing moment conditions, for which

$$\langle f, \varphi_{jk} \rangle = f(2^{-j}k) + O(2^{-Mj}) \qquad (53)$$

holds, where the integer M is related to the number of vanishing moments (such scale functions are often referred to as "coiflets"). Note that the preceding approx-

---

[14] Recall that $\langle \cdot, \cdot \rangle$ denotes the inner product in L_2.
[15] Recall that $\bar{h}$ denotes the complex conjugate of h.

imation is at the same time good and very easy to compute. Alternative techniques to get simple approximations similar to (53) are proposed in [54, 55].

Because QMF pairs are known to allow exact reconstruction of filtered-and-decimated signals [52, 53], Eqs. (50) and (51) can be "inverted" to yield the synthesis equation

$$\alpha_{jk} = \sum_l h_{k-2l}\alpha_{j-1,l} + g_{k-2l}\beta_{j-1,l}. \tag{54}$$

For $f \in V_{j_0}$, we have, by definition of this space,

$$f = \sum_k \alpha_{j_0 k}\varphi_{j_0 k}, \tag{55}$$

and, because $V_{j_0} = V_0 \oplus W_0 \oplus W_1 \oplus \cdots \oplus W_{j_0}$,

$$f = \sum_k \alpha_{0k}\varphi_{0k} + \sum_{j,k} \beta_{jk}\psi_{jk}. \tag{56}$$

Formulas (50) and (51) allow us to switch from representation (55) to representation (56). The latter one is generally much more compact because, when f is smooth, most β_{jk} are negligible. In the multidimensional case, $f \in L_2(\mathbf{R}^d)$, formula (56) generalizes as follows:

$$f = \sum_k \alpha_{0k}\Phi_{0k} + \sum_{j=0}^{\infty}\sum_{k \in \mathbf{Z}^d}\sum_{l=1}^{2^d-1} \beta_{jk}^{(l)}\Psi_{jk}^{(l)},$$

$$\alpha_{jk} = \langle f, \Phi_{jk}\rangle, \qquad \beta_{jk}^{(l)} = \langle f, \Psi_{jk}^{(l)}\rangle, \tag{57}$$

where the Φ_{0k}'s and $\Psi_{jk}^{(l)}$'s are the basis functions defined in (48).

C. WAVELETS AND FUNCTIONAL SPACES

We first state a result [5, 56] concerning functions that satisfy Hölder-type conditions. This result then motivates introducing Besov functional spaces. Recall that a function f is called Hölder continuous with exponent s at point x_0, written $f \in C_{x_0}^s$, if there is a polynomial P of degree at most $\lfloor s \rfloor$ such that[16]

$$\left|f(x) - P(x - x_0)\right| \le C|x - x_0|^s.$$

---

[16]Recall that $\lfloor s \rfloor$ denotes the largest integer $\le s$.

If f is Hölder continuous, with exponent s at x_0, then there exists $C < \infty$ such that, for $j > 0$,

$$\max_{\{k:\, x_0 \in \text{supp}\, \psi_{jk}\}} \langle f, \psi_{jk} \rangle \leq C 2^{-j(s+d/2)}. \tag{58}$$

Conversely, if (58) holds and f is known to be $C^{\varepsilon}_{x_0}$ for some $\varepsilon > 0$, then

$$\left| f(x) - P(x - x_0) \right| \leq C |x - x_0|^s \log \frac{2}{|x - x_0|}.$$

This result states that local smoothness of Hölder type can be characterized with the vanishing rate of the wavelet coefficients in the neighborhood of this point. This property is specific to the wavelet transform, and does not hold for other orthogonal bases. This remark also motivates introducing Besov spaces of functions.

1. Besov Spaces as Spaces of Smooth Functions with Localized Singularities

Smooth functions with sparse singularities are typically encountered in nonlinear systems, for example, in mechanical and chemical systems. As we shall see, Besov spaces are spaces

- of smooth functions with possibly localized singularities,
- in which norms are easily evaluated using wavelet coefficients.

For the sake of clarity we consider only compactly supported functions $f \colon \text{supp}\, f \subseteq [0, 1]^d$, though all of the following definitions can be generalized for the noncompact and multidimensional case (we recomend [57, 58] as extremely complete presentations of the current state of the theory of functional spaces).

For $f \in L_1$ and $M \in \mathbf{N}$ we define the local oscillation of order M (or M-oscillation for short) at the point $x \in [0, 1]$ by

$$\text{osc}_M f(x, t) \overset{\Delta}{=} \inf_P \frac{1}{t^d} \int_{|x-y|<t} \left| f(y) - P(y) \right| dy, \tag{59}$$

where the infimum is taken over all polynomials P of degree less than or equal to M. This quantity measures the quality of the local fit of f by polynomials on balls of radius t.

Select $p, q > 0$, $s > d(p^{-1} - 1)$, and take $M = \lfloor s \rfloor$. The following set of functions:

$$\mathcal{B}^s_{pq} = \left\{ f \in L_{1 \wedge p} \colon \|f\|_{\mathcal{B}^s_{pq}} = \|f\|_p \right.$$
$$\left. + \left(\sum_{j=1}^{\infty} (2^{js} \|\mathrm{osc}_M f(x, 2^{-j})\|_p)^q \right)^{1/q} < \infty \right\} \tag{60}$$

(with the usual modification for p or $q = \infty$) is identical to the *Besov spaces* of functions [59], and it is shown in [57] that $\| \cdot \|_{\mathcal{B}^s_{pq}}$ is equivalent to the classical Besov norm.

Comments

1. The triple parametrization using s, p, and q provides a very accurate characterization of the smoothness properties. As usual for Hölder or Sobolev spaces, the index s indicates how many derivatives are smooth. Then, for larger p, $\|f\|_{\mathcal{B}^s_{pq}}$ is more sensitive to details. Finally, the index q has no useful practical interpretation, but it is a convenient instrument that serves to compare Besov spaces with the more usual Sobolev spaces $\mathcal{W}^s_p$, as indicated next. It is interesting to notice that the indicator functions of intervals belong to the spaces $\mathcal{B}^s_{s-1 \infty}$ for all $s > 0$, this illustrates our claim in the title of this subsection.

2. It can be shown that (cf. [57]) for $s \geq 0$, $0 \leqslant p, q \leqslant \infty$:

- The family of Besov spaces includes some more classical spaces. For s noninteger, Hölder classes $\mathcal{C}^s = \mathcal{B}^s_{\infty\infty}$, and Sobolev spaces $\mathcal{W}^s_2 = \mathcal{B}^s_{22}$;
- $\mathcal{B}^s_{pq} \subset \mathcal{B}^{s'}_{p'q'}$ if $p' \geqslant p$, $q' \geqslant q$, $s' \leq s - d/p + d/p'$ (strict inequality if $p = \infty$);
- $\mathcal{B}^0_{pq} \subseteq L_p \subseteq \mathcal{B}^0_{pq'}$ where $q = 2 \wedge p$ and $q' = 2 \vee p$;
- $\mathcal{B}^s_{pp} \subset \mathcal{W}^s_p \subset \mathcal{B}^s_{p2}$ for $p \leqslant 2$;
- $\mathcal{B}^s_{p2} \subset \mathcal{W}^s_p \subset \mathcal{B}^s_{pp}$ for $p \geqslant 2$.

In particular, if $s > d/p$, then $\mathcal{B}^s_{pq} \subset \mathcal{C}$.

2. Approximation in Besov Spaces: Some General Results

We consider the d-dimensional case and supp $f \subseteq [0, 1]^d$. *Free-knot spline approximations* have been analyzed in [60, Theorems 7.3 and 7.4] using Besov spaces. Recall that a function f_n is called the spline function on $[0, 1]$ of order k with n knots if $f_n \in \mathcal{C}^{k-2}$ and there exist points (knots) $0 = x_0 < x_1 \leq x_2 \leq \cdots \leq x_{n-1} \leq x_n = 1$ such that f_n is an algebraic polynomial of degree $k - 1$

in each interval (x_{i-1}, x_i). Therefore, a spline is a smooth piecewise-polynomial function. One can also consider a d-dimensional spline which is the natural generalization of the one-dimensional one.

We now state the so-called *Jackson inequality* for spline approximations. Consider $f \in \mathcal{B}_{pq}^s$, $p, q > 0$. Then there exists a spline function with n free knots f_n such that the following bound holds:

$$\| f_n - f \|_u \leq C(s, p, q) n^{-s/d} \| f \|_{\mathcal{B}_{p\infty}^s}, \tag{61}$$

where u satisfies $s - d/p + d/u > 0$. The converse bound is provided by the *Bernstein inequality*: For any $f \in L_u$, $s - d/p + d/u = 0$, $u < \infty$,

$$\| f \|_{\mathcal{B}_{pp}^s} \leq C(s, p, q) \left(1 + n^{s/d} \inf_{f_n} \| f - f_n \|_u \right),$$

where the infimum ranges over the set of spline functions f_n of order $k \geq s + 2$ with n free knots. A similar result holds for n-order *rational fraction approximations*; see [60, Theorem 8.3].

In contrast, *linear approximations* perform poorly in Besov spaces. Consider some increasing family $(\mathcal{L}_n)$ of n-dimensional linear subspaces of L_u, $u > p$. Let f_n denote the linear projection of $f \in \mathcal{B}_{pq}^s$ on $\mathcal{L}_n$ using the L_u-norm. Then, for any such family $(\mathcal{L}_n)$, there exists a least favorable f such that the following lower bound holds:

$$\| f - f_n \|_u \geq C n^{-s'/d} \| f \|_{\mathcal{B}_{uu}^{s'}}, \tag{62}$$

where $s' = s - d/p + d/u$. Consider again the example of the indicator function $f(x) = 1_{\{0 \leq x < a\}}$. Recall that $f \in \mathcal{B}_{s^{-1}\infty}^s$ for any $s > 0$. On the one hand, (61) shows that f is approximated using rational fractions with an L_u-error of order $O[\exp(-C\sqrt{n})]$, where n is the order of the rational fraction [60]. Thus rational approximations are very efficient for such a function, and the same is true for splines with free knots. On the other hand, by (62), linear approximations of the same function have an L_u-error of order $O(n^{-1/u})$, where n is the dimension of the linear subspace, which is extremely poor for large u. This remark would make rational approximations or splines with free knots very attractive for approximation in Besov spaces. Unfortunately, such approximations are very hard to compute, for example, the optimal positioning of the knots of the spline approximation is very hard to find. It is amazing that *wavelet approximations are as good as spline or rational ones, but are much more easily constructed.* We discuss this next.

3. Wavelets and Besov Spaces: Mathematically Efficient and Practically Effective

Let φ be a piecewise-continuous scale function satisfying the following conditions:

$$\exists a > 0: \; \text{supp}\, \varphi \in \{|x| \leq a\}, \tag{63}$$

$$\exists r > s: \; \varphi \in \mathcal{B}_{u\infty}^r. \tag{64}$$

We have the following result (cf. [61, Theorem 4]):

THEOREM 6 (Besov norms and wavelet decompositions).   Let $s > d(1/u - 1)$ and let φ be a scale function satisfying conditions (63) and (64). For any $f \in \mathcal{B}_{pq}^s$, define

$$\|f\|_{spq} = \left(\sum_k |\alpha_k|^p \right)^{1/p} + \left(\sum_{j=0}^{\infty} [2^{j(s+d/2-d/p)} \|\beta_{j\cdot}\|_p]^q \right)^{1/q} \tag{65}$$

and $\|\beta_{j\cdot}\|_p = (\sum_{i,k} |\beta_{jk}^{(l)}|^p)^{1/p}$; see (49) and (57) for the definition of coefficients $\alpha_k = \alpha_{0k}$ and $\beta_{jk}^{(l)}$. Then (65) is equivalent to the norm of Besov space $\mathcal{B}_{pq}^s$; that is, there exist constants C_1 and C_2, independent of f, such that

$$C_1 \|f\|_{\mathcal{B}_{pq}^s} \leq \|f\|_{spq} \leq C_2 \|f\|_{\mathcal{B}_{pq}^s}. \tag{66}$$

Theorem 6 states that norms in Besov spaces are suitably evaluated using orthonormal wavelet decompositions. This fact can be used to obtain very efficient approximations.

We now indicate how such a wavelet approximation of f can be constructed. Consider the full wavelet decomposition of f:

$$f(x) = \sum_{k \in \mathbf{Z}} \alpha_{0k} \Phi_{0k}(x) + \sum_{j=0}^{\infty} \sum_{k \in \mathbf{Z}^d} \sum_{l=1}^{2^d - 1} \beta_{jk}^{(l)} \Psi_{jk}^{(l)}(x). \tag{67}$$

1. Keep the projection of f on the subspace V_0; this corresponds to the leftmost sum in (67). When f and Φ are both compactly supported, this requires computing only a fixed amount of coefficients, say m.
2. Select in the second (triple) sum those coefficients β_λ, $\lambda = (i, j, k)$, with largest absolute value; denote by Λ the set of the $n - m$ so selected wavelet coefficients.
3. Add $n - m$ detail terms $\beta_\lambda \Psi_\lambda$ to the sum taken in step 1.

This procedure yields the approximation

$$
w_n(x) = \underbrace{\sum_k \alpha_{0k} \Phi_{0k}(x)}_{\substack{m \text{ coeffs.} \neq 0 \\ (f, \Phi \text{ compact. supp.})}} + \underbrace{\sum_{j=0}^{\infty} \sum_{k \in \mathbf{Z}^d} \sum_{l=1}^{2^d-1} \beta_{jk}^{(l)} \Psi_{jk}^{(l)}(x)}_{\text{keep the largest } n-m \text{ coeffs.}}
\tag{68}
$$

and the following theorem provides corresponding approximation bounds.

THEOREM 7 (DeVore *et al.* [62]). *Consider* $f \in \mathcal{B}_{pp}^s$, $s, p > 0$ *and* $s - d/p + d/u \geq 0$. *Let* w_n *denote the approximation* (68) *of* f. *If the scale function satisfies conditions* (63) *and* (64), *then*

$$
\| f - w_n \|_u \leq C(s, p) n^{-s/d} \| f \|_{\mathcal{B}_{pp}^s}
$$

holds. If, in addition, u satisfies $s - d/p + d/u = 0$, $u < \infty$, *and it is a priori known that* $f \in L_u$, *then the following converse bound holds:*

$$
\| f \|_{\mathcal{B}_{pp}^s} \leq C(s, p, q)\big(1 + n^{s/d} \| f - w_n \|_u\big).
$$

This result is very interesting to us. It implies that, in the wavelet decomposition of a function $f \in \mathcal{B}_{pq}^s$, $p < 2$, only a small number of coefficients are important, and the other ones can be neglected. Consider once more our example $f(x) = 1_{\{0 \leq x < a\}}$. Consider the wavelet decomposition of this function using a compactly supported wavelet $\psi(x)$ such that $\int \psi(x) \, dx = 0$. It is evident that the coefficient β_{jk} vanishes for any wavelet $\psi_{jk}(x)$ which does not cross the (local) singularities of f. Thus, if we consider the projection of f on the subspace V_j, only $O(j)$ coefficients of the decomposition significantly differ from zero (among 2^j potential candidates).

Discussion

At this point, we have the requested background for understanding how to perform wavelet-based estimation. Roughly speaking, the crux is the following. The function $f \in \mathcal{B}_{pq}^s$ to be estimated can be *approximated* using the expansion w_n in (68) with n terms. This is achieved with a rate of $O(n^{-s/d})$. Then the coefficients α_k and β_λ in (68) are estimated via empirical means based on N noisy observations, exactly as for the projection estimates in Section II, formula (28). The mean square error on the estimate of each coefficient is $O(1/N)$. Thus the total mean square error of the estimate will be, as usual, the sum of the stochastic part and of the bias due to the approximation error: this yields $O(n/N) + O(n^{-2s/d})$. The optimal choice for n balances these two terms: $n = N^{1/(2s+d)}$. This choice for n yields a quadratic error of order $N^{-2s/(2s+d)}$ (independent of p, q). As we shall see, this is the typical minimax rate of convergence on Besov spaces. Thus we might be ready to deduce that wavelet estimators are minimax optimal in Besov spaces. Unfortunately, the set Λ of "important" coefficients in truncation (68) is *not* known *a priori* when noisy data sets are at hand for estimation. Thus some kind of hypothesis-testing problem must be solved to obtain the optimal approximation. This adds to the estimation problem a nice stochastic flavor. We address this point in the next section.

IV. WAVELETS: THEIR USE IN NONPARAMETRIC ESTIMATION

We consider here some simple results concerning the estimation of a regression function or a density $f: \mathbf{R}^d \to \mathbf{R}$, and we assume f to be compactly supported (supp $f \subseteq [0, 1]^d$). For the sake of simplicity, we measure the estimation error in the L_2-norm. Similar results were proved for a general d-dimensional case and a variety of error measures, which includes, for instance, L_p-norms for $0 < p \leq \infty$ (see the references at the end of the section). We successively discuss the problems of nonparametric regression and density estimation.

A. WAVELET SHRINKAGE ALGORITHMS

Nonparametric Regression

Assume an N-sample of input–output observations of the following system are available:

$$Y_i = f(X_i) + w_i,$$

where (X_i) and (w_i) are i.i.d. sequences of random variables, X_i is *uniformly distributed* on $[0, 1]^d$, and $Ew_i = 0$, $Ew_i^2 \leq \sigma_w^2$. These assumptions are introduced for the sake of simplicity. They can be weakened, in particular, the (unusual) assumption that X is uniformly distributed can easily be relaxed; see [63]. This would introduce additional burden to our presentation, however.

For $f \in L_2$, recall the wavelet expansion

$$f(x) = \sum_{k \in \mathbf{Z}} \alpha_{0k} \Phi_{0k}(x) + \sum_{j=0}^{\infty} \sum_{k \in \mathbf{Z}^d} \sum_{l=1}^{2^d-1} \beta_{jk}^{(l)} \Psi_{jk}^{(l)}(x), \tag{69}$$

where

$$\alpha_{0k} = \int f(x)\Phi_{0k}(x)\,dx \quad \text{and} \quad \beta_{jk}^{(l)} = \int f(x)\Psi_{jk}^{(l)}(x)\,dx. \tag{70}$$

To construct an estimate of f, a first idea consists of using the law of large numbers and replacing, in expansion (69), the coefficients α_k and $\beta_{jk}^{(l)}$ by their empirical estimates

$$
\hat{\alpha}_{0k}(N) = \frac{1}{N} \sum_{i=1}^{N} Y_i \Phi_{0k}(X_i) \quad \text{and} \quad \hat{\beta}_{jk}^{(l)}(N) = \frac{1}{N} \sum_{i=1}^{N} Y_i \Psi_{jk}^{(l)}(X_i). \quad (71)
$$

Note that the assumption that input X is uniformly distributed has been used at this point.

Density Estimation

Assume independent observations $X_1, \ldots, X_N$ of some random variable X with unknown density $f(x)$ are available. Again f can be expanded using (69) and (70). However, it turns out that

$$
\alpha_{0k} = \int f(x) \Phi_{0k}(x) \, dx = \mathbf{E}_f \Phi_{0k}(X_i),
$$

where $\mathbf{E}_f$ denotes expectation with respect to density f, and the same holds for the β's. Thus empirical estimates of the wavelet coefficients α_k and β_{jk} are given by

$$
\hat{\alpha}_{0k} = \frac{1}{N} \sum_{i=1}^{N} \Phi_{0k}(X_i) \quad \text{and} \quad \hat{\beta}_{jk}^{(l)} = \frac{1}{N} \sum_{i=1}^{N} \Psi_{jk}^{(l)}(X_i). \quad (72)
$$

Thus both nonparametric regression and density estimation are faced with the same issue: in formulas (71) and (72), there may not even be available X_i's within the support of many of the Φ's and Ψ's! We shall now discuss this key point for the case of density estimation.

Obviously, to compute the empirical coefficient $\hat{\beta}_{jk}^{(l)}$, we need that at least several observations X_i hit the support of $\Psi_{jk}^{(l)}(x)$. Statistical laws of loglog type guarantee that this would generically hold for scales that are not too fine. More specifically, for $j \leq j_{\max}$, where

$$
\frac{N}{\ln N} \leq 2^{d j_{\max}} \leq \frac{2N}{\ln N}.
$$

Thus, using brute force, we set $\hat{\beta}_{jk}^{(l)} = 0$ for $j > j_{\max}$. At this point, we have built an estimator of the linear projection type, as in the case of Fourier series in Section II.A. Because these estimators are linear, we cannot expect them to be efficient for Besov spaces [64].

First Proposal

Our first attempt to construct an "interesting estimate" is, following the intuition at the end of the previous section, to keep a properly chosen number of coefficients with largest absolute values and set the others to zero. More precisely, let us consider the set $\widehat{\Lambda}_n$ of pairs $\lambda = (j, k)$ corresponding to the n estimated wavelet coefficients $\hat{\beta}_{jk}^{(l)}$ with largest absolute values. We construct the estimate $\hat{f}_N$ as follows:

$$\hat{f}_N(x) = \underbrace{\sum_k \hat{\alpha}_{0k} \Phi_{0k}(x)}_{\substack{m \text{ coeffs. } \neq 0 \\ (f, \Phi \text{ compact. supp.})}} + \underbrace{\sum_{j=0}^{\infty} \sum_{k \in \mathbf{Z}^d} \sum_{l=1}^{2^d-1} \hat{\beta}_{jk}^{(l)} \Psi_{jk}^{(l)}(x)}_{\text{keep the largest } n-m \text{ coeffs.}}. \tag{73}$$

The following result can be proved about estimate (73) [see (60) for the definition of the Besov spaces]:

THEOREM 8.   *Let $f \in \mathcal{B}_{p\infty}^s$ with $s \geq d/p$, $\|f\|_\infty < \infty$. If $n = N^{1/(2s+d)}$ is selected in (73), then*

$$E\|\hat{f}_N - f\|_2^2 = O\left(\frac{\ln N}{N}\right)^{2s/(2s+d)}. \tag{74}$$

The idea of the proof of Theorem 8 is quite intuitive and typical for wavelet estimators. We follow the argument at the end of the previous section with the only following difference: because no information is available about the distribution of the error $|\hat{\beta}_\lambda - \beta_\lambda|$ for $\lambda \in \widehat{\Lambda}_n$, we take a cautious upper bound for it:

$$\mathbf{E}|\hat{\beta}_\lambda - \beta_\lambda|^2 \mathbf{1}_{\{\hat{\beta} \neq 0\}} \leq \mathbf{E} \sup_{i,j,k} |\hat{\beta}_{jk}^{(l)} - \beta_{jk}^{(l)}|^2 = O\left(\frac{\ln N}{N}\right),$$

which explains the extra logarithmic factor in (74).

Final Solution

Note that n in Theorem 8 depends on s, which is generally unknown. Hence, to complete the estimation algorithm, we need a method to estimate our model order n. Though generalized cross-validation techniques could be used, we prefer a somewhat different estimation approach developed by Donoho *et al.* (see the following references). It uses simple thresholding rules[17]:

$$\tilde{\beta}_{jk}^{(l)} = \hat{\beta}_{jk}^{(l)} \mathbf{1}_{\{|\hat{\beta}| \geq \lambda_j\}}, \tag{75}$$

[17]We consider here the so-called "hard thresholding"; meanwhile, other rules can also be studied, for example, "soft thresholding" [65]. See also the discussion in [66].

where λ_j is a threshold parameter, so we set

$$\hat{f}(x) = \sum_k \hat{\alpha}_{0k} \Phi_{0k}(x) + \sum_{j=0}^{\infty} \sum_{k \in \mathbf{Z}^d} \sum_{l=1}^{2^d-1} \hat{\beta}_{jk}^{(l)} 1_{\{|\hat{\beta}_{jk}^{(l)}| \geq \lambda_j\}} \Psi_{jk}^{(l)}(x). \quad (76)$$

In other words, in expansion (69), we keep those empirical estimates of wavelet coefficients which exceed some properly selected threshold. How this threshold should be selected is provided by the following result:

THEOREM 9 (Donoho *et al.* [41, 67]). *Let* $f \in \mathcal{B}_{p\infty}^s$ *with* $s \geq d/p$, $\|f\|_\infty < \infty$. *Select* $\lambda_j = \lambda = \sqrt{C \ln N / N}$, *with an appropriate* $C < \infty$. *Then*

$$E\|\hat{f}_N - f\|_2^2 = O\left(\frac{\ln N}{N}\right)^{2s/(2s+d)}.$$

The constant C in the expression for the threshold parameter λ is a sort of a "hyperparameter" of the procedure, which can easily be estimated; see [66, 67] for related discussions. Note that the estimator $\hat{f}_N$ is adaptive because *it does not require prior knowledge of the regularity parameter.*

Discussion

• Theorem 9 has the following intuitive explanation. As already mentioned, Besov classes $\mathcal{B}_{pq}^s$ for $p < 2$ have a special structure: a relatively small number of "important" wavelet coefficients are sufficient for obtaining a good function approximation. In the wavelet decomposition $(\hat{\alpha}_k, \hat{\beta}_{jk}^{(l)})$ using noisy data, all coefficients are "contaminated" by noise. A central limit theorem argument suggests that this noise is approximately Gaussian with zero mean and variance $O(1/N)$. Thus loglog law implies that the maximal error in the estimates has magnitude given by

$$\max_{j,k} |\hat{\beta}_{jk}^{(l)} - \beta_{jk}^{(l)}| \approx \sqrt{\frac{2 \ln N}{N}}.$$

Thus, when small (according to the threshold λ in Theorem 9) coefficients are shrinked to zero, noise is canceled with very high probability. On the other hand, coefficients exceeding this threshold are likely to be significantly different from zero. This property of thresholding explains another useful feature of the estimator: the estimate $\hat{f}_N$ has the same regularity as the unknown function f to be estimated (cf. the discussion in [41]).

• Let us now consider again our example of estimating the regression function or density $f(x) = 1_{\{0 \leq x < a\}}$. Theorem 9 states that the mean square rate of convergence of the wavelet estimator for any bounded function $f \in \mathcal{B}_{s^{-1}\infty}^s$ is very close to $O(N^{-1})$, which is nearly as good as the "parametric" rate of convergence, though the function we estimate is not even continuous. Let us compare the preceding results with the lower rate of convergence for this problem obtained in [68]. Using Comment 2 of Section III.C.1, the following lower bound is a direct corollary of the results of [68] which were originally formulated in terms of Sobolev spaces:

$$\inf_{\hat{f}_N} \sup_{f \in \mathcal{B}_{pq}^s} E\|\hat{f}_N - f\|_2 \geq CN^{-2s/(2s+d)} \qquad (77)$$

for any estimator $\hat{f}_N$. As compared to (77), there is an extra logarithmic factor in the upper bound of Theorem 9. In the more subtle construction presented in [46], this logarithmic factor is eliminated (and even a precise minimax constant is obtained) in the case of Gaussian noises and deterministic design (observations are $x_i = i/N$, $i = 1, \ldots, N$). In [69] a cross-validation procedure is proposed to adapt the optimal algorithm to unknown smoothness. Finally, in [66] the authors of this paper showed that properly selecting the threshold λ for shrinking provides the optimal rate of convergence (without a logarithmic factor). An adaptive version of this algorithm is developed in [70].

B. PRACTICAL IMPLEMENTATION OF WAVELET ESTIMATORS

We now move to the practical implementation of wavelet estimators. We propose two versions of it which differ in the way the empirical estimates of the wavelet coefficients $\hat{\alpha}_{jk}$ and $\hat{\beta}_{jk}$ are computed. The first one, called it *direct realization*, is based on the explicit formulas (71) and (72) for empirical coefficients. The second one, called the *fast realization* procedure, relies on the quadrature mirror filters (QMFs) presented in Section III.B.2.

Direct Wavelet Estimation Procedure for an N-Sample Length (Put $Y_i \equiv 1$ for Density Estimation): The Wavelet Shrinkage Algorithm Procedure

(Recall that the assumption that X is uniformly distributed is required for the case of regression.)

1. Select j_{max} scales for the wavelet expansion, where

$$\frac{N}{\ln N} \leq 2^{d\,j_{max}} \leq \frac{2N}{\ln N}.$$

2. For $j \leq j_{\max}$, compute the empirical estimates

$$\hat{\alpha}_k = \frac{1}{N} \sum_{i=1}^{N} Y_i \Phi_{0k}(X_i), \qquad \hat{\beta}_{jk}^{(l)} = \frac{1}{N} \sum_{i=1}^{N} Y_i \Psi_{jk}^{(l)}(X_i). \tag{78}$$

3. Shrink these estimates according to

$$\tilde{\beta}_{jk}^{(l)} = \hat{\beta}_{jk}^{(l)} 1_{\{|\hat{\beta}_{jk}^{(l)}| \geq \lambda_j\}}, \tag{79}$$

where λ_j is a properly selected threshold (cf. Theorem 9).
4. The final estimate is given by

$$\hat{f}_N(x) = \sum_k \hat{\alpha}_k \Phi_{0k}(x) + \sum_{i,j,k} \tilde{\beta}_{jk}^{(l)} \Psi_{jk}^{(l)}(x). \tag{80}$$

This procedure for nonparametric regression can be extended to the case in which X is *not* uniformly distributed over $[0, 1]^d$, and has density $g(x)$. In this case, we have

$$\hat{\alpha}_k = \frac{1}{N} \sum_{i=1}^{N} Y_i \Phi_{0k}(X_i) \approx \int f(x) \Phi_{0k}(x) g(x)\, dx = \int [fg](x) \Phi_{0k}(x)\, dx$$

and similarly for the $\hat{\beta}_{jk}^{(l)}$'s. Thus applying the WSA to estimate the regression function f (with the Y_i in the empirical estimates) as if X was uniformly distributed yields, in fact, an estimate $\widehat{[fg]}_N$ of $[fg]$. From this remark the following procedure follows:

1. apply the WSA to estimate the density g (without the Y_i in the empirical estimates); this yields $\hat{g}$;
2. apply the WSA to estimate the regression function f (with the Y_i in the empirical estimates) as if X was uniformly distributed; this yields $\hat{f}_{\text{uniform}}$;
3. the final estimate is $\hat{f} = \hat{f}_{\text{uniform}}/\hat{g}$.

Comment

The preceding *direct* estimate has some drawbacks (we consider only the computational aspect for a moment). First, we know that there is no closed form for the scale function Φ or wavelet Ψ. Thus, to compute $\hat{\alpha}_{jk}$ and $\hat{\beta}_{jk}$, we would have to compute and store the values of Φ and Ψ on a fine grid, which is prohibitive. Second, we would like to take advantage of the fast QMF algorithms of Section III.B.2 for computing orthonormal wavelet decompositions. We cannot apply these algorithms directly on the data, because the available observations $X_1, \ldots, X_N$ are randomly sampled and do not form a regular grid. To circumvent this difficulty, we preprocess the observations to obtain the empirical coefficients

$\hat{\alpha}_{j_{\max},k}$ at the finest resolution level $j_{\max}$; then we can apply the QMF algorithms of Section III.B.2 to compute the coefficients at coarser scales. The proposed procedure is close to the *empirical wavelet transform* or *hybrid transform*, studied in [71, Sect. 5]; mathematical details can be found in [55]. We assume that the function f is supported on $[0, 1]^d$.

Fast Wavelet Estimator (X Does Not Need to Be Uniformly Distributed)

1. *Preprocessing.* Select again $j_{\max}$ such that

$$\frac{N}{\ln N} \le 2^{d j_{\max}} < \frac{N}{\ln N}.$$

Let $k = (k_1, \ldots, k_d)^T$ be a multi-index, and consider the bin

$$\Delta_k = \left[2^{-j_{\max}} k_1, 2^{-j_{\max}}(k_1 + 1)\right] \times \cdots \times \left[2^{-j_{\max}} k_d, 2^{-j_{\max}}(k_d + 1)\right].$$

For density estimation, we first take the empirical probability of bin Δ_k (recall that Δ_k has volume $2^{-d j_{\max}}$); this yields

$$\tilde{f}_{N,k} = 2^{d j_{\max}} \frac{1}{N} \sum_{i=1}^{N} 1_{\{X_i \in \Delta_k\}}$$

and then

$$\hat{\alpha}_{j_{\max},k} = 2^{d j_{\max}/2} \tilde{f}_{N,k}.$$

For nonparametric regression, similarly, compute

$$\tilde{f}_{N,k} = \frac{\sum_{i=1}^{N} Y_i 1_{\{X_i \in \Delta_k\}}}{\sum_{i=1}^{N} 1_{\{X_i \in \Delta_k\}}}$$

and then

$$\hat{\alpha}_{j_{\max},k} = 2^{d j_{\max}/2} \tilde{f}_{N,k}.$$

At this point, we have constructed synthetic input–output pairs, where the input is the considered bin and the output is the associated $\hat{\alpha}_{j_{\max},k}$ estimate. Getting the full wavelet expansion is then performed by applying to these synthetic data the QMF fast formulas (50) and (51).

2. *QMF filtering.* Use the multidimensional version of filters (50) and (51) to compute $\hat{\alpha}_{jk}$, $\hat{\beta}_{jk}^{(l)}$, $j = 0, \ldots, j_{\max} - 1, l = 1, \ldots, 2^d - 1$:

$$\hat{\alpha}_{jk} = \sum_i \bar{h}_{i-2k} \hat{\alpha}_{j+1,i},$$

$$\hat{\beta}_{jk}^{(l)} = \sum_i \bar{g}_{i-2k}^l \hat{\alpha}_{j+1,i}.$$

3. Shrink the estimates $\hat{\beta}_{jk}^{(l)}$ according to

$$\tilde{\beta}_{jk}^{(l)} = \hat{\beta}_{jk}^{(l)} 1_{\{|\hat{\beta}_{jk}^{(l)}| \geq \lambda_j\}},$$

where λ_j is a properly selected threshold (cf. Theorem 9).

4. Use the "inverse" filter (54) to obtain $\tilde{\alpha}_{j\max,k}$:

$$\tilde{\alpha}_{jk} = \sum_{il} h_{k-2i} \tilde{\alpha}_{j-1,i} + g_{k-2i}^l \tilde{\beta}_{j-1,i}^{(l)}. \tag{81}$$

5. Finally, set

$$\hat{f}_N(2^{-j\max}k) \triangleq \hat{f}_{N,k} = 2^{-dj\max/2} \tilde{\alpha}_{jk}.$$

In this way we obtain estimates of $f(2^{-j\max}k)$. If this accuracy is not sufficient, it is possible to interpolate $\hat{f}_N$ at a finer grid by applying upsampling (81), using the filters that are biorthogonal to those associated with the Haar basis (see [6, Chap. 8, 71]).

V. WAVELET NETWORK FOR PRACTICAL SYSTEM IDENTIFICATION

The estimation procedure described in the previous section may not be effective for X of higher dimension and for sparse input data sets for training. In this section we attempt to cope with highly dimensional problems and bad data sampling using an alternative technique of wavelet estimation. We present here a method for constructing estimators with nonorthogonal wavelets; the corresponding software is available [72]. We investigate Problem 1 of Section I.B in the case of additive noise; that is, we suppose that the pair of random variables X, Y satisfies

$$Y = f(X) + e, \tag{82}$$

where $f(x): \mathbf{R}^d \mapsto \mathbf{R}$ and e is some noise of zero mean and independent of X. We want to estimate f based on a sample of size N that we shall refer to as the *training data set*: $\mathcal{O}_1^N = \{(X_1, Y_1), \ldots, (X_N, Y_N)\}$. We are particularly interested in training with sparse data sets. Sparse data often occur in classification problems and in the modeling of control systems, where available data can be relatively few as compared to the dimension of input X. Throughout this section, φ shall denote a radial wavelet as defined in Theorem 3; thus we are *not* using orthonormal wavelets.

A. ADAPTIVE DILATION/TRANSLATION SAMPLING

We present here a result which can be regarded as a theoretical justification of the techniques in this section. Note that in the orthonormal wavelet expansion

$$f(x) = \sum_k \alpha_{0k} \Phi_{0k}(x) + \sum_{ljk} \beta_{jk}^{(l)} \Psi_{jk}^{(l)}(x),$$

the dilation and translation parameters -2^{-dj} and k do not depend on the function to expand and only the linear weights α_{jk} and $\beta_{jk}^{(l)}$ depend on f. Suppose that we construct a wavelet "basis" with dilations and translations depending on the function f. The wavelet expansion of f using these basis functions is expected to use less wavelets, and thus we expect it to be more convenient for estimation purposes. To obtain such a basis, *we discretize the continuous wavelet transform* (42) (see Section III.A).

We first recall the following algorithm proposed in [73]. Consider the continuous wavelet transform (42), which we rewrite as

$$f(x) = \int u(a, t) \varphi(a(x - t)) a^{d-1/2} \, da \, dt$$

$$= \int \varphi(a(x - t)) \operatorname{sign}(u(a, t)) a^{(d-1)/2} |u(a, t)| \, da \, dt$$

$$= \frac{1}{C} \int \varphi(a(x - t)) \operatorname{sign}(u(a, t)) w(a, t) \, da \, dt,$$

where we have renormalized $u(a, t)$ by a constant factor C so that the function $w(a, t) = Ca^{(d-1)/2} |u(a, t)|$ can be considered as a probability density. Then we draw n independent random samples $(a_i, t_i)_{i=1,\dots,n}$ from the distribution with density $w(a, t)$. Then we build

$$f_n(x) = \frac{1}{n} \sum_{i=1}^n a_i^{d/2} \varphi(a_i(x - t_i)) \operatorname{sign}(u(a_i, t_i)), \qquad (83)$$

which, owing to the law of large numbers, converges to the true wavelet transform. Some faster implementations of this algorithm are given in [73]. Improving this estimate by some "bootstrapping" technique yields the following approximation result:

THEOREM 10 (Delyon *et al.* [73]). *φ is any radial wavelet function such that there exists a related radial function ψ which satisfies condition (40). Let p, μ, l, ρ be real numbers satisfying*

$$1 < p < \left(1 - \frac{\rho - l}{d}\right)^{-1}, \qquad \mu = \min\left(1 - \frac{1}{p}, \frac{1}{2}\right)$$

and let f be a function of the Sobolev space $\mathcal{W}_1^\rho(\mathbf{R}^d)$; then, for any $n > 0$, there exists a function f_n of the form

$$f_n(x) = \sum_{i=1}^n u_i \varphi\big(a_i(x - t_i)\big) \tag{84}$$

such that

$$\|f_n - f\|_{\mathcal{W}_p^l} \leq Cn^{-\mu}\|f\|_{\mathcal{W}_1^\rho}.$$

In particular, if $\rho > d/2$ then

$$\|f_n - f\|_2 \leq n^{-1/2}C\|f\|_1^\rho .$$

Comment

Theorem 10 provides us with an upper bound for the rate of approximation when adaptive dilation/translation sampling is used to discretize the continuous wavelet transform. We should compare this rate with rates of convergence for approximations based on *fixed* dilation/translation sampling. For example, the following theorem is proved in [73]:

THEOREM 11. *Let* $p = 2$ *and* $\rho = d/2 + \varepsilon$, $\varepsilon > 0$. *For a collection* $h_1, \ldots, h_n$ *of basis functions,*[18] *consider the error*

$$V_n = \inf_{h_1,\ldots,h_n} \sup_{\|f\|} \|f - \operatorname{span}\{h_1, \ldots, h_n\}\|_2,$$

where $\operatorname{span}\{\cdots\}$ *denotes the linear space spanned by the listed functions, and the supremum is taken over the unit ball* $\mathcal{B} = \{f: \|f\|_1^\rho \leq 1\}$ *of the Sobolev space* $\mathcal{W}_1^\rho$. *Then there exists a universal constant* C *such that, for any fixed basis* $h_1, \ldots, h_n$,

$$V_n \geq Cn^{-\varepsilon/d}.$$

The result of the theorem implies that for any *fixed* basis $h_1, \ldots, h_n$ and any set of $\alpha_1, \ldots, \alpha_n$, there are "worst functions" f for which a projection approximation $f_n^h(x)$ of the form

$$f_n^h(x) = \sum_{i=1}^n \alpha_i h_i$$

[18]One can take, for instance, the trigonometric basis on $[0, 1]^d$, or a truncated wavelet basis with fixed dilation and translation sampling.

converges much slower than the approximation (84). Note that this is not in contradiction with the optimality of wavelet shrinkage procedures, because shrinking coefficients in the wavelet expansion make the estimator nonlinear.

B. Wavelet Network and Its Structure

Though the preceding adaptive dilation/translation sampling algorithm provides us with a good basis, its implementation using a Monte Carlo technique is of prohibitive computational cost. We rather implement adaptive sampling in a different way, by combining regressor selection and backpropagation algorithms to find good dilations and translations. The resulting estimator is called the *wavelet network*. Related works have been reported in [74–76]. We refer the reader to [74] for heuristic comparisons between neural and wavelet networks. For any wavelet function $\varphi\colon \mathbf{R}^d \to \mathbf{R}$, the wavelet network is written as follows:

$$f_n(x) = \sum_{i=1}^{n} u_i \varphi\big(a_i \star (x - t_i)\big), \tag{85}$$

where $u_i \in \mathbf{R}$, $a_i \in \mathbf{R}^d$, $t_i \in \mathbf{R}^d$, and "$\star$" denotes the component-wise product of two vectors. Note that we could have used scalar dilation parameters a_i, but we prefer vectorial dilation parameters because they considerably increase the flexibility of network (85) at a reasonable price. The structure of the wavelet network is depicted in Fig. 3.

In this section we present an efficient comprehensive method for wavelet network training. The following is an outline of this method:

1. Construct a library W of dilated/translated versions of a given wavelet φ. This library W is adapted to the available training data set, by selecting a subset from all dilated/translated versions of φ on a regular grid. This technique makes it feasible to build the library W even for significantly large input dimension when the training data are sparse.

2. Not all wavelets from library W are useful in fitting f from noisy data, however. This leads to the problem of selecting the best wavelet regressors among W. Three heuristic methods will be proposed for this. When the regressors are conveniently selected, fitting model (85) amounts to identifying the u_i coefficients, which is a standard least squares estimation problem.

3. Steps 1 and 2 yield a fast training procedure. The result can still be further improved by subsequently applying an iterative backpropagation algorithm with steps 1 and 2 as fast initialization. In fact, because initialization was good, a faster Newton procedure can be used.

More details of each step are given in the following discussion.

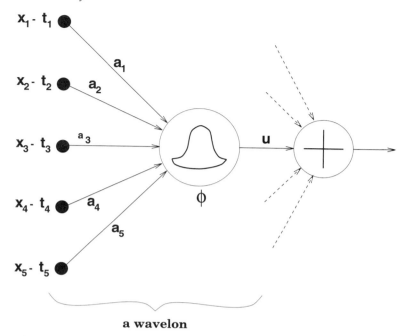

$x_1 - t_1$
$x_2 - t_2$ a_2
a_1
$x_3 - t_3$ $a\,_3$
u
$x_4 - t_4$ a_4
a_5
ϕ
$x_5 - t_5$

a wavelon

Figure 3 Wavelet network. A wavelon is shown, which corresponds to one term $\varphi(a_i \star (x - t_i))$. Dashed arrows figure output connections to other wavelons.

C. CONSTRUCTING THE WAVELET LIBRARY W

First, we should build a library W of wavelets which will be considered as candidates of regressors. We have to restrict ourselves to a finite set of regressor candidates, in order to apply regressor selection algorithms. Naturally W is chosen to be a subset of the continuously parameterized family $\{\varphi(a(x - t)): a \in \mathbf{R}^+, \ t \in \mathbf{R}^d\}$. The choice of W is in principle the same problem as discretizing the continuous wavelet reconstruction (42) to obtain the discrete reconstruction (43). The standard discretization is a regular lattice:

$$\{\varphi(a_0^n x - m t_0): \ n \in \mathbf{Z}, \ m \in \mathbf{Z}^d\}, \tag{86}$$

where $a_0, t_0 > 0$ are two scalar constants defining the discretization step sizes for dilation and translation, respectively. Typically we take a *dyadic lattice*. Now the countable family (86) should be truncated into a finite set. Usually we only want to estimate $f(x)$ on a compact domain $D \subset \mathbf{R}^d$ and the wavelet function $\varphi(x)$ is chosen to have compact or rapidly vanishing support. Therefore, we can replace in (86) $m \in \mathbf{Z}^d$ by $m \in S_t$ with a finite set $S_t \subset \mathbf{Z}^d$; on the other hand,

$n \in \mathbf{Z}$ should be replaced by $n \in S_a$ with a finite set $S_a \subset \mathbf{Z}$ corresponding to the "desired" resolution levels of the estimation. In practice, four or five consecutive dilation levels are usually sufficient, with the largest wavelet scale corresponding to the size of D, the compact domain on which f is to be estimated. After such a truncation is performed, the family (86) is replaced by

$$\{\varphi(a_0^n x - mt_0): n \in S_a, \ m \in S_t(n)\}. \tag{87}$$

Note that the cardinality of this wavelet library grows exponentially with the dimension d. The following procedure is used to overcome this curse of dimensionality when the training data are sparse: scan the training data set $\mathcal{O}_1^N$; for each sample point in $\mathcal{O}_1^N$, determine the wavelets in (87) whose supports[19] contain this data point; and add these wavelets to W if they have not figured in it. With this method, the dimension d is not a critical factor of complexity, because the family (87) does not need to be actually created. For a sparse training data set, this method allows us to handle problems of relatively large input dimension d. In particular, if the supports of the wavelets are approximated by hypercubes in $\mathbf{R}^d$, this method is easily implemented. From now on we denote by W the resulting *library of wavelet regressor candidates*. For computational convenience, we normalize the wavelets and get the library W composed of the wavelets:

$$\varphi_i(x) = \alpha_i \varphi(a_i(x - t_i)), \qquad i = 1, \ldots, L,$$

$$\alpha_i = \left(\sum_{k=1}^{N} [\varphi(a_i(x_k - t_i))]^2 \right)^{-1/2},$$

where L is the number of elements in W, a_i, t_i correspond to the dilation and translation parameters a_0^n and $a_0^{-n} mt_0$ of the wavelet φ_i, and α_i is the normalizing factor. The numbering order with i is arbitrary.

D. SELECTING THE BEST WAVELET REGRESSORS

The problem of regressor selection is to select a number $M \leqslant L$ of wavelets, which are the "best" ones from W for building the regression

$$f_M(x) = \sum_{i \in I} u_i \varphi_i(x), \tag{88}$$

where I is an M-element subset of the index set $\{1, 2, \ldots, L\}$. This is a classical problem in regression analysis [77]. Let $\mathcal{I}_M$ be the set of all the M-element subsets of $\{1, 2, \ldots, L\}$. For any $I \in \mathcal{I}_M$, the optimal linear weights u_i of (88) are found

[19]For noncompactly supported but rapidly vanishing wavelets, the term "support" should be interpreted in an approximate way as some domain around the center of the wavelet.

using the least squares method. Then the question is how to choose $I \in \mathcal{I}_M$ which minimizes the averaged square residuals

$$J(I) = \min_{\{u_i : i \in I\}} \frac{1}{N} \sum_{k=1}^{N} \left(Y_k - \sum_{i \in I} u_i \varphi_i(X_k) \right)^2. \tag{89}$$

Determining the optimal number M should be performed using generalized cross validation; cf. Section II.A.2. For given M, selecting the M optimal regressors from W must be performed via exhaustive search which may involve massive computations. To overcome this difficulty, three different heuristics are proposed instead; details can be found in Section IX.

Residual-Based Selection

The idea of residual-based selection (RBS) is to select, for the first stage, the wavelet in W that best fits the observations $\mathcal{O}_1^N$. Then repeatedly select the wavelet that best fits the residual of the fitting of the previous stage. In the literature of the classical regression analysis, it is considered as a simple, but not quite effective method, for example, in [77] where it is called the *stagewise regression procedure*. For classical regressions the number of regressor candidates is usually small; hence alternative more complicated and more effective procedures are preferred. In our situation the number of regressor candidates may reach several hundreds or even more, the computational efficiency becomes more important, and the simple residual-based selection should be a first choice. Recently it has also been used in the matching pursuit algorithm of Mallat and Zhang [78] and the adaptive signal representation of Qian and Chen [79]. This procedure is described in Section IX.A

Stepwise Selection by Orthogonalization

The idea of stepwise selection by orthogonalization (SSO) is to select, for the first stage, the wavelet in W which best fits the observations $\mathcal{O}_1^N$, then repeatedly select the wavelet that best fits $\mathcal{O}_1^N$ while working together with the previously selected wavelets. This method has been used in radial basis function (RBF) networks and other nonlinear modeling problems by Chen *et al.* [80, 81]. This procedure is described in Section IX.B.

Backward Elimination

In contrast to the previous methods, the backward elimination (BE) method starts building the regression (88) by using all wavelets in W, then eliminates one wavelet per stage, while trying to increase as little as possible the residual at each stage. This procedure is described in Section IX.C.

E. Combining Regressor Selection and Backpropagation

Any of the procedures mentioned previously can be used to initialize the wavelet network (85). This network is then further trained using a backpropagation procedure. Note that in (85) we use vectorial dilation parameters a_i, but for the regressor selection procedures the dilation parameters a_i in W are scalars. Before applying any backpropagation procedure, change the scalar dilation parameters resulting from the regressor selection procedures into vectors with identical components. Standard backpropagation is a stochastic gradient procedure; a quasi-Newton algorithm is, however, preferred for training the wavelet network, owing to the good performance of the initialization procedures. Finally, to better capture linear properties in regressions, we replace (85) by

$$f_n(x) = \sum_{i=1}^{n} u_i \varphi\big(a_i \star (x - t_i)\big) + c^T x + b, \tag{90}$$

with the additional parameters $c \in \mathbf{R}^d$, $b \in \mathbf{R}$. The initialization procedures are slightly modified accordingly.

VI. FUZZY MODELS: EXPRESSING PRIOR KNOWLEDGE IN NONLINEAR NONPARAMETRIC MODELS

A. Fuzzy Rules and Prior Knowledge in Nonparametric Models

We first begin by introducing fuzzy models such as typically used in fuzzy control [7]. Several presentations are possible; see, for instance, [82]. The presentation we give now is slightly heterodox, but is simple and consistent.

1. Input variables are scalar and are written as $x_1, \ldots, x_d$. Input locations are encoded via fuzzy set membership functions, that is, functions $\mu_A(x_i)$ with values in $[0, 1]$ where the symbol A is just a label; the fuzzy set membership function μ_A is the mathematical meaning of "fuzzy set A." Thus, for each actual value of x_i, the statement "x_i is A" has a value equal to $\mu_A(x_i)$, such statements are premises of so-called "fuzzy rules." Be careful that a typical form of such statements is "x_i is large," which does not convey as much information as formula $\mu_A(x_i)$ does, because the function μ_A is not explicitly specified by this statement.

2. Fuzzy sets can be combined using the "and, or, not" operators of first-order predicate logic. For instance,

(x_1 is A_1) and (x_2 is A_2) ... and (x_d is A_d)

is a fuzzy set involving the vector $(x_1, \ldots, x_d)$. The keyword "and" is a combinator of fuzzy sets which must be defined formally in terms of combination of membership set functions. Several choices have been proposed by various authors [83]. The most widely used ones are

$$
\begin{aligned}
\text{and}(u, v) &= \min(u, v), & \text{or}(u, v) &= \max(u, v), \\
\text{and}(u, v) &= uv, & \text{or}(u, v) &= u + v - uv, \\
\text{and}(u, v) &= \max(0, u + v - 1), & \text{or}(u, v) &= \min(1, u + v)
\end{aligned} \tag{91}
$$

(corresponding definitions for and and or are written on the same line) and $\text{not}(u) = 1 - u$. Then, as usual in logic, the implication "(x is A) implies (y is B)," also written as

if x is A then y is B

is a macro which expands into[20]

(y is B) or not(x is A)

In the sequel, we shall encode the " and" as the product: $\text{and}(u, v) = uv$, *with corresponding codings for the " not, or." Finally the implication is expanded as stated previously.*

3. Fuzzy rules are statements of the form

if x is A then y is B

Note that more complex premises can be used, using and, or, not. Here we state the mathematical translation of the classical "modus-ponens" mechanism, which can be written as

Rule:	if	x is A	then	y is B
Fact:		x is A'		
Conclusion:				y is ?B

Modus ponens is a mechanism which combines membership functions and yields a membership function. It can be viewed as a mechanism to express interpolation. Denote by $\mu_A(x)$ the membership function associated with the fuzzy set x is A, and denote by $\mu_{A \Rightarrow B}(x, y)$ the membership function of "if x is A then

---

[20]This is the point where we deviate from the usual presentation: in the fuzzy literature, implication is often encoded as an "and," and the modus-ponens mechanism is modified accordingly. We preferred this presentation, because it is fully consistent and in accordance with the usual predicate calculus.

y is B." We now state the mathematical translation of the modus ponens [83]. It is defined as

$$\mu_{?B}(y) = \text{proj}_u \{\mu_{A'}(u) \text{ and } \mu_{A \Rightarrow B}(u, y)\}$$
$$\triangleq \max_u \{\mu_{A'}(u) \text{ and } \mu_{A \Rightarrow B}(u, y)\}, \tag{92}$$

where elimination of component u has been performed via maximization. We now consider the particular case in which the fact is a *crisp* statement, that is, has the standard form "x is x," where x is an ordinary value. In this case, we have $\mu_{A'}(u) = 1$ if $u = x$, and $\mu_{A'}(u) = 0$ otherwise. Hence, for such a case, the modus-ponens mechanism (92) reduces to

$$\mu_{?B}(y) = \mu_{A \Rightarrow B}(x, y) = 1 - \mu_A(x)(1 - \mu_B(y)), \tag{93}$$

where we have used the formulas $u \Rightarrow v = v$ or not $u = v + (1-u) - v(1-u) = 1 - u(1 - v)$. To conclude, because we only consider crisp facts, the fuzzy rule

                    if x is A then y is B

represents fuzzy set (93).

    4. A "fuzzy rule basis" is a collection of fuzzy rules of the form, say,

            if (x_1 is A_1_1) and (x_2 is A_1_2) ...
            and (x_d is A_1_d) then (y is B_1)
            ....
            if (x_1 is A_p_1) and (x_2 is A_p_2) ...
            and (x_d is A_p_d) then (y is B_p)

where the $A_{j,i}$ are doubly indexed labels, i is the index of the input coordinate, and j is the index of the rule. The mathematical translation of this rule basis is now given. We assume that the fuzzy sets form a *fuzzy partition* of the space, that is,

$$\sum_{j=1}^{p} \prod_{i=1}^{d} \mu_{A_{j,i}}(x_i) \equiv 1. \tag{94}$$

Then, *combining fuzzy rules within our fuzzy rule basis is interpreted as taking the "and" of their conclusions.* Thus, using the notation of item 3, the preceding fuzzy rule basis represents the fuzzy set ?B equal to

        y is ?B_1 and ... and y is ?B_p

where the ?B_j's are defined according to (93). Expressing the and combinator as the product of membership functions, we get

$$\mu_{?B}(y) = \prod_{j=1}^{p} \mu_{?B_j}(y),$$

$$= \prod_{j=1}^{p}\left(1 - \prod_{i=1}^{d}\mu_{A_{j,i}}(x_i)\left(1 - \mu_{B_j}(y)\right)\right) \quad \text{[by (93)]}$$

$$\approx 1 - \sum_{j=1}^{p}\left(1 - \mu_{B_j}(y)\right)\prod_{i=1}^{d}\mu_{A_{j,i}}(x_i),$$

$$= \sum_{j=1}^{p}\mu_{B_j}(y)\prod_{i=1}^{d}\mu_{A_{j,i}}(x_i), \quad \text{[by (94)]} \tag{95}$$

where we have used the property (94) of fuzzy partition, and approximation $\prod_{j=1}^{p}(1 - u_j) \approx 1 - \sum_{j=1}^{p} u_j$, which is valid for u_j small and p large. Next, we also assume that sets B_j are *crisp*; that is, they are of the form "y is y_j." Thus $\mu_{B_j}(y) = 1$ if $y = y_j, = 0$ otherwise. Hence, assuming that both the consequences of the rules and the facts are crisp statements, we get for the conclusion the fuzzy set "y is ?B," where

$$\mu_{?B}(y) = \sum_{j=1}^{p} 1_{\{y=y_j\}}\prod_{i=1}^{d}\mu_{A_{j,i}}(x_i). \tag{96}$$

At this point, setting $x = (x_1, \ldots, x_d)$, formula (96) defines a function mapping points $x \in \mathbf{R}^d$ into fuzzy sets. To get a function in the usual setting $\mathbf{R}^d \mapsto \mathbf{R}$, we perform *defuzzification* of $\mu_{?B}(y)$ in (96). That is, we replace $\mu_{?B}$ by its center of gravity, using again fuzzy partition property (94); see [7, 83]. This finally yields the ordinary function

$$y = \sum_{j=1}^{p} y_j \left(\prod_{i=1}^{d}\mu_{A_{j,i}}(x_i)\right) \stackrel{\Delta}{=} \sum_{j=1}^{p} y_j w_j(x), \tag{97}$$

where $x = (x_1, \ldots, x_d)$; this defines the weights $w_j(x)$. If property (94) does not hold, that is, if our fuzzy rule basis is sparse so that the range of each coordinate x_i is not covered by a fuzzy partition, then the preceding defuzzification formula is modified accordingly:

$$y = \frac{\sum_{j=1}^{p} y_j w_j(x)}{\sum_{j=1}^{p} w_j(x)}. \tag{98}$$

Usually, fuzzy set membership functions are parametrized functions of the form

$$\mu_A(x) = \mu\big(a(x - t)\big),\qquad(99)$$

where $\mu(x)$ is a given function with values in $[0, 1]$, a is a dilation factor, and t is a translation factor, and the pair (a, t) encodes the fuzzy set A. Mostly used is the piecewise-linear function μ such that $\mu(1) = 1$ and $\mu(x) = 0$ for x outside the interval $[0, 2]$, that is, a spline of order 1. In this case, the defuzzification mechanism (97) just performs interpolation. If the fuzzy partition is fixed and not adjustable, then we get a particular case of the kernel estimate (15). Obviously, fuzzy models such as (97) or (98) are amenable to identification because they have some unknown parameters for tuning, namely, the y's, a's, and t's. Identified fuzzy models are often referred to as "neuro-fuzzy models" in the AI literature [84], because standard backpropagation (i.e., stochastic gradient) can be used for their training, exactly as for neural networks. It is also proved that fuzzy models are universal approximants [85], which is not surprising.

To summarize, fuzzy models are described by fuzzy rule bases, plus some additional parameters which make vague statements such as "large," "small," etc., to be precise in terms of fuzzy set membership functions. The fuzzy rule basis exhibits the structure of the model, plus some coarse features related to the location of the elementary functions in the decomposition (97) or (98). Thus *fuzzy models are just particular instances of the kind of nonlinear nonparametric model we consider here, with the advantage of providing the fuzzy rules as a way to describe some possibly available prior knowledge.* In the experiments reported in Section VII.A.2, neuro-fuzzy modeling is used in this sense.

B. FUZZY RULE BASES
FOR WAVELET-BASED ESTIMATORS

In this short section, we briefly discuss a proposal for blending the practical advantages of fuzzy models with the mathematical quality of wavelet-based identification techniques. Further development of this proposal will be the subject of future work and will be reported elsewhere.

Requirements

Formulas (94) and (97) reveal that fuzzy models can be viewed as interpolation procedures: interpolation is performed between points where the set membership function takes value 1, with associated y value. Thus fuzzy models cannot reflect hierarchical or multiresolution approximations of a function such as performed by wavelet-based identification techniques. So the following natural question can be considered: *how does one provide fuzzy rule bases for wavelet-based estimators?*

Thus what we need is to abstract wavelet networks, say, of the form (85), in the form of syntax similar to fuzzy rule bases. Such syntax would not specify the considered wavelet network exactly, but should capture some essential features of it. Objectives would be to use such a syntax for a rough but easy description of a wavelet network based on some qualitative prior knowledge on the system, or to use it as an initial guess for some iterative identification procedure based on recorded data from the system.

Reflecting the notion of multiresolution or hierarchy within rules calls for a syntactic notion of *context*. For instance, in the context "x is large," we may want to write "x is small" to express that x is not too large, and "x is large" again to insist that x is very large indeed. This calls for logics handling context-dependent statements. Such logics are studied under different frameworks independently in the AI and theoretical computer science communities. The notion of a "conditional object" proposed and studied by Dubois and Prade [86] in the AI community is a candidate model for such "context-dependent rules." In [86] various definitions are investigated for such "conditional objects," based on some reasonable requirements accepted as axioms. On the other hand, "structured operational semantics" (SOS) was introduced by Plotkin [87] in theoretical computer science. SOS rules describe the legal transitions of a considered program *for a given context*. SOS rules are used to specify primitives as well as the various combinators for program contruction. We shall not elaborate any further on possible theoretical models for the kind of context-dependent statements we shall take the liberty to write in the sequel.

Let us propose the following syntax we call *hierarchical fuzzy models*.

1. Standard fuzzy rules are hierarchical fuzzy rules. Thus we can still write

```
if (x_1 is A_1) and (x_2 is A_2) ...
and (x_d is A_d) then (y is B)
```

with the same mathematical meaning as before.

2. Let us give names to the fuzzy rule bases, for example,

```
                RULE_BASE is
if (x_1 is A_1_1) and (x_2 is A_1_2) ...
and (x_d is A_1_d)
    then (y is B_1)
..........
if (x_1 is A_p_1) and (x_2 is A_p_2) ...
and (x_d is A_p_d)
    then (y is B_p)
end
```

Then the following statement:

```
    if (x_1 is C_1) and (x_2 is C_2) ...
    and (x_d is C_d) then RULE_BASE applies
```

is a *hierarchical fuzzy rule*. Its premise is an ordinary fuzzy statement

```
    (x_1 is C_1) and (x_2 is C_2) ... and (x_d is C_d)
```

as before. The second part of this statement, namely, "then RULE_BASE applies," has "then" and "applies" as keywords and RULE_BASE as a parameter. This hierarchical fuzzy rule has the following interpretation:

(a) The reference space for input x, which was, say, $[0, 1]^d$, is now stretched down, to enforce validity of the statement

```
    (x_1 is C_1) and (x_2 is C_2) ...
    and (x_d is C_d)
```

Thus all premises of RULE_BASE are stretched down accordingly.

(b) Since RULE_BASE was a standard fuzzy rule basis, our new rule is a hierarchical fuzzy rule.

3. Collections of hierarchical fuzzy rules are termed *hierarchical fuzzy rule bases*. Hierarchical fuzzy rules can call hierarchical fuzzy rule bases; this captures multiresolution. Obviously, in doing so, the question of *recursivity* in the computer science setting occurs: does it happen that a rule recursively calls itself? Recursion may or may not be accepted. Anyway, simple syntactic constraints in writing rule bases would prevent recursivity.

"Down Stretching" Mechanism

The key issue in this informal discussion is the precise mathematical meaning of the "down stretching" mechanism. We assume for convenience that the default context is $[0, 1]^d$. Consider a fuzzy partition satisfying condition (94), which we recall now

$$\sum_{j=1}^{p} \prod_{i=1}^{d} \mu_{A_j}(x_i) \equiv 1. \tag{100}$$

Down stretching this fuzzy partition to a given membership function $\mu_C(x_i)$ consists of building a collection $\mu_{(A_{j,i}|C)}$, $j = 1, \dots, p$, of membership functions which satisfy

$$\sum_{j=1}^{p} \prod_{i=1}^{d} \mu_{(A_{j,i}|C)}(x_i) \equiv \mu_C(x_i) \tag{101}$$

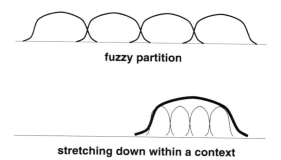

fuzzy partition

stretching down within a context

Figure 4 Down stretching mechanism.

and, in addition, preserve the "geometry" of the original fuzzy partition. This is illustrated in Fig. 4. A possible procedure achieving this is described now.

We first need to define the notion of a fuzzy set more accurately. A fuzzy set A is a triple $A = (\mu_A(x), a, b)$, where

$$\mu_A: [a, b] \to [0, 1] \text{ is the membership function and } -\infty < a \leqslant b < +\infty.$$

The interval $[a, b]$ is the context of the fuzzy set A. For example, when we define a fuzzy set "small," we must specify its context interval $[a, b]$ in addition to its membership function. This "small" label means that the μ_A membership function is mainly concentrated on the small values of this context interval. Note that this set may not be "small" within other context intervals.

Now consider a fuzzy set $C = (\mu_C(x), a', b')$. We consider its left and right boundaries defined by

$$l_C = \inf_{\mu_C(x) > 0} x, \qquad r_C = \sup_{\mu_C(x) > 0} x;$$

that is, $[l_C, r_C]$ is the support of μ_C. Consider a pair (A, C) of fuzzy sets, and define the *contextual fuzzy set* $(A|C)$ as follows:

$$(A|C) = (\mu_{(A|C)}, l_C, r_C),$$

$$\mu_{(A|C)}(x) = \mu_A \left(\frac{b - a}{r_C - l_C}(x - l_C) + a \right) \mu_C(x). \tag{102}$$

Hence $(A|C)$ has the support of C as context, and its membership function is obtained by mapping the interval $[a, b]$ onto $[l_C, r_C]$ and then multiplying by μ_C. With this definition of contextual fuzzy sets, a fuzzy partition having the default context, that is, satisfying property (100), is down stretched to a fuzzy partition satisfying property (101).

Mathematical Implementation of the Hierarchy

Here we formalize what it means for a rule base to be called within a given context. As an example, we give the meaning of the hierarchical statement

```
if (x_1 is C_1) and (x_2 is C_2) ... and (x_d is C_d)
then (y = y_o)
if (x_1 is C_1) and (x_2 is C_2) ... and (x_d is C_d)
then RULE_BASE applies
```

where RULE_BASE has been defined before. We may also rewrite this as

```
if (x_1 is C_1) and (x_2 is C_2) ... and (x_d is C_d)
then (y = y_0) and RULE_BASE applies
```

First, we have to combine two rules, and this is performed using the general formula (97). Then we must recall that RULE_BASE is called within the context of $C_1 \times \cdots \times C_d$; hence we use definition (102) of contextual fuzzy sets. This yields the following mathematical interpretation of the previous hierarchical rule base:

$$y = y_o \left(\prod_{i=1}^{d} \mu_{C_i}(x_i) \right) + \left[\sum_{j=1}^{p} y_j \left(\prod_{i=1}^{d} \mu_{(A_{j,i}|C_i)}(x_i) \right) \right]. \tag{103}$$

This shows that the value y_o can be interpreted as a "first-order approximant," whereas the y_j's, $j = 1, \ldots, p$, are increments corresponding to a refinement of our modeled function. Thus *truncating such an approximation is simply performed by truncating the tree of the nested calls of rule bases.*

Thus what we have at this point is a flexible way to associate syntax with multiresolution expansions of functions. If, in addition, *we carefully choose our membership functions μ_A to be derived from scale functions φ associated with wavelets, we now have a way to abstract wavelet networks in the form of hierarchical fuzzy rule bases.* See Section III.A for scale functions which are nonnegative and bounded, and thus satisfy the requirements for being prototypes of membership functions. The "call" mechanism provides some kind of genericity, because the same rule base can be called within different contexts. This genericity is expected to be useful mainly when adjustable parameters, which are hidden inside fuzzy rules, are identified from data. On the other hand, for fuzzy models specified based on the prior knowledge of the user, it is not expected that the same rule base will be called under different contexts.

VII. EXPERIMENTAL RESULTS

In this section we consider the application examples introduced in Section I.A. We provide detailed results obtained with the wavelet networks and the fuzzy network. For the gas turbine example, we also compare them with alternative semiphysical models which were developed in [88] for the purpose of monitoring and diagnostics.

A. MODELING THE GAS TURBINE SYSTEM

1. Using the Wavelet Network

In the gas turbine system we introduced in Section I.A.1, the temperature profile at the exhaust of the turbine is considered as the output. We need a model which predicts this temperature profile from available measurements. For the semiphysical model we mentioned in Section I.A.1, the temperature profile is predicted from the mean temperature in the combustion chambers T_e, the mean temperature at the exhaust of the turbine T_s, and the rotation velocity of the turbine N. The velocity N is directly measured, T_s is given by the average of a set of thermocouples installed at the exhaust of the turbine, and T_e is computed from T_s and the compression rate π of the compressor [9, 10]. By substituting T_e, the temperature profile at the exhaust of the turbine depends on T_s, π, and N. As suggested by this semiphysical model, we assume that the temperature measured by each of the thermocouples installed at the exhaust of the turbine is a function of T_s, π, and N, which are all measured. Therefore, we can try to construct, for each of the thermocouples, a wavelet network with T_s, π, and N as its input variables, and train it to predict the temperature measured by the thermocouples.

We have experimented with this approach on the data taken from a gas turbine of European Gas Turbine SA. The training data were collected during about 48 hours. We have resampled the data and kept only 1000 measurement points. This gas turbine system is equipped with 18 thermocouples at its exhaust. For the sake of brevity, we show only the results concerning the first thermocouple. The resampled data are depicted in Fig. 5 where the plots correspond to T_s, π, and N and $y = t_1 - T_s$, where t_1 is the measurement of the first thermocouple. These 1000 measurement points, which we refer to as the *training data*, are used for training models whose input vector is $x = (T_s, \pi, N)^T$. The obtained models are tested on another set of measured data, which we refer to as the *test data* set and depict in Fig. 6.

We have chosen the radial wavelet function $\varphi(x) = (d - x^T x) \exp(-\frac{1}{2} x^T x)$ with $d = \dim(x)$. The number of wavelets used in the networks is set to 40. Note that there are 18 thermocouples.

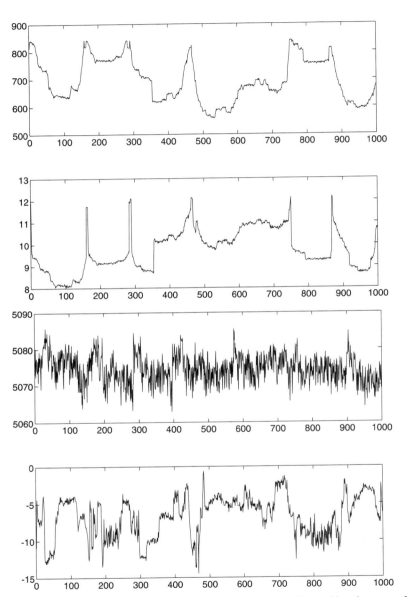

Figure 5   Training data. The plots correspond to, from top to bottom, T_s, π, N, and $y = t_1 - T_s$.

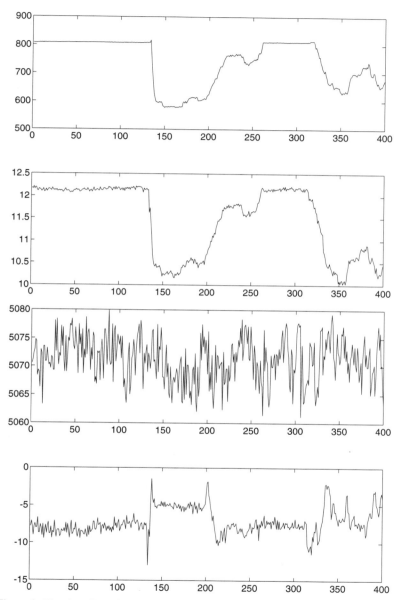

Figure 6 Test data. The plots correspond to, from top to bottom, T_s, π, N, and $y = t_1 - T_s$.

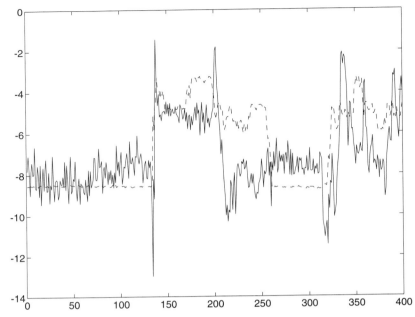

Figure 7   Result with the semiphysical model on the test data set. The solid line represents the true measurement and the dashed line represents the output of the model.

We initialize the wavelet networks with each of the proposed (RBS, SSO, BE) procedures and train them with the Gauss–Newton procedure.

To show the performance of the resulting models, we compare their results with those of the semiphysical model and a third-order polynomial model. In Figs. 7 and 8 the results obtained with the semiphysical model and the third-order polynomial model are, respectively, shown. The results obtained with the wavelet networks initialized with procedures RBS, SSO, and BE, and the results after 10 iterations of the Gauss–Newton procedure are given in Figs. 9–11. In Table I we list the mean of square errors (MSE) of these models on the training data set as well as on the test data set. For each of these networks, we give the result of its initialization (init. MSE) and the result after 10 iterations of the Gauss–Newton procedure (final MSE). The time of computation for building these models is also listed in Table I, based on our programs in MATLAB 4.1 language executed on a Sun Sparc-2 workstation. Because the execution time of the programs is perturbed by other processes on the workstation, another figure of merit is provided, namely the the MATLAB's Flop which measures the computational complexity of a program.

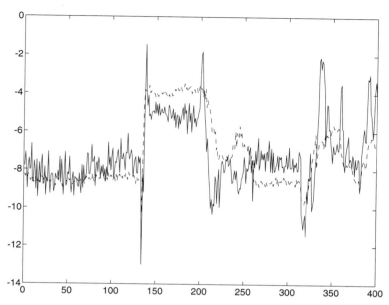

Figure 8 Result with the third-order polynomial model on the test data set. The solid line represents the true measurement and the dashed line represents the output of the model.

The following observations can be made:

- The semiphysical model performs quite poorly in predicting the output of the system.
- The system is truly nonlinear; in addition, the results obtained with the polynomial model are quite poor.
- The wavelet networks do improve the performance on prediction. Recall, however, that we get in turn increasing computational complexity and loss of the physical meaning of the model parameters.

2. Using the Fuzzy Network

We also applied the (classical) neuro-fuzzy network as briefly introduced in Section VI for modeling the gas turbine system. Similarly to the wavelet network, we train the fuzzy network using the training data set, and then evaluate it on the test data set.

To build the network, we have taken a fuzzy partition of the state space using triangular membership functions (i.e., first-order splines); this divides the varia-

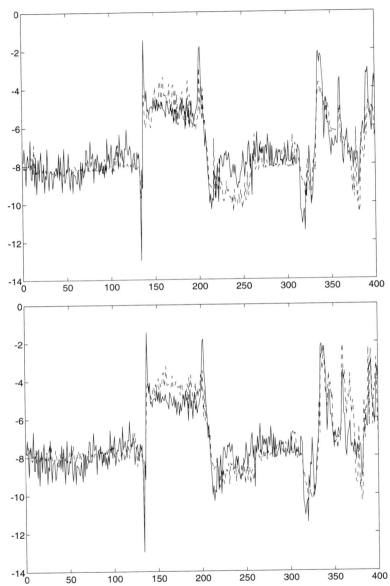

Figure 9 Results with wavelet network initialized by procedure RBS (top) and after 10 iterations of the Gauss–Newton procedure (bottom). The solid lines represent the true measurement and the dashed lines represent the output of the model.

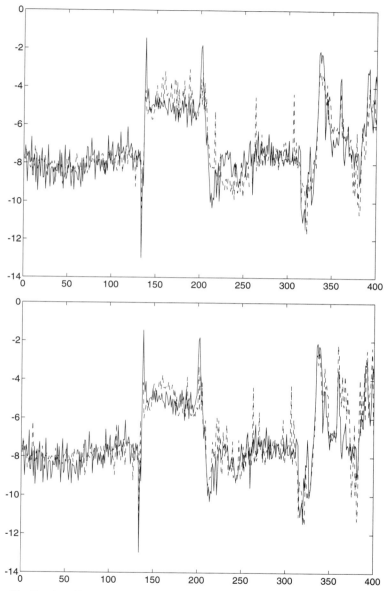

Figure 10   Results with wavelet network initialized by procedure SSO (top) and after 10 iterations of the Gauss–Newton procedure (bottom). The solid lines represent the true measurement and the dashed lines represent the output of the model.

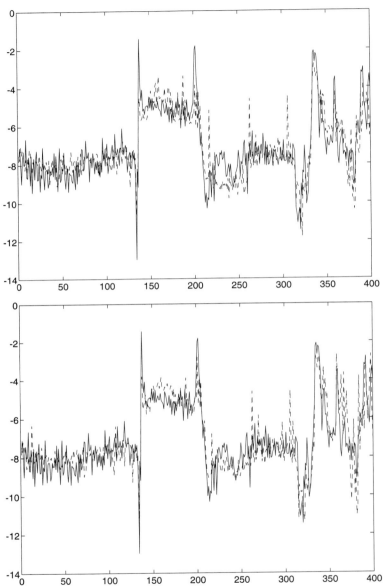

Figure 11 Results with wavelet network initialized by procedure BE (top) and after 10 iterations of the Gauss–Newton procedure (bottom). The solid lines represent the true measurement and the dashed lines represent the output of the model.

Table I

Performance Evaluation of the Models

Models	RBS net	SSO net	BE net	Semiphysical	Polynomial
Train. init. MSE	1.2656	1.0453	1.0381		
Train. final MSE	0.5395	0.4239	0.4503	3.5268	2.8438
Test. init. MSE	1.2368	1.1229	1.1576		
Test. final MSE	1.1886	1.2348	1.0898	2.8914	2.1135
Init. flops	2.0718×10^7	4.3714×10^8	7.5143×10^7		
Train. flops	1.5365×10^9	1.5365×10^9	1.5365×10^9	9.8041×10^8	4.7056×10^5
Init. time (s)	41.6	251.2	87.2		
Train. time (s)	2461.8	2383.8	2456.5	2265.0	1.5362

tion domain of each input into five equal parts. Following Section VI, the mathematical translation for both conjunction and implication operators is taken to be the product.

Before learning, we have initialized the network using a simple interpolation procedure. Consider the "defuzzification" formula (97) which we recall now:

$$y = \sum_{j=1}^{p} y_j \left(\prod_{i=1}^{d} \mu_{A_{j,i}}(x_i) \right) \triangleq \sum_{j=1}^{p} y_j w_j(x), \qquad (104)$$

where the index j labels the rules. For each rule j, select the training input data point X_{n_j} closest to the center of the corresponding fuzzy set, that is, $w_j(x)$ is maximal for $x = X_{n_j}$. Then take $y_j = Y_{n_j}$ where Y_{n_j} is the output value corresponding to X_{n_j}. Results of this procedure are shown in Fig. 12.

The second stage consists of performing a least squares fit of the parameters θ_j in the function $f_\theta(x) = \sum_{j=1}^{P} \theta_j w_j(x)$, where $\theta = (\theta_1, \ldots, \theta_p)$ based on the whole training data sample $O_1^N = \{(X_1, Y_1), \ldots, (X_N, Y_N)\}$. A brute-force implementation of least squares would be difficult, due to the need for inverting the Hessian of the least squares functional. Thus an iterative stochastic gradient procedure has been preferred instead, using the preceding simple initialization technique. Training was stopped after only three successive scannings of the learning set.

The identified fuzzy network is then evaluated on the test data set. The output of the identified fuzzy network is plotted in Fig. 12, and is compared to the actual one. The solid line represents the true measurement and the dashed line represents the output of the model. The mean of square errors (MSE) on the test data set is 1.5860.

388

A. Juditsky et al.

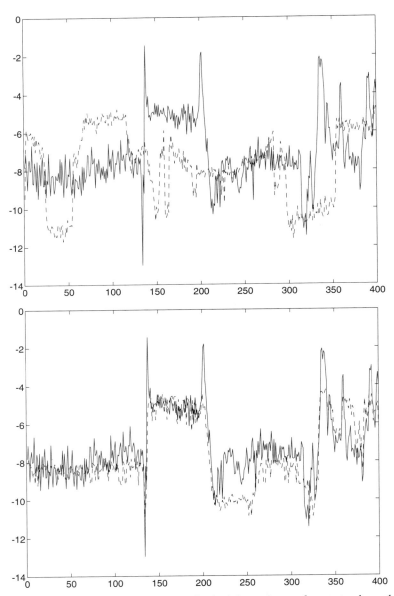

Figure 12 Results with the initialized (top) and trained (bottom) neuro-fuzzy networks on the test data set. The solid lines represent the true measurement and the dashed lines represent the output of the model.

B. MODELING THE HYDRAULIC ACTUATOR OF THE ROBOT ARM

Let us denote by $u(t)$ and $p(t)$ the position of the valve and the oil pressure at time t, respectively. A sample of 1024 pairs of $(u(t), p(t))$ was registered.[21] We divide it into two equal parts for training and testing the models. The training data are depicted in Fig. 13, and the test data in Fig. 14.

We first tried to model this system with linear autoregressive exogenous (ARX) models. More precisely, we tried to use models of the following form:

$$p(t) = a_1 p(t-1) + a_2 p(t-2) + \cdots + a_n p(t-n)$$
$$+ b_1 u(t - \tau - 1) + b_2 u(t - \tau - 2) + \cdots + b_m u(t - \tau - m) + e(t),$$

where the pure time delay τ is assumed to be an integer and $e(t)$ is some noise independent of $u(t)$ and past values of $p(t)$. After the identification of the model parameters a_i, b_j, τ, we plot the output of the following system to visually evalu-

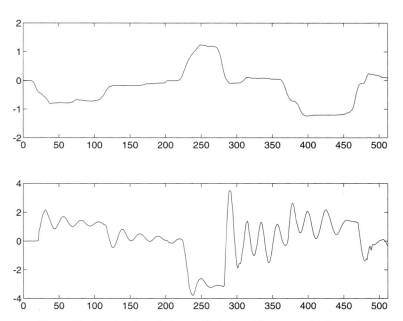

Figure 13   Training data: the input $u(t)$ (top) and the output $p(t)$ (bottom).

[21] We gratefully acknowledge Jonas Sjöberg and Svante Gunnarsson from Linköping University for providing the data.

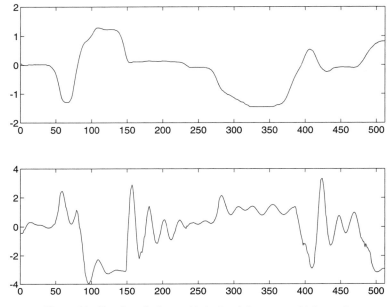

Figure 14 Test data: the input $u(t)$ (top) and the output $p(t)$ (bottom).

ate the quality of the model:

$$\hat{p}(t) = a_1 \hat{p}(t - 1) + a_2 \hat{p}(t - 2) + \cdots + a_n \hat{p}(t - n)$$
$$+ b_1 u(t - \tau - 1) + b_2 u(t - \tau - 2) + \cdots + b_m u(t - \tau - m).$$

We processed the data with Ljung's System Identification Toolbox, Version 3.0a. It turns out that the ARX model that gives the best simulation result on the test data set has the model order with $n = 3$, $m = 2$, $\tau = 0$. This result is shown in Fig. 15. It does not seem to be satisfactory. The wavelet networks as defined in (90) are then considered as candidates of nonlinear models.

In analogy with the linear ARX model, we build models of the following form:

$$p(t) = \hat{f}(p(t - 1), p(t - 2), p(t - 3), u(t - 1), u(t - 2)) + e(t),$$

where the nonlinear estimator $\hat{f}$ is a wavelet network composed of six wavelets, and $e(t)$ represents the modeling error. To train the network, compose its input and output vectors with the training data $\{u(t), p(t)\}$:

$$x(t) = \left[p(t - 1), p(t - 2), p(t - 3), u(t - 1), u(t - 2) \right]^T,$$
$$y(t) = p(t).$$

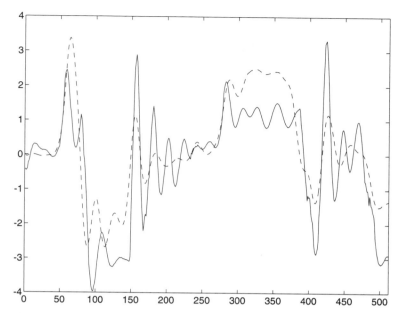

Figure 15 Result with the linear ARX model on the test data set. The solid line represents the true measurement and the dashed line represents the simulated output.

Then apply the initialization algorithms and the Gauss–Newton procedure. Again we take

$$\varphi(x) = (d - x^T x) \exp\left(-\tfrac{1}{2} x^T x\right),$$

with $d = \dim(x)$ as the wavelet function. It happens that for this example the Gauss–Newton procedure does not significantly improve the performance of the wavelet models, so we only show the results obtained with the initialized networks.

We then simulate the output $\hat{p}(t)$ on the test data set with the wavelet models, in a similar way as with the linear ARX model:

$$\hat{p}(t) = \hat{f}\big(\hat{p}(t-1),\, \hat{p}(t-2),\, \hat{p}(t-3),\, u(t-1),\, u(t-2)\big).$$

The simulation results obtained with the wavelet networks initialized with algorithms RBS, SSO and BE are depicted in Figs. 16–18.

Clearly, the wavelet models significantly improve the results of the simulation. Although the results obtained with initialization algorithms SSO and BE are very similar, the result of algorithm RBS is obviously not as good.

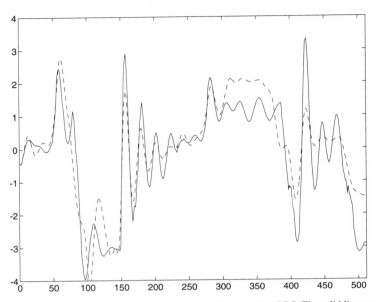

Figure 16   Result with the wavelet network initialized with algorithm RBS. The solid line represents the true measurement and the dashed line represents the simulated output.

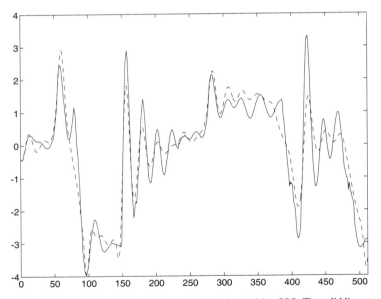

Figure 17   Result with the wavelet network initialized with algorithm SSO. The solid line represents the true measurement and the dashed line represents the simulated output.

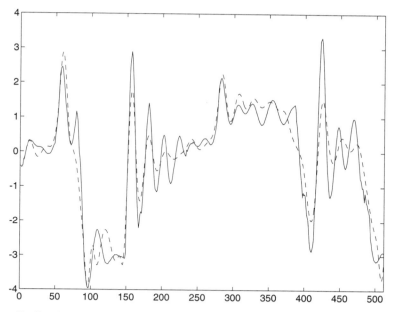

Figure 18 Result with the wavelet network initialized with algorithm BE. The solid line represents the true measurement and the dashed line represents the simulated output.

C. PREDICTIVE FUZZY MODELING OF GLYCEMIC VARIATIONS

1. Variables of Interest and Their Qualitative Labels

Diabetologists' knowledge is expressed under the form of "rule of thumb" advice. We have used this knowledge to build a two-hour-ahead predictive model of glycemic variations. This predictive model will be subsequently used in a control system. We have restricted our model to six inputs as shown in Table II (current instant t is omitted for simplicity).

The ouput is the predicted variation of glycemia at time $t + 2$ hours, DG(t+2) $\in$ {PVB, PB, PM, PS, Z, NS, NM, NB, NVB}, where P means "Positive," N "Negative," S "Small," B "Big," etc. Figure 19 shows the membership functions of glycemia, where the $(g_i)_{i=0}^4$ parameters must be determined by learning because their optimal value depends on the patient. Membership functions have been represented by simple first-order splines with free knots.

Table II

Fuzzy Variables for Glycemic Variation Modeling

Item	Symbol	Fuzzy values				
Glycemia	Gl	Very Low (VL)	Low (L)	Normal (N)	High (H)	Very High (VH)
Basis insulin injection rate	Ba		Low,	Normal,	High	
Flash insulin injection rate	Bo		Low,	Normal,	High	
Elapsed time since previous meal	Dr	Far Before,	Near,	Just After,	Far	
Diet	Nr		Fiber, Normal, Glucidic			
Expected future activity	Ac		Low, Normal, High			

Our method follows the following two steps:

1. start with an initial guess of the model, based on available (qualitative) prior knowledge;
2. tune this model to the particular patient under consideration, by performing learning from data.

2. Expressing Prior Knowledge

Combining all possible qualitative values for the different inputs yields 1620 different cases, corresponding to the same amount of candidate fuzzy rules. In fact, only 50 rules were considered for our prior model, thus reflecting the actual domain for the input variables where meaningful knowledge exists. Example of such rules are

```
if (GL(t) is VL) and (Nr(t) is N) then DG(t+2) is PB
if (GL(t) is  L) and (Ba(t) is L) then DG(t+2) is NS
```

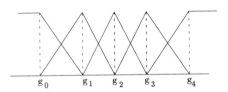

Figure 19   Fuzzy partition for glycemia.

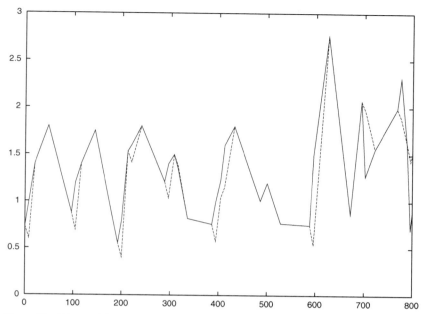

Figure 20 Prior model: two-hour-ahead prediction (dashed line) vs. actual (solid line) glycemia.

Figure 20 shows predicted glycemia at $t + \delta$ from glycemia at time t, with $\delta = 2$ hours, *before learning*, that is, with only use of the prior model. The solid line shows the actual glycemia and the dashed line the predicted one. The doctor's rules are quite efficient in predicting the effect of insulin injections. Still some spikes occur in the prediction error. The prediction error has mean $\mu = -0.20$ and standard deviation $\sigma = 0.38$.

3. Tuning the Model for Each Patient

Using data from the patient's notebook, we divided the data file into two parts, one for learning and the other for generalization (i.e., testing). Figure 21 shows predicted glycemia at $t + 2$ from glycemia at time t, *after learning*, that is, sub-sequent learning of the g_i parameters on the data. A simple stochastic gradient was used. The prediction error has mean $\mu = -0.0003$ and standard deviation $\sigma = 0.29$. Some improvement is seen; note that such an improvement is likely to be patient dependent. The errors around time steps 700 and 800 are due to catheter changes (as marked in the notebook) which usually lead to the injection of more insulin than expected.

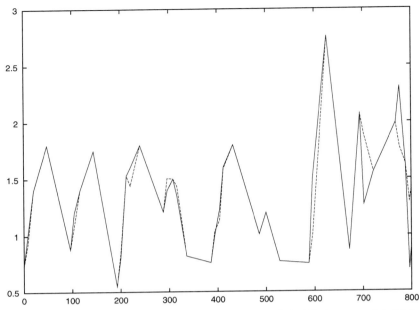

Figure 21 Model after learning: two-hour-ahead prediction (dashed line) vs. actual (solid line) glycemia.

4. Comments and Conclusions about This Example

The following conclusions can be drawn from this case study:

• Fuzzy rules turned out to be a convenient way to express prior knowledge from doctors, in part because this prior knowledge is mainly qualitative. It is important to notice that this fuzzy rule basis was far from being equivalent to an exhaustive table describing the input–output map, because about only a few percent (50/1620) of this table was described by the rules. This restriction is by itself a useful prior information about the range of validity of the modeling.

• Subsequent tuning of the prior model was performed while preserving the structure of the model; that is, the fuzzy rules were not modified, only the g_i parameters hidden in the splines were adjusted. It would also be possible to use our prior model as an initial guess but allow other "rules" (i.e., additional splines) to be introduced via learning; corresponding experiments are under progress.

• Another advantage of describing the model via fuzzy rules is the possibility to "decompile" the model after learning, again in the form of fuzzy rules, for return to the user (doctor or patient). Returning a mathematical model would be of little use for the average user having no training in mathematics.

• In this application, high accuracy was not a key point. For other cases where model accuracy is more important, replacing fuzzy membership functions in the form of first order splines by more efficient wavelets could be easily performed.

• On the practical side, on can notice from both Figs. 20 and 21 that human control of glycemia injection performs quite poorly. The desired range would be, say, about 1 ± 0.3, which is far from being accessible to human control. Thus nonlinear fuzzy control design is now under progress for this application.

VIII. DISCUSSION AND CONCLUSIONS

In this chapter we have discussed the wide area of nonparametric nonlinear estimation from the point of view of system identification. We have seen that a huge amount of work has been pursued in the statistics community. We also know from numerous press releases that, in parallel, the AI community revitalized the same area by advertising neural networks, fuzzy models, and neuro-fuzzy models. In addition, AI scientists and engineers packaged these techniques with user-oriented software and even hardware. It is not until recently that the AI community become interested in the mathematical developments and algorithms from statistics. At the same time, statisticians became involved in the mathematical study of the methods advertized by the AI community and engineering practice. In parallel, the control community recognized those models and estimation algorithms as possible candidates for nonlinear black-box system identification. In this chapter we have tried to put together material—both classical and very modern—from different areas, and have discussed both mathematical and practical issues. Here is a summary of our tentative conclusions and suggestions for future work.

Practical Issues

• *Models for prediction and simulation.* As reflected by the reported experiments, our experience has been that nonlinear nonparametric models are very good at predicting behaviors, provided that the training data set reflects all actual operating conditions that can occur. This is especially true for models that are multiresolution in nature, for example, wavelet-based models. More interesting, prediction is still efficient even for a sparse training data set—a situation which is almost unavoidable for high-dimensional input data. The quality of prediction can rapidly vanish outside the range of the training data set, however, but this is not really surprising.

• *System monitoring and diagnostics.* The reported experiments on the gas turbine case study show that the data fit is much better for our wavelet network (and even for the neuro-fuzzy network) than for our semiphysical model. Accordingly, one may expect a better performance in change detection and diagnostics by using the wavelet network. Designing a change detection procedure based on the wavelet network can be performed by applying the general asymptotic local approach discussed in [10, 89]. However because the parameters of the network have no useful interpretation, diagnostics would require learning the failure modes from training data sets: this is unrealistic because real data corresponding to failure modes are (fortunately) seldom. Thus diagnostics requires a combination of data *and* prior knowledge, preferably in the form of a (semi)-physical model: data are the current data (from safe or failure mode), and the model is used to describe prior knowledge about failure modes. In fact, gas turbine monitoring and diagnostics were successfully performed using our seemingly poor semiphysical model; see [10, 88] for an account of the results.

• *Describing prior knowledge.* Fuzzy models and their associated rules can be used to describe prior knowledge for nonlinear nonparametric models. Now if it is desired to blend the style of fuzzy rules with the mathematical quality of modern nonparametric models, we are faced with the need for a notion of "multiresolution" or "hierarchical" fuzzy rule bases. We have discussed a possible proposal toward this objective. This has to be further explored. In addition, it would be interesting to develop statistical methods checking for violation of a particular subset of fuzzy rules; this would blend methods from artificial intelligence and statistics model-based diagnostics.

• *Software support.* Our current experience can be summarized as follows. There are three different kinds of needs for nonlinear black-box identification: low-dimensional input (say, 1, 2, 3), medium-dimensional input (in the range of tens), and large-dimensional input (in the range of hundreds or thousands). The first case typically corresponds to curve fitting and is useful in signal or image processing and sometimes in control. High-performance algorithms based on wavelets are available today, which outperform others in both accuracy and computational cost (see Section IV), and software is available, such as Taswell's WavBox in the Matlab language [90]. The second case has its main applications in system identification and control. There, RBF (radial basis function) networks, which provide fast noniterative training procedures, are preferred; theoretical studies and experiments suggest that wavelet networks [72], such as discussed in this chapter, are likewise more efficient candidates. Finally, sigmoid-based neural networks with their iterative backpropagation algorithm, both simple and time consuming, are still effective for very large dimensional cases such as encountered in some pattern recognition applications. We have seen that alternative models with much more efficient iterative training procedures can also work well, such as Breiman's hinge functions [50]; Breiman's hinging hyperplane algorithm fits piecewise-linear models on nonlinear systems in a very efficient way.

Mathematical Issues

• *Assessing the quality of an approximation.* What is the convenient figure of merit for the estimation error $\| f - \hat{f} \|$? We have emphasized in this chapter the central role played by *Besov spaces*: this is a triply parametrized family of spaces of functions that are generally smooth but may have sparse singularities. *Being smooth outside localized singularities is a common feature of most of the nonlinear systems encountered in practice; thus Besov spaces are suitable to assess the quality of an estimator.*

• *Quality of fit from noisy data, and "Cramer–Rao bounds."* Maximal risks and lower rates of convergence provide adequate frameworks; they have to be used in combination with Besov spaces. And we have shown that wavelet-based estimators are optimal for systems in Besov spaces.

• *How efficient identification algorithms really are in terms of computational cost and quality of conditioning.* When orthonormal wavelet librairies can be efficiently built (this is feasible for low-dimensional input, say, up to 4 or slightly more), wavelet estimators from Section IV are the fastest ones. For very large dimensions, wavelet librairies cannot be built today, and standard sigmoid-based neural networks are preferred; Breiman's hinging hyperplane models are very promising alternative candidates. In the medium-range situation, wavelet networks using partial wavelet librairies seem to be efficient alternatives to RBF networks.

Research Directions

Based on the material of this chapter, we can suggest the following three major challenges for future research.

• Providing wavelet-based identification methods for higher-dimensional inputs. The central question here lies in the efficient construction of wavelet librairies in higher dimensions.
• Taking advantage of multiresolution in both *time and space* is a major challenge for dynamical system identification. Functional nonlinear autoregressions of the form $Y_k = f(Y_{k-1}, \ldots, Y_{k-p}) + e_k$, or their state-space counterparts, are naturally used with both neural and wavelet networks. These models do not allow playing with multiresolution for time, however, because discretization is fixed and rigid. Thus a new framework would be needed for this purpose.
• Investigating the interplay between the syntax of fuzzy modeling and modern nonparametric models certainly is a topic of major practical

interest. It would provide the user with ways of describing prior knowledge within nonparametric models.

IX. APPENDIX: THREE METHODS FOR REGRESSOR SELECTION

Recall that $W = \{\varphi_i \colon i = 1, \ldots, L\}$ is the library of the wavelet regressor candidates. Introduce the following notation:

$$
v_i = \begin{bmatrix} \varphi_i(x_1) \\ \vdots \\ \varphi_i(x_N) \end{bmatrix}, \tag{105}
$$

where $\varphi_i \in W$ and $x_1, \ldots, x_N$ are input observations in the training data set

$$
\mathcal{O}_1^N = \{(x_1, y_1), \ldots, (x_N, y_N)\}.
$$

φ_i has been normalized so that v_i is unitary:

$$
v_i^T v_i = 1, \qquad i = 1, \ldots, L.
$$

Now collect all the v_i, $i = 1, \ldots, L$, in a set V:

$$
V = \{v_1, \ldots, v_L\}. \tag{106}
$$

We also define the output observation vector

$$
y = \begin{bmatrix} y_1 \\ \vdots \\ y_N \end{bmatrix}, \tag{107}
$$

where $y_1, \ldots, y_N$ are output observations in $\mathcal{O}_1^N$.

Let span$\{v_i \colon i \in I\}$ be the space linearly spanned by the vectors v_i, $i \in I$, and let $\mathcal{I}_M$ be the set of all the M-element subsets of the index set $\{1, 2, \ldots, L\}$. Using this notation, selecting $I \in \mathcal{I}_M$ so that the corresponding M wavelets in W minimize the mean square residual $J(I)$ in (89) is equivalent to selecting the M vectors v_i from V which minimize the Euclidean distance from the vector y to the space span$\{v_i \colon i \in I\}$. Such an optimal solution requires an exhaustive examination of all the M-element subsets of W, which may not be feasible in practice, because of its massive computational burden. Some suboptimal and heuristic solutions have to be considered. Here we present three heuristic procedures.

A. RESIDUAL-BASED SELECTION: DETAILS

Define the initial residual $\gamma_0(k) = y_k$, $k = 1, \ldots, N$, with y_k the output observations in $\mathcal{O}_1^N$. Set $f_0(x) \equiv 0$.

At stage i, $i = 1, \ldots, M$, search among W the wavelet φ_j that minimizes

$$J(\varphi_j) = \frac{1}{N} \sum_{k=1}^{N} \left(\gamma_{i-1}(k) - u_j \varphi_j(x_k) \right)^2,$$

where

$$u_j = \left(\sum_{k=1}^{N} \left(\varphi_j(x_k) \right)^2 \right)^{-1} \sum_{k=1}^{N} \varphi_j(x_k) \gamma_{i-1}(k)$$

and $\gamma_{i-1}(k)$, $k = 1, \ldots, N$, are the residuals of stage $i - 1$. Note

$$l_i = \arg \min_{1 \leq j \leq L} J(\varphi_j).$$

Then φ_{l_i} is the wavelet selected at stage i. Update f_i and γ_i:

$$f_i(x) = f_{i-1}(x) + u_{l_i} \varphi_{l_i}(x),$$
$$\gamma_i(k) = \gamma_{i-1}(k) - u_{l_i} \varphi_{l_i}(x_k), \qquad k = 1, \ldots, N.$$

This procedure can be more conveniently described with the aid of vectorial notation as follows. Define the initial residual vector $\gamma_0 = y$ with y as defined in (107) and set $f_0(x) \equiv 0$. At stage i, $i = 1, \ldots, M$, search among V the vector v_j that minimizes

$$J(v_j) = (\gamma_{i-1} - u_j v_j)^T (\gamma_{i-1} - u_j v_j),$$

with

$$u_j = (v_j^T v_j)^{-1} v_j^T \gamma_{i-1} = v_j^T \gamma_{i-1},$$

where the last equality is due to the normality $v_j^T v_j = 1$.

Substituting u_j into $J(v_j)$ yields

$$J(v_j) = \left(\gamma_{i-1} - v_j^T \gamma_{i-1} v_j \right)^T \left(\gamma_{i-1} - v_j^T \gamma_{i-1} v_j \right) \tag{108}$$
$$= \gamma_{i-1}^T \gamma_{i-1} + (v_j^T \gamma_{i-1})^2 v_j^T v_j - 2 \left(v_j^T \gamma_{i-1} \right)^2 \tag{109}$$
$$= \gamma_{i-1}^T \gamma_{i-1} - \left(v_j^T \gamma_{i-1} \right)^2. \tag{110}$$

It turns out that minimizing $J(v_j)$ at stage i is equivalent to maximizing $(v_j^T \gamma_{i-1})^2$.

The algorithm is summarized as follows.

Regressor Selection Algorithm for Residual-Based Selection

Step 0. Set $\gamma_0 = y$ and $f_0(x) \equiv 0$.

Step i, i = 1, ..., M. Let $I_i = \{j: j = 1, ..., L \text{ and } j \neq l_1, ..., l_{i-1}\}$. Find

$$l_i = \arg \max_{j \in I_i} \left(v_j^T \gamma_{i-1}\right)^2$$

and set

$$u_{l_i} = v_{l_i}^T \gamma_{i-1},$$
$$f_i(x) = f_{i-1}(x) + u_{l_i}\varphi_{l_i}(x),$$
$$\gamma_i = \gamma_{i-1} - u_{l_i}v_{l_i}.$$

It is easy to prove (see [78])

$$\gamma_i^T \gamma_i = \gamma_{i-1}^T \gamma_{i-1} - \left(v_{l_i}^T \gamma_{i-1}\right)^2,$$

so $\gamma_i^T \gamma_i$ monotonically decreases as i increases. It also means that the ith term added to $f_M(x)$ has a contribution to the minimization of $\gamma_M^T \gamma_M$ measured by $(v_{l_i}^T \gamma_{i-1})^2$.

B. STEPWISE SELECTION
BY ORTHOGONALIZATION: DETAILS

At stage i of this procedure, assume that the $i - 1$ already selected wavelets correspond to the vectors $v_{l_1}, ..., v_{l_{i-1}}$. To select the ith wavelet, we have to compute the distance from y to the space span$(v_{l_1}, ..., v_{l_{i-1}}, v_j)$ for each $j = 1, ..., L$ and $j \neq l_1, ..., l_{i-1}$. For computational efficiency, we orthogonalize the later selected vectors v_j to the earlier selected ones. Assume that $v_{l_1}, ..., v_{l_{i-1}}$ are already orthonormalized and renamed as $w_{l_1}, ..., w_{l_{i-1}}$, then span$(v_{l_1}, ..., v_{l_{i-1}}, v_j) =$ span$(w_{l_1}, ..., w_{l_{i-1}}, v_j)$. For each $j = 1, ..., L$ and $j \neq l_1, ..., l_{i-1}$, compute

$$p_j = v_j - \left((v_j^T w_{l_1})w_{l_1} + \cdots + (v_j^T w_{l_{i-1}})w_{l_{i-1}}\right), \tag{111}$$
$$q_j = (p_j^T p_j)^{-1/2} p_j. \tag{112}$$

Then we should search the v_j, or equivalently the q_j, that minimizes

$$
\begin{aligned}
J(v_j) &= J(q_j) \\
&= \left[y - \left(\tilde{u}_{l_1}w_{l_1} + \cdots + \tilde{u}_{l_{i-1}}w_{l_{i-1}} + \tilde{u}_j q_j\right)\right]^T \\
&\quad \times \left[y - (\tilde{u}_{l_1}w_{l_1} + \cdots + \tilde{u}_{l_{i-1}}w_{l_{i-1}} + \tilde{u}_j q_j)\right] \\
&= [y - W_j U_j]^T [y - W_j U_j],
\end{aligned}
$$

with the matrix $W_j = (w_{l_1}, \ldots, w_{l_{i-1}}, q_j)$ and the vector

$$U_j = (\tilde{u}_{l_1}, \ldots, \tilde{u}_{l_{i-1}}, \tilde{u}_j)^T = (W_j^T W_j)^{-1} W_j^T y = W_j^T y, \qquad (113)$$

where the last equality is due to the orthonormality of $w_{l_1}, \ldots, w_{l_{i-1}}, q_j$. Continue the computation:

$$\begin{aligned} J(v_j) &= y^T y + U_j^T W_j^T W_j U_j - 2U_j^T W_j^T y \\ &= y^T y + U_j^T U_j - 2U_j^T W_j^T y. \end{aligned}$$

By (113) we have $U_j = W_j^T y$. Therefore,

$$\begin{aligned} J(v_j) &= y^T y + U_j^T U_j - 2U_j^T U_j \\ &= y^T y - U_j^T U_j \\ &= y^T y - (\tilde{u}_{l_1}^2 + \cdots + \tilde{u}_{l_{i-1}}^2 + \tilde{u}_j^2). \end{aligned}$$

Consequently, minimizing $J(v_j)$ is equivalent to maximizing $\tilde{u}_{l_1}^2 + \cdots + \tilde{u}_{l_{i-1}}^2 + \tilde{u}_j^2$. By (113) we have

$$\begin{aligned} \tilde{u}_{l_k} &= w_{l_k}^T y, \qquad k = 1, \ldots, i-1, \\ \tilde{u}_j &= q_j^T y, \end{aligned}$$

so $\tilde{u}_{l_1}^2 + \cdots + \tilde{u}_{l_{i-1}}^2$ is independent of q_j. We conclude that minimizing $J(v_j)$ is equivalent to maximizing $\tilde{u}_j^2 = (q_j^T y)^2$.

After M iterations, the values of $l_1, \ldots, l_M$ are determined, as well as $\tilde{u}_{l_1}, \ldots, \tilde{u}_{l_M}$. We still need to determine the values of $u_{l_1}, \ldots, u_{l_M}$ in

$$f_M(x) = \sum_{i=1}^{M} u_{l_i} \varphi_{l_i}(x).$$

By the definitions of w_l and $\tilde{u}_l$,

$$y = [w_{l_1}, \ldots, w_{l_M}][\tilde{u}_{l_1}, \ldots, \tilde{u}_{l_M}]^T + \gamma_M.$$

On the other hand,

$$y = [v_{l_1}, \ldots, v_{l_M}][u_{l_1}, \ldots, u_{l_M}]^T + \gamma_M.$$

Therefore,

$$[w_{l_1}, \ldots, w_{l_M}][\tilde{u}_{l_1}, \ldots, \tilde{u}_{l_M}]^T = [v_{l_1}, \ldots, v_{l_M}][u_{l_1}, \ldots, u_{l_M}]^T. \qquad (114)$$

In (111) and (112) let $j = l_i$. Then combining them yields

$$\begin{aligned} v_{l_i} &= ((v_{l_i}^T w_{l_1})w_{l_1} + \cdots + (v_{l_i}^T w_{l_{i-1}})w_{l_{i-1}}) + p_{l_i} \\ &= ((v_{l_i}^T w_{l_1})w_{l_1} + \cdots + (v_{l_i}^T w_{l_{i-1}})w_{l_{i-1}}) + (p_{l_i}^T p_{l_i})^{1/2} q_{l_i} \end{aligned}$$

$$= \left((v_{l_i}^T w_{l_1}) w_{l_1} + \cdots + (v_{l_i}^T w_{l_{i-1}}) w_{l_{i-1}} \right) + \left(p_{l_i}^T p_{l_i} \right)^{1/2} w_{l_i}$$
$$= (\alpha_{1i} w_{l_1} + \cdots + \alpha_{i-1i} w_{l_{i-1}}) + \alpha_{ii} w_{l_i},$$

with

$$\alpha_{ki} = v_{l_i}^T w_{l_k}, \qquad k = 1, \ldots, i - 1,$$
$$\alpha_{ii} = \left(p_{l_i}^T p_{l_i} \right)^{1/2}.$$

Consequently,

$$[w_{l_1}, \ldots, w_{l_M}] A = [v_{l_1}, \ldots, v_{l_M}], \tag{115}$$

where A is the triangular matrix

$$A = \begin{bmatrix} \alpha_{11} & \alpha_{12} & \alpha_{13} & \cdots & & \alpha_{1M} \\ 0 & \alpha_{22} & \alpha_{23} & \cdots & & \alpha_{2M} \\ 0 & 0 & \alpha_{33} & \cdots & & \alpha_{3M} \\ \vdots & \vdots & & \ddots & \ddots & \vdots \\ 0 & 0 & \cdots & 0 & \alpha_{M-1M-1} & \alpha_{M-1M} \\ 0 & 0 & \cdots & & 0 & \alpha_{MM} \end{bmatrix}.$$

Then u_{l_i} can be obtained by solving the triangular system of equations obtained by combining (114) and (115):

$$A[u_{l_1}, \ldots, u_{l_M}]^T = [\tilde{u}_{l_1}, \ldots, \tilde{u}_{l_M}]^T. \tag{116}$$

Let us summarize the algorithm as follows.

Regressor Selection Algorithm for Stepwise Selection by Orthogonalization

Step 1.   Find

$$l_1 = \arg \max_{1 \leqslant j \leqslant L} \left(v_j^T y \right)^2.$$

Set

$$\tilde{u}_{l_1} = v_{l_1}^T y, \qquad w_{l_1} = v_{l_1}, \qquad \alpha_{11} = 1.$$

Step i, $i = 2, \ldots, M$.   Let $I_i = \{j : j = 1, \ldots, L$ and $j \neq l_1, \ldots, l_{i-1}\}$. For each $j \in I_i$, compute

$$p_j = v_j - \left((v_j^T w_{l_1}) w_{l_1} + \cdots + (v_j^T w_{l_{i-1}}) w_{l_{i-1}} \right),$$
$$q_j = (p_j^T p_j)^{-1/2} p_j.$$

Find

$$l_i = \arg\max_{j \in I_i} \left(q_j^T y \right)^2$$

and set

$$\tilde{u}_{l_i} = q_{l_i}^T y,$$
$$w_{l_i} = q_{l_i},$$
$$\alpha_{ki} = v_{l_i}^T w_{l_k}, \qquad k = 1, \ldots, i-1,$$
$$\alpha_{ii} = \left(p_{l_i}^T p_{l_i} \right)^{1/2}.$$

Step M + 1. Solve (116) to obtain u_{l_i}, $i = 1, \ldots, M$, and build

$$f_M(x) = \sum_{i=1}^{M} u_{l_i} \varphi_{l_i}(x).$$

C. BACKWARD ELIMINATION: DETAILS

The regression with all the wavelets of W is written as

$$f_L(x) = \sum_{i=1}^{L} u_i \varphi_i(x),$$

where u_i are determined by the least squares algorithm:

$$(u_1, \ldots, u_L)^T = \left[(v_1, \ldots, v_L)^T (v_1, \ldots, v_L) \right]^{-1} (v_1, \ldots, v_L)^T y. \quad (117)$$

Note that inverting the matrix $(v_1, \ldots, v_L)^T (v_1, \ldots, v_L)$ may cause a problem when it is singular. This situation rarely occurs with the set V of vectors corresponding to the wavelet library W. Whenever it happens, the two previously presented regressor selection algorithms should be used.

The residuals

$$\gamma_L(k) = y_k - f_L(x_k), \qquad k = 1, \ldots, N,$$

can be written in their vectorial form as

$$\gamma_L = y - (v_1, \ldots, v_L)(u_1, \ldots, u_L)^T. \quad (118)$$

Combining (117) and (118), we get

$$\gamma_L^T \gamma_L = y^T y^T - y^T V_0 (V_0^T V_0)^{-1} V_0^T y,$$

where the matrix $V_0 = (v_1, \ldots, v_L)$.

If we remove one wavelet, say φ_j, from $f_L(x)$, the same computation can be repeated to get a similar result

$$\gamma_{L-1}^T \gamma_{L-1} = y^T y^T - y^T C(v_j|V_0)\big(C(v_j|V_0)^T C(v_j|V_0)\big)^{-1} C(v_j|V_0)^T y,$$

where the operator C means the complement of a matrix; that is, if a matrix $U = [U_1, U_2, U_3]$, then $C(U_2|U) = [U_1, U_3]$. Hence the increment of the sum of square residual caused by removing φ_j from $f_L(x)$ is

$$
\begin{aligned}
J(\varphi_j) &= \gamma_{L-1}^T \gamma_{L-1} - \gamma_L^T \gamma_L \\
&= y^T V_0 (V_0^T V_0)^{-1} V_0^T y - y^T C(v_j|V_0) \\
&\quad \times \big(C(v_j|V_0)^T C(v_j|V_0)\big)^{-1} C(v_j|V_0)^T y.
\end{aligned}
\tag{119}
$$

Removing from $f_L(x)$ the wavelet φ_j that minimizes (119) yields $f_{L-1}(x)$. Repeat the same procedure to remove another wavelet from $f_{L-1}(x)$, and so on. This results in the following algorithm.

Regressor Selection Algorithm for Full Backward Elimination

Step 0. Set $V_0 = (v_1, \ldots, v_L)$.

Step i, $i = 1, \ldots, L - M$. Let $I_i = \{j : j = 1, \ldots, L \text{ and } j \neq l_1, \ldots, l_{i-1}\}$. Find

$$l_i = \arg\max_{j \in I_i} y^T C(v_j|V_{i-1})\big(C(v_j|V_{i-1})^T C(v_j|V_{i-1})\big)^{-1} C(v_j|V_{i-1})^T y.$$

Set $V_i = C(v_{l_i}|V_{i-1})$.

Step $L - M + 1$. Let $I_{L-M+1} = \{j : j = 1, \ldots, L \text{ and } j \neq l_1, \ldots, l_{L-M}\}$. Build

$$f_M(x) = \sum_{j \in I_{L-M+1}} u_j \varphi_j(x),$$

with u_j the components of the vector u given by

$$u = \big(V_{L-M}^T V_{L-M}\big)^{-1} V_{L-M}^T y.$$

The computation required by this procedure is quite heavy. For instance $L - i + 1$ matrices need to be inverted at step i. The computation for inverting the matrices can be reduced in the following way.

For any matrix $U = [U_1, U_2, U_3]$ where U_1, U_2, U_3 are subblocks of U,

$$U^T U = \begin{bmatrix} U_1^T U_1 & U_1^T U_2 & U_1^T U_3 \\ U_2^T U_1 & U_2^T U_2 & U_2^T U_3 \\ U_3^T U_1 & U_3^T U_2 & U_3^T U_3 \end{bmatrix}.$$

Assume $(U^T U)^{-1}$ is already calculated and partitioned in the same way as $U^T U$:

$$(U^T U)^{-1} = \begin{bmatrix} \Lambda_{11} & \Lambda_{12} & \Lambda_{13} \\ \Lambda_{21} & \Lambda_{22} & \Lambda_{23} \\ \Lambda_{31} & \Lambda_{32} & \Lambda_{33} \end{bmatrix}.$$

Then the following formula can be easily verified:

$$\begin{aligned} ([U_1, U_3]^T [U_1, U_3])^{-1} &= \begin{bmatrix} U_1^T U_1 & U_1^T U_3 \\ U_3^T U_1 & U_3^T U_3 \end{bmatrix}^{-1} \\ &= \begin{bmatrix} \Lambda_{11} & \Lambda_{13} \\ \Lambda_{31} & \Lambda_{33} \end{bmatrix} - \Lambda_{22}^{-1} \begin{bmatrix} \Lambda_{12} \\ \Lambda_{32} \end{bmatrix} [\Lambda_{21} \quad \Lambda_{23}]. \quad (120) \end{aligned}$$

In this way only $V_0^T V_0$ needs to be actually inverted using the conventional method. Using (120), $(C(v_j|V_i)^T C(v_j|V_i))^{-1}$ can be obtained from subblocks of $(V_i^T V_i)^{-1}$.

This procedure can be further simplified as follows.

Assume that $f_L(x)$ is built with all the wavelets of W as before. Now eliminate one wavelet from $f_L(x)$, say φ_j, but keep the values of u_l unchanged, $l = 1, \ldots, L$. The residual becomes

$$\gamma_{L-1}(k) = y_k - \big(f_L(x_k) - u_j \varphi_j(x_k)\big) = \gamma_L(k) + u_j \varphi_j(x_k), \qquad k = 1, \ldots, N,$$

so

$$\gamma_{L-1} = \gamma_L + u_j v_j.$$

Then

$$\begin{aligned} \gamma_{L-1}^T \gamma_{L-1} &= \gamma_L^T \gamma_L + u_j^2 v_j^T v_j + 2u_j \gamma_L^T v_j \\ &= \gamma_L^T \gamma_L + u_j^2 + 2u_j \gamma_L^T v_j. \end{aligned}$$

The last term of this equation can be neglected under the assumptions that γ_L is close to zero mean and independent of v_j. Therefore,

$$\gamma_{L-1}^T \gamma_{L-1} - \gamma_L^T \gamma_L \approx u_j^2.$$

This means that removing φ_j from $f_L(x)$ will cause an increment of the sum of square residuals approximatively equal to u_j^2. Repeating the same reasoning on $f_{L-1}(x)$, $f_{L-2}(x)$, etc. yields the following procedure.

Regressor Selection Algorithm for Backward Elimination

Step 0.   Set $V_0 = (v_1, \ldots, v_L)$.

Step i, $i = 1, \ldots, L - M$.   Let $I_i = \{j: j = 1, \ldots, L \text{ and } j \neq l_1, \ldots, l_{i-1}\}$ and compute

$$u = (V_{i-1}^T V_{i-1})^{-1} V_{i-1}^T y,$$

where u is a vector composed of u_j, $j \in I_i$. Find

$$l_i = \arg \min_{j \in I_i} u_j^2.$$

Set $V_i = C(v_j | V_{i-1})$.

Step $L - M + 1$.   Let $I_{L-M+1} = \{j: j = 1, \ldots, L \text{ and } j \neq l_1, \ldots, l_{L-M}\}$. Build

$$f_M(x) = \sum_{j \in I_{L-M+1}} u_j \varphi_j(x),$$

with u_j the components of the vector u given by

$$u = (V_{L-M}^T V_{L-M})^{-1} V_{L-M}^T y.$$

Note that Eq. (120) is used for inverting $V_i^T V_i$, $i > 0$; only $V_0^T V_0$ is inverted using the conventional algorithm. Alternatively, if the mother wavelet function φ is chosen to have compact support, then the matrices V_i and $V_i^T V_i$ are sparse. $V_i^T V_i$ is symmetric and usually has diagonal dominance. In such situations, and for large matrices $V_i^T V_i$, instead of directly computing

$$u = (V_i^T V_i)^{-1} V_i^T y,$$

iterative methods [91] should be used for solving

$$(V_i^T V_i) u = V_i^T y.$$

The preceding RBS, SSO, and BE algorithms have been implemented in the Matlab 4.1 language. The full BE algorithm has not been implemented due to its high computational cost.

REFERENCES

[1] L. Ljung. Perspectives on the process of identification. In *Proceedings of the 12th IFAC World Congress* (Sydney), 1993.

[2] T. Poggio and F. Girosi. Networks for approximation and learning. *Proc. IEEE* 78:1481–1497, 1990.

[3] K. Hunt, D. Sbarbaro, R. Zbikowski, and P. Gawthrop. Neural networks for control systems—a survey. *Automatica* 28:1083–1112, 1992.

[4] K. Narendra and K. Parthasarathy. Identification and control of dynamical systems using neural networks. *IEEE Trans. Neural Networks* 1:4–27, 1990.

[5] Y. Meyer. *Ondelettes et Opérateurs.* Hermann, Paris, 1990.

[6] I. Daubechies. *Ten Lectures on Wavelets. CBMS–NSF Regional Series in Applied Mathematics,* 1992.

[7] C. Lee. Fuzzy logic in control systems, 1 and 2. *IEEE Trans. Systems Man Cybernet.* 20:1990.

[8] E. Sontag. Nonlinear regulation: the piecewise linear approach. *IEEE Trans. Automat. Control* 26:346–358, 1981.

[9] Q. Zhang. Contribution à la surveillance de procédés industriels. Ph.D. Thesis, Université de Rennes I, 1991.

[10] Q. Zhang, M. Basseville, and A. Benveniste. Early warning of slight changes in systems. *Automatica* 30:95–113, 1994.

[11] L. Devroye and L. Györfi. *Nonparametric Density Estimation L_1 View.* Wiley, New York, 1985.

[12] J. Van Ryzin. Bayes risk consistency of classification procedures using density estimation. *Sankhyā* 28:261–270, 1966.

[13] C. Wolverton and T. Wagner. Asymptotically optimal discriminant functions for pattern classifications. *IEEE Trans. Inform. Theory* 15:258–265, 1969.

[14] S. Csibi. *Stochastic Processes with Learning Properties.* Springer-Verlag, Berlin, 1975.

[15] E. Nadaraya. On estimating regression. *Theory Probab. Appl.* 9:141–142, 1964.

[16] G. Watson. Smooth regression analysis. *Sankhyā Ser. A* 26:359–372, 1969.

[17] E. Parzen. On estimation of probability density function and the mode. *Ann. Math. Statist.* 33:1065–1076, 1962.

[18] M. Rosenblatt. Remarks on some nonparametric estimates of density functions. *Ann. Math. Statist.* 27:832–835, 1956.

[19] M. Rosenblatt. Curve estimation. *Ann. Math. Statist.* 42:1815–1842, 1971.

[20] M. Duflo. *Recursive Stochastic Methods.* Springer-Verlag, Berlin, 1993.

[21] G. Oppenheim and B. Portier. Commande adaptative du processus de Markov $x_{t+1} = f_t + u_t + x_t$, $t \in n$. Technical Report 90-18, Université d'Orsay, 1990.

[22] B. Portier. Estimation non paramétrique et commande adaptative de processus Markoviens non linéaires. Ph.D. Thesis, Université Paris Sud, Orsay, 1992.

[23] G. Wahba. *Spline Functions for Observational Data.* SIAM, Philadelphia, 1991.

[24] C. Stone. Optimal global rates of convergence for nonparametric regression. *Ann. Statist.* 10:1040–1053, 1982.

[25] N. Cencov. Statistical decision rules and optimal inference. *Amer. Math. Soc. Transl.* 53:1982.

[26] I. Ibragimov and R. Khasminskij. *Statistical Estimation Asymptotic Theory.* Springer-Verlag, Berlin, 1981.

[27] J. Rice. Bandwidth choice for nonparametric regression. *Ann. Statist.* 12:1215–1230, 1984.

[28] W. Härdle and J. Marron. Optimal bandwidth selection in nonparametric regression function estimation. *Ann. Statist.* 13:1465–1481, 1985.

[29] K. Li. Asymptotic optimality of c_L and generalized cross-validation in ridge regression and application to the spline smoothing. *Ann. Statist.* 14:1101–1112, 1986.

[30] P. Craven and G. Wahba. Smoothing noisy data with spline functions. *Numer. Math.* 31:337–403, 1979.

[31] B. Polyak and A. Tsybakov. Asymptotical optimality of c_p criterion for projection regression estimates. *Theory Probab. Appl.* 35:305–317, 1990.

[32] K. Li. Asymptotic optimality of c_L and generalized cross-validation: discrete index set. *Ann. Statist.* 15:958–975, 1987.

[33] C. Mallows. Statistical predictor identification. *Technometrics* 15:661–675, 1973.

[34] H. Akaike. Statistical predictor identification. *Ann. Inst. Math. Statist.* 22:203–217, 1970.

[35] S. Efroimovich and M. Pinsker. Estimation of square-integrable spectral density based on a sequence of observations. *Problems Inform. Transmission* 182–196, 1982 (in Russian).

[36] S. Efroimovich and M. Pinsker. Estimation of square-integrable probability density of a random variable. *Problems Inform. Transmission* 175–189, 1983 (in Russian).

[37] S. Efroimovich and M. Pinsker. A learning algorithm for nonparametric filtering. *Avtomat. i Telemekh.* 11:58–65, 1984 (in Russian).

[38] L. Devroye. Any discrimination rule can have an arbitrary bad probability of error for final sample size. *IEEE Trans. Pattern Anal. Machine Intell.* 4:154–157, 1982.

[39] L. Devroye and T. Wagner. Distribution free consistency result in nonparametric discrimination and regression function estimation. *Ann. Statist.* 8:231–239, 1980.

[40] A. Korostelev and A. Tsybakov. *Minimax Theory of Image Reconstruction.* Springer-Verlag, Berlin, 1981.

[41] D. Donoho, I. Johnstone, G. Kerkyacharian, and D. Picard. Wavelet shrinkage: asymptopia. Technical Report, 1993. Available at ftp playfair.stanford.edu.

[42] J. Friedman and W. Stuetzle. Projection pursuit regression. *J. Amer. Statist. Assoc.* 76:817–823, 1981.

[43] L. Breiman, J. Friedman, J. Olshen, and C. Stone. *Classification and Regression Trees.* Wadsworth, Belmont, CA, 1984.

[44] J. Friedman. Multivariate adaptive regression splines (with discussion). *Ann. Statist.* 19:1–141, 1991.

[45] H.-G. Müller and U. Stadtmüller. Variable bandwidth kernel estimators of regression curves. *Ann. Statist.* 15:182–201, 1987.

[46] D. Donoho and I. Johnstone. Minimax estimation via wavelet shrinkage. Technical Report, Department of Statistics, Stanford University, 1992. Available at ftp playfair.stanford.edu.

[47] P. Huber. Projection pursuit (with discussion). *Ann. Statist.* 13:435–475, 1985.

[48] J. Morgan and J. Sonquist. Problems in the analysis of survey data, and a proposal. *J. Amer. Statist. Assoc.* 58:415–434, 1963.

[49] L. Ljung. Neural networks in identification, a tutorial. In *Proceedings of the 10th IFAC Symposium on Identification and System Parameter Estimation* (Copenhagen), 1994.

[50] L. Breiman. Hinging hyperplanes for regression, classification and function approximation. *IEEE Trans. Inform. Theory* 39:999–1013, 1993.

[51] A. Barron. Universal approximation bounds for superpositions of a sigmoidal function. *IEEE Trans. Inform. Theory* 39:1993.

[52] A. Benveniste. *Digital Signal Processing Techniques and Applications.* Academic Press, San Diego, 1993.

[53] P. Vaidyanathan. Quadrature mirror filters banks, m-band extensions and perfect reconstruction techniques. *IEEE-ASSP Mag.* 4:4–20, 1987.

[54] D. Donoho. Interpolating wavelet transforms. Technical Report, Department of Statistics, Stanford University, 1993. Available at ftp playfair.stanford.edu.

[55] B. Delyon. Orthogonal and biorthogonal wavelets. Technical Report 732, IRISA, 1993.

[56] S. Jaffard and P. Laurentçot. *Wavelets: A Tutorial.* Academic Press, San Diego, 1989.

[57] H. Triebel. *Theory of Function Spaces.* Birkhäuser, Berlin, 1983.

[58] H. Triebel. *Theory of Function Spaces II*. Birkhäuser, Berlin, 1993.

[59] O. Besov. On a family of functional spaces: embedding theorems and applications. *Dokl. Akad. Nauk SSSR* 126:1163–1165, 1959.

[60] P. Petrushev and V. Popov. *Rational Approximation of Real Functions*. Cambridge University Press, 1987.

[61] W. Sickel. Spline representations of functions in Besov–Triebel–Lizorkin spaces on $\mathbf{r}^n$. *Forum Math.* 2:451–476, 1990.

[62] R. DeVore, B. Jawerth, and V. Popov. Compression of wavelet decompositions. *Amer. J. Math.*, to appear.

[63] B. Delyon and A. Juditsky. Optimal estimators for functional autoregression. Technical Report, IRISA, in preparation.

[64] G. Kerkyacharian and D. Picard. Density estimation in Besov spaces. *Statist. Probab. Lett.* 13:15–24, 1992.

[65] D. Donoho and I. Johnstone. Minimax risk over l_p-balls. Technical Report, Department of Statistics, Stanford University, 1992. Available at ftp playfair.stanford.edu.

[66] B. Delyon and A. Juditsky. Wavelet estimators, global error measures revisited. Technical Report 782, IRISA, 1993.

[67] D. Donoho, I. Johnstone, G. Kerkyacharian, and D. Picard. Density estimation by wavelet thresholding. Technical Report, Department of Statistics, Stanford University, 1993. Available at ftp playfair.stanford.edu.

[68] A. Nemirovskij. Nonparametric estimation of smooth regression functions. *Izv. Akad. Nauk SSSR Techn. Kibern.* 3:50–60, 1985 (in Russian).

[69] D. Donoho and I. Johnstone. Adapting to unknown smoothness via wavelet shrinkage. Technical Report, Department of Statistics, Stanford University, 1993. Available at ftp playfair.stanford.edu.

[70] A. Juditsky. Adaptive wavelet estimators. Technical Report 815, IRISA, 1994.

[71] D. Donoho. Smooth wavelet decompositions with blocky coefficient kernels. Technical Report, Department of Statistics, Stanford University, 1993. Available at ftp playfair.stanford.edu.

[72] Q. Zhang. Wavenet. Public domain Matlab toolbox, 1993. Available at ftp://ftp.irisa.fr/local/wavenet/wnet2.1.tar.Z.

[73] B. Delyon, A. Juditsky, and A. Benveniste. Accuracy analysis for wavelet approximations. *IEEE Trans. Neural Networks* 6:332–348, 1995.

[74] Q. Zhang and A. Benveniste. Wavelet networks. *IEEE Trans. Neural Networks* 3:889–898, 1992.

[75] Q. Zhang. Wavelet networks: the radial structure and an efficient initialization procedure. Technical Report LiTH-ISY-I-1423, Linköping University, 1992.

[76] Q. Zhang. Regressor selection and wavelet network construction. Technical Report 709, Inria, 1993.

[77] N. Draper and H. Smith. *Applied Regression Analysis*, 2nd ed. Wiley, New York, 1981.

[78] S. Mallat and Z. Zhang. Matching pursuit with time-frequency dictionaries. Technical Report 619, Computer Science Department, New York University, 1993.

[79] S. Qian and D. Chen. Signal representation using adaptive normalized Gaussian functions. *Signal Process.* 36:1994.

[80] S. Chen, S. Billings, and W. Luo. Orthogonal least squares methods and their application to non-linear system identification. *Internat. J. Control* 50:1873–1896, 1989.

[81] S. Chen, C. Cowan, and P. Grant. Orthogonal least squares learning algorithm for radial basis function networks. *IEEE Trans. Neural Networks* 2:302–309, 1991.

[82] L. Zadeh. Fuzzy logic, neural networks, and soft computing. *Comm. ACM* 37:77–86, 1994.

[83] D. Dubois and H. Prade. Fuzzy sets in approximate reasoning, 1. *Fuzzy Sets Systems* 40:1992.

[84] P. Glorennec. A general class of fuzzy inference systems. In *Proceedings of the CES2 Conference* (Prague), 1993.

[85] L. Wang. Fuzzy systems are universal approximators. In *Proceedings of the First IEEE Conference on Fuzzy Systems* (San Diego), pp. 1163–1169, 1992.

[86] D. Dubois and H. Prade. *Conditional Logic in Expert Systems*, pp. 115–158. North-Holland, Amsterdam, 1991.

[87] G. Plotkin. *A Structural Approach to Operational Semantics*. Lecture Notes, Aarhus University, 1981.

[88] M. Basseville, A. Benveniste, G. Mathis, and Q. Zhang. Monitoring the combustion set of a gas turbine. In *Proceedings of SAFEPROCESS'94* (Helsinki), 1994.

[89] A. Benveniste, M. Basseville, and G. Moustakides. The asymptotic local approach to change detection and model validation. *IEEE Trans. Automat. Control* 32:583–592, 1987.

[90] C. Taswell. Wavbox. Public domain Matlab toolbox, 1993. Available at anonymous ftp: simplicity.stanford.edu:/pub/taswell.

[91] D. M. Young. *Iterative Solution of Large Linear Systems*. Academic Press, San Diego, 1971.

Index